Mobile
Telecommunications
Networking with IS-41

Other McGraw-Hill Communications Books of Interest

To order or receive additional information on these or any other McGraw-Hill titles, in the United States please call 1-800-722-4726. In other countries, contact your local McGraw-Hill representative.

Mobile Telecommunications Networking with IS-41

Michael D. Gallagher
Randall A. Snyder

McGraw-Hill

New York   San Francisco   Washington, D.C.   Auckland   Bogotá
Caracas   Lisbon   London   Madrid   Mexico City   Milan
Montreal   New Delhi   San Juan   Singapore
Sydney   Tokyo   Toronto

Library of Congress Cataloging-in-Publication Data

Gallagher, Michael D.
    Mobile telecommunications networking with IS-41 / Michael D.
Gallagher, Randall A. Snyder.
       p.   cm.
    Includes bibliographical references and index.
    ISBN 0-07-063314-2 (alk. paper)
    1. Personal communication service systems. 2. Mobile
communication systems. I. Snyder, Randall A. II. Title.
TK5103.485.G35 1997
621.3845′6′02187—dc21   96-51017
                                  CIP

McGraw-Hill

A Division of The McGraw-Hill Companies

    4 5 6 7 8 9 0 DOC/DOC 9 0 2 1 0 9

ISBN 0-07-063314-2

*The sponsoring editor for this book was Stephen S. Chapman, the editing
supervisor was Paul Sobel, and the production supervisor was Suzanne
W. B. Rapcavage. It was set in Century Schoolbook by Victoria Khavkina
of McGraw-Hill's Professional Book Group composition unit.*

Printed and bound by R. R. Donnelley & Sons Company.

McGraw-Hill books are available at special quantity discounts to use
as premiums and sales promotions, or for use in corporate training pro-
grams. For more information, please write to the Director of Special
Sales, McGraw-Hill, 11 West 19th Street, New York, NY 10011. Or
contact your local bookstore.

 This book is printed on recycled, acid-free paper containing a
minimum of 50% recycled, de-inked fiber.

*From **Michael D. Gallagher**: I would like to dedicate this book to Halina, Samuel, and Alexandra—for their love.*

*From **Randall A. Snyder**: I would like to dedicate this book to my wife, Cecilia, whom I love dearly and continues to encourage me; to my parents who instilled in me the concept of personal success through the examples of their successes and to my brother Brad, who will always live in my heart.*

Contents

Part 1 Introduction to Mobile Telecommunications, Network Architecture and Functions

Part 2 IS-41 Revision C Explained

Part 3 IS-41 Network Implementations

Foreword

[handwritten inscription: "Peter, Enjoy the "read". Mike 1/16/01"]

People are still amazed today when they call a seven-digit phone number to a cellular phone and find out that the person that answers the call is not in the city they thought they'd be in. The magic of being able to call a person—not a place, and have that person answer the call anywhere in North America, Europe and eventually the world, is simply that—magic.

To those in the industry, this magic is called mobility management. It is based on a standard known as IS-41. It is a revolution that sits in the shadow of the TDMA, CDMA and GSM debate. Tremendous effort has gone into the creation, development and implementation of this standard. This vision of person-to-person communication, anytime, anywhere has been promoted by the industry leaders like Craig McCaw (then CEO of McCaw Cellular), John Stupka (then President of Southwestern Bell Mobile Systems) and Nick Kauser (then CTO of Cantel) to name a few. Others not so visible also played a tremendous part, but none as great as Kirk Carlson (still with Synacom Technology).

The future direction of telecommunication is set on an irrevocable course. Both wireline and wireless service providers will need to provide a healthy dose of "magic" that their customers will demand. Read on as Randy Snyder and Mike Gallagher unlock the magic. I found this book unlocked the "magic," providing the secrets of the trade for all who want to know.

Mike Buhrmann
Vice President, Wireless Strategy
AT&T Wireless Services

Preface

Who Should Read This Book

The subject matter of Interim Standard-41 (IS-41) for mobile telecommunications has typically fallen into the purview of telecommunications system engineers and equipment software developers only. This book attempts to serve many purposes, the primary one being to expand the understanding of IS-41 beyond this select group. For system engineers and software developers, this book can serve as a reference to the IS-41 standard, as well as explaining some of the more complex technical features of the standard. For wireless telecommunications technical and marketing managers, as well as executives, this book is intended to also be a learning tool. Although the IS-41 standard itself is the focus of the book, there is enough information included about wireless and cellular telecommunications networking to provide a better understanding of mobile network architecture and functionality.

This book is a direct result of our involvement as engineers in the wireless industry participating in the creation of the IS-41 standard, as well as many others. The bulk of the material is derived from the IS-41 standard itself, however there is valuable information provided to expand one's understanding beyond the standard. Information about the application of IS-41 in the *real world*, deployment strategies, insights into network architectures and the standards-making process are provided so that this book serves as more than just a reference for a single industry standard. The intent of this book is to provide a comprehensive introduction to IS-41, as well as a reference for those already familiar with the protocol.

The most recent published revision of the IS-41 standard, TIA/EIA IS-41-C, has been raised to a full national ANSI standard known as ANSI/TIA/EIA-41. There were a few very minor protocol changes made to IS-41 as a result of being advanced to a full ANSI standard. However, these changes do not affect the content of this book and should not affect overall implementations of IS-41. The differences between IS-41 and ANSI-41 are marked with change bars in the ANSI-41 specification.

All standards are open to interpretation and IS-41 is no different.

Although we have taken part in the development of the IS-41 standard and the design of systems that use IS-41, the substance of this book reflects our understanding and interpretation of the standard. This understanding certainly should not be taken as a unique view. The IS-41 specification and related standards are the primary references for this book and we disclaim any responsibility for any usage of the interpretations described in this book.

Graphical Map of This Book

This book is divided into three parts. The subject matter of each part logically builds on the information presented in the previous part. Part 1 provides an introduction to the concept of mobile telecommunications networking, network architecture and functions that enable mobility within the telecommunications network. Part 2 provides a full description of IS-41 including revisions, operational details and protocol usage with a focus on IS-41 Revision C (IS-41-C). Part 3 provides IS-41 network implementations including network interconnections and practical information about network interoperability and roaming.

A Few Words about Terminology

Terminology can be a major stumbling block in the understanding of technical subject matter, and mobile telecommunications networks are no exception. Seemingly familiar terms develop new meanings and engineers begin to drown in a sea of acronyms. Also, informal usage of synonyms for many of the technical terms can lead to confusion and semantic arguments. This section presents an explanation of some commonly misunderstood and confusing terms that should help with the understanding of this book. Of course, anyone familiar with these terms can still find some ambiguity in our descriptions and definitions; however, it is our intention to clarify some of these semantics, provide distinctions between similar terms and provide a basis for consistent usage within this book.

Wireless versus cellular

These terms can be very confusing based on the context in which they are used. Many standards bodies have vehemently argued the meaning of these terms from a technical, political and commercial perspective. The term *wireless* can encompass nearly any type of communications without the use of wires. Good examples are radio and infrared communications. Cellular systems, paging, wireless local area networks and even remote control units are considered to be types of wireless communications.

Part 1

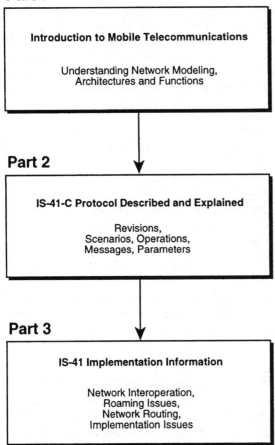

Part 2

Part 3

The term *cellular* is used in the United States to specifically refer to the current operating mobile telecommunications system and is the type of system referred to by IS-41. Cellular systems can be considered a subset of wireless communications employing certain unique characteristics, such as high extensible capacity, frequency reuse and mobility management that controls communications to and from specific cells. The term cellular is sometimes used politically and commercially to differentiate first generation mobile telecommunications systems from next generation personal communications services (PCS) systems. Technically, however, many systems considered to be PCS actually do employ cellular technology. IS-41 can support any system based on this cellular concept and the term wireless can often times refer to both cellular and PCS networks. The term *wireless* is used within this book to refer to systems employing cellular technology.

Intersystem versus Internetwork

The IS-41 standard specifies cellular network *intersystem* operations. A system is generally defined as a single mobile switching center (MSC) system and its peripheral functionality (i.e., its associated HLR, VLR, etc.). The term *intersystem* in this context implies operations between MSC systems used to support wireless telecommunications. The term *internetwork* sometimes implies conversion of network messages between non-homogeneous networks.

Operation versus message

An IS-41 operation refers to an entire operation process which includes all components of the information exchange used to effect that operation. An IS-41 message refers to each specific component of a given operation. For example, the Registration Notification operation consists of the entire information exchange process including the "Registration Notification Invoke" component as well as the "Registration Notification Return Result" component. Each of these components can be considered an individual IS-41 message.

Subscriber versus end-user

These terms are generally used synonymously. *Subscriber* is used throughout the IS-41 specification to mean the end-user of the wireless network services. The subscriber is the human using the mobile station (MS).

Feature versus service

These terms are generally used synonymously. The term *service* is somewhat all encompassing and represents any function that can be supported in the wireless network. In the context of IS-41, the term feature specifically refers to a specific supplementary service provided to a subscriber, such as call waiting or call forwarding.

CDMA and TDMA

Code division multiple access (CDMA) and time division multiple access (TDMA) are general categories of digital radio technology. There are many mobile radio standards (both national and international) that are based on these technologies. However, in reference to IS-41-based networks, the terms have specific meaning. In this book, CDMA specifically refers to the *TIA/EIA IS-95* family of related standards and TDMA specifically refers to the *TIA/EIA IS-54* and *TIA/EIA IS-136* family of related standards. The IS-136 TDMA family of standards is sometimes referred to as digital AMPS or D-AMPS (refer to the *Bibliography* for these standards).

CCITT versus ITU-T

The original CCITT (International Telegraph and Telephone Consultative Committee) which published the well known *Blue Book* technical standards, officially changed its name to ITU-T (International Telecommunications Union-Telecommunications) in 1993. International standards published from this organization before 1993 carry the name CCITT, while those published after 1992 carry the name ITU-T. The appropriate term is used within this book depending on when the actual standard referred to was published.

Cell

An individual geographic area that is the topological component of a cellular or personal communications services system. The area is defined by the telecommunications coverage of the radio equipment located at the cell site. A *cell sector* is a geographic portion of a cell (typically a third) that is served by directional antennas dividing the coverage area of the original cell. The aggregate of the cell sectors that form an entire cell increases capacity by providing frequency reuse among the sectors. A cell site is the physical location of a cell's radio equipment and supporting systems. This term also refers to the equipment located at the cell site.

Protocol

A protocol is simply a set of rules or conventions governing the interactions of processes, applications and components in a communications system. IS-41 is a standard protocol governing the communications among cellular telecommunications network elements.

IS-41 parts, chapters, and sections

IS-41 is one of the few Telecommunications Industry Association (TIA) specifications that is divided into multiple *parts* (i.e., IS-41.1, IS-41.2, IS-41.3, IS-41.4, IS-41.5, and IS-41.6). These parts are sometimes referred to as *chapters*. Within each part of IS-41 is a Table of Contents where topics are divided into *sections* (e.g., Section 5.4.3 of Part IS-41.3).

Revision 0

The term *Revision 0* is an informal term used to refer to the initial version of a standard or specification published by the TIA. This revision number is not officially listed as an actual revision number by the TIA. Subsequent revisions to an initial specification are officially denoted as *Revision A, Revision B, Revision C*, etc. Revision 0 documents carry no revision designator in their title.

Michael D. Gallagher

Randall A. Snyder

Acknowledgments

We gratefully acknowledge the contributions of the following organizations:

- Synacom Technology, Inc., for allowing us to use company resources for the preparation of this book and for giving us the opportunity to be part of a very special company providing engineering products and services in a very exciting field.
- Telecommunications Industry Association, Inc. (TIA), for allowing us to use their published standards as source material for this book.
- *Cellular Networking Perspectives* Industry Newsletter, written by Mr. David Crowe, who continues to provide valuable insights into the wireless industry and provided us with many excellent ideas that added to the quality of this book.

We would also like to thank the following individuals:

- Mr. Kirk Carlson, without whom IS-41 would not be the high-quality standard that it is, and for reviewing our entire manuscript.
- Dr. Ming Lee and Mr. Vilnis Grencions who founded Synacom Technology and gave us the encouragement to write this book.
- Dr. Yick-Man Chan, who provided very useful comments on the manuscript of this book.
- Ms. Aparna Shute, who provided valuable engineering assistance.

Finally, we would like to thank the following group of individuals—the contributors to the IS-41 Revision C standard: Jean Alphonse, Phil Audino, Cheryl Blum, Jane Bissonnette, Alain Boudreau, Terri Brooks, Arzu Calis, Kirk Carlson, Lynn Carlson, Sharat Chander, Ben Chen, Jeff Crollick, David Crowe, Erkin Cubukcu, Amy Dwyer, Robert Ephraim, A. Gains Gardner, Thomas Ginter, Anthony Golka, Ed Hall, Huel Halliburton, Alejandro Holcman, Michel Houde, Chuck

Ishman, Terry Jacobson, Stephen Jones, Joe Juliano, Irfan Khan, Barry Kratz, Bill Krehl, Sunit Lohtia, Paxton J. Louis, John Marinho, Eileen McGrath-Hadwen, Charlene Meins, Robert Montgomery, David Munsinger, C. Peter Musgrove, Peter Oldfield, Gustavo Pavon, Gary Pellegrino, Tom Richter, Raul Rodriguez, Asim Roy, Todd Shingler, Jay Tang, Ed Tiedemann, Dennis Velte, Tim Verdonk, Terry Watts, John Willse, and Mike Youngberg. Our apologies to those contributors that we have missed, including the "behind the scenes" contributors who did not attend the TR45.2 meetings.

Mobile
Telecommunications
Networking with IS-41

Introduction to Mobile Telecommunications, Network Architecture, and Functions

1

Basics of Mobile Telecommunications

To begin to understand the IS-41 signaling protocol, it is necessary to understand some basics about mobile telecommunications. This chapter provides a general overview of the most important concepts in mobile and cellular telecommunications that apply to IS-41. For some readers, this will be a high-level review; for others, it may clarify some misconceptions and provide an overall understanding of cellular technology that will be useful in understanding IS-41.

What Are Mobile Telecommunications?

The concept of mobile telecommunications can be viewed from two perspectives: the mobile subscriber's and the mobile network's. From the subscriber's perspective, *mobile telecommunications* is a service that allows telephone calls to be made or received while the telephone equipment is moved from place to place or is in motion. From this perspective, the telephone handset (known as the *mobile station*) is wireless and affords the ability to be mobile.

From the network's perspective, *mobile telecommunications* is a service provided to end users. Mobile telecommunications, within the context of IS-41, is a service based on a set of functions internal to the network known as *mobility management*. Mobility management functions enable the network to maintain location and subscriber status information so that end users can make and receive calls while they move from place to place.

Origin of Advanced Mobile Phone System

Mobile telecommunications can be considered both a system and a service. The network equipment including antennas, radios, switches, databases, and all hardware and software within the network represents a mobile telecommunications system that provides mobile telecommunications service to subscribers.

The first mobile telecommunications system based on cellular technology was (and is) known as *AMPS*—the Advanced Mobile Phone System, a technology developed by Bell Laboratories in 1947. The term *cellular* refers to a network of small *cells,* or radio transceivers, each providing a limited range of radio coverage, that are linked by a computer-controlled switching system which manages subscriber mobility and interfaces to the fixed wire line telephone network. The technology is based on cellular frequency reuse (described below), providing a high-capacity system and allowing network access using low-power mobile stations (typically less than 6 W). The radio transceiver modulation methods used are based on analog frequency modulation (FM) signals similar to those used for commercial radio, but at a higher frequency range and lower bandwidth.

The first commercial cellular system in the United States became operational in Chicago in 1983. However, other countries around the world provided operational cellular systems several years earlier.

Today, cellular systems based on AMPS technology are implemented in well over 100 countries. It is interesting to note, however, that there is no single worldwide standard for the implementation of these cellular systems. The different systems deployed generally represent differing radio technologies, each based on the concepts of AMPS. The networking technologies that are used to link the cells are also quite different. These technologies are considered supplemental to the defining characteristics of AMPS (i.e., cells, frequency reuse, etc.). In fact, AMPS can operate within a variety of networking schemes.

Some Basic Cellular Concepts

Basic radio technology

Cellular radio technology allows a subscriber to send and receive telephone calls wherever compatible cellular radio coverage is provided. A cell is an individual radio coverage area controlled by a radio base station (BS) system. Individual calls within a single cell use different frequencies. These frequencies can be *reused* by other cells provided that there is no interference with the other cells. The frequency reuse pattern of the cells is dependent on the distance between the cells and the radio transmission power.

First-generation radio technologies (i.e., AMPS-based) use signals based on analog FM for speech transmission. Subsequent generations of radio technology for wireless systems include NAMPS (narrowband AMPS), which is also based on analog FM, and a variety of sophisticated digital technologies based on TDMA (time division multiple access) and CDMA (code division multiple access).

The cellular radio spectrum (range of allowable and available radio frequencies) used for these cellular system technologies is regulated by government agencies in different countries. In the United States, cellular service providers are categorized by one of two sets of noncontiguous 25-MHz radio-frequency bandwidths.

A-side and B-side carriers

The two sets of bandwidths that are licensed for cellular radio service are known as the *A side* and the *B side*. A-side carriers are cellular service providers that were originally termed the *nonwire line* licensees. These original licensees are companies that provide cellular service and are not associated with any local wire line telephone company.

B-side carriers are cellular service providers that were originally termed the *wire line* licensees. These licensees are companies that provide cellular service and are associated with the local wire line telephone company (i.e., the Regional Bell Operating Company, or RBOC) in the area where they provide cellular service.

The concept of A-side and B-side carriers was devised as part of the Modification of Final Judgement (MFJ) consent decree in 1982 that broke up the AT&T/Bell System monopoly in 1984. The AMPS technology originally developed by Bell Laboratories was given up to the seven RBOCs as part of the compromise to divest them from AT&T. The mandated provision to allow two cellular service provider licenses in a given geographic area was designed to provide competition between an independent cellular carrier and the cellular carrier owned by a wire line carrier.

Frequency reuse

Basic AMPS technology in the United States provides for each mobile station to occupy 60 kHz of bandwidth (30 kHz for transmission and 30 kHz for reception) within an entire radio-frequency (RF) allocation of 25 MHz for each of the two cellular carriers (A and B) in a given area (that is, 12.5 MHz for transmit and 12.5 MHz for receive for each carrier). This system permits each carrier access to 416 channels, 21 of which are set aside for control functions. The remaining 395 channels are available to support subscriber speech.

Cellular systems use a technique known as *frequency reuse.* A particular available channel frequency is transmitted from one base station at a power level that supports communications within a moderate cell radius around that base station (anywhere from a few hundred feet to about 50 mi!). Because this transmitted signal power is controlled to serve only a limited range, the same frequency can be transmitted simultaneously, or *reused,* by another base station, provided there is no interference between it and any other base station using that same frequency.

Figure 1.1 depicts a typical cellular frequency reuse model using a seven-cell pattern that provides uniform distances for channel reuse. In the model, each base station is considered to be located at the center of a hexagon with the hexagons (or cells) labeled A through G, representing seven channel sets. The frequencies used for the channel sets in the A cells are the same, as are the frequencies in the B cells, C cells, etc. There are many possible frequency reuse patterns. Since the total number of channels for AMPS is fixed, the selection of a reuse pattern and the cell size determine how many subscribers can be supported in a given service area. Available capacity, however, is much greater than the actual number of channels accessible in a given cell. This is due to the nature of end-user calling behavior; i.e., not every-

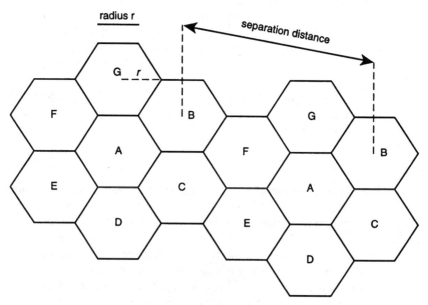

Figure 1.1   Seven-cell frequency reuse pattern.

one wants to talk at the same time. A cell can typically serve 10 to 20 times more subscribers than the total number of channels supported.

Digital radio

Digital radio technologies have been developed for use on the air (or radio) interface that can dramatically increase the number of subscribers supported on the range of frequencies used for mobile telecommunications systems (note that the analog NAMPS technology also provides an increase in the number of subscribers supported). The two basic types of digital technology are time division multiple access and code division multiple access. Many standards exist for the use of these basic technologies. TDMA is based on the use of time-interleaving of multiple signals to provide an apparently simultaneous transmission of those signals on a single radio frequency. CDMA is based on a technique known as *direct sequence spread spectrum* that digitally codes a base signal and employs different signal encoding patterns and frequency hopping to redistribute that base signal across a broad range of frequencies. Logical channels for each signal are created through the use of unique code sequences.

TDMA and CDMA can provide many benefits over AMPS-based systems. Examples are better voice quality, increased capacity, less noise and interference, and the ability to provide digital services such as data and messaging. The IS-41 networking protocol is designed to support versions of these newer-generation digital technologies as well as the original analog systems.

Handoff

Handoff encompasses a set of functions supported between a mobile station (MS) and the network that allows the MS to move from one cell to another (or one radio channel to another, within or between cells) while a call is in progress. The handoff function requires sophisticated coordination between the network and the MS to smoothly transfer the MS from one radio channel to another during a call. There are two types of handoff: *intra*system handoff and *inter*system handoff.

Intrasystem handoff (see Fig. 1.2) is a handoff between two cells or radio channels that subtend the same mobile switching center (MSC). In this case, no coordination is required between MSCs to support the movement of an MS between cells. Intersystem handoff (see Fig. 1.3) is a handoff between two cells subtending two different MSCs. This type of handoff requires specialized signaling between the two MSCs to coordinate the movement of the MS between the cells. Since the IS-41 protocol is concerned with intersystem operations, it provides the oper-

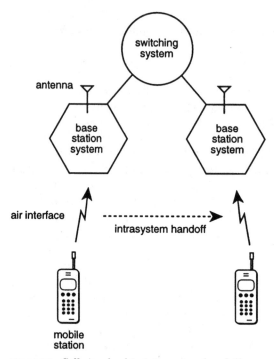

Figure 1.2 Cells involved in *intra*system handoff.

ations necessary to support intersystem handoff. Intrasystem handoff
is not within the scope of IS-41 and is handled via proprietary methods
at the MSC.

There are three strategies for performing a handoff:

1. MS-controlled

2. Network-controlled

3. MS-assisted

These strategies differ mainly in which side of the radio interface
determines when to hand off the MS to another channel. MS-controlled handoff is a technique where the MS itself continuously monitors the radio signal's strength and quality. When predefined criteria
are met, the MS checks the best-candidate cell for an available traffic
channel and requests the handoff to occur. Network-controlled handoff is a technique where the radio base station, MSC, or both monitor
the radio signal. When the signal's strength and quality deteriorate
below a predefined threshold, the network arranges for a handoff to
another channel. MS-assisted handoff (MAHO) is a variant of net-

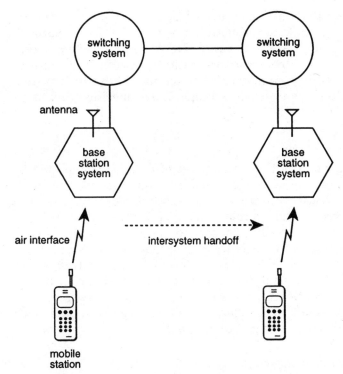

Figure 1.3   Cells involved in *inter*system handoff.

work-controlled handoff and is a technique where the network directs the MS to measure signals from surrounding cells and to report those measurements back to the network. The network then uses these measurements to determine where a handoff is required and to which channel. The IS-41 protocol supports both network-controlled and MS-assisted strategies for intersystem handoff.*

Network systems

In the context of mobile telecommunications systems, the network consists of entities not directly related to the radio interface between the mobile station and the radio base station. Network systems consist of switching functions, service logic functions, database functions, and all mobility management functions that enable subscribers to be

---

*Examples of MS-controlled handoff are in the Personal Access Communications System (PACS) and Digital European Cordless Telephone (DECT) protocols.

mobile. These are the functions provided by the MSCs, by the control and database systems (known as *location registers*), and by other logical functional entities in the network. Network intersystem operations are required to provide communications between these entities to enable the mobility management functions. These intersystem operations, including intersystem handoff, are specified and standardized by IS-41.

Mobility management

Mobility management is the primary set of functions supported by the network to enable subscriber mobility. Mobility management enables the network to keep track of the subscriber's status and location for the delivery of calls to that subscriber. It also enables the network to authorize a subscriber for service in a given cellular service area. The key component to mobility management is the control of the subscribers' *service profile*. The service profile is simply a database record in the network that contains information about each subscriber. This information includes temporary data, such as current location and status of a subscriber, as well as permanent data, such as the subscribed features (e.g., call waiting).

Basic network system architecture

Mobile telecommunications networks are primarily comprised of the following four basic network elements (see Fig. 1.4):

1. Radio systems
2. Switching systems
3. Data-based systems (i.e., location registers)
4. Operations, administration, and maintenance (OA&M) systems

Radio systems consist of low-power mobile stations and base station systems. The radio system provides the communications path between the mobile station and the cellular base station. The base station system includes the antennas, transceivers, and controller systems that provide radio access into the network.

Switching systems provide interfaces for subscriber traffic between the cellular network and other public switched networks and within the cellular network. The switching systems coordinate the establishment of calls to and from the mobile subscriber. These systems are directly responsible for managing transmission facilities, subscriber mobility, and call processing.

Data-based systems in the mobile telecommunications network are known as *location registers*. Location registers provide a database

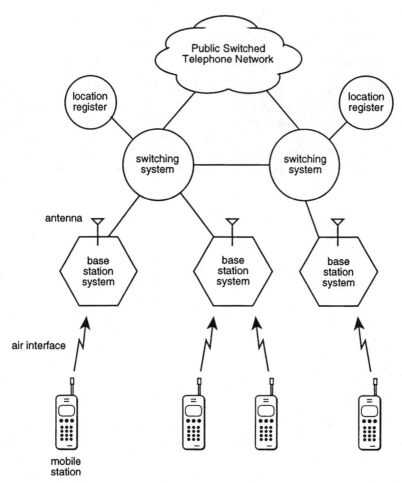

Figure 1.4 Basic mobile telecommunications system architecture.

(hence, the term *data-based*) along with service logic to actively control the mobile services provided to subscribers. The database provides information about subscribers to the network. This information includes the subscriber's identification, directory number (phone number), current location, subscribed features (such as call forwarding to voice mail), and call-routing information, as well as many other types of data.

Operations, administration, and maintenance are a set of functions that enable the service provider to monitor and control the network. The OA&M functions allow the service provider to perform the following general actions:

1. Observe and record operational characteristics of the network.

2. Modify and configure the network equipment and functions.

3. Identify and correct failures and defects within the network.

The IS-41 specification provides a standard protocol for the operations that enable subscriber mobility between MSC serving areas. IS-41 specifies the signaling communications that occur between MSCs, network location registers, and some specialized network nodes (such as short message service centers and authentication centers) to allow subscriber movement between networks based on the standard.

2

Mobile Telecommunications Standards

IS-41 is a technical standard for mobile network signaling. This chapter explains the importance of standards; it also provides an overview of the organizations that govern the development of IS-41 and how those organizations relate to others in the world that also define standards.

What Is a Standard?

Technical standards are evident in everyday life all around us. There are basically two types of technical standards: those that are *prescribed* and those that are *de facto*. An example of a prescribed standard is the design and function of electric appliance plugs and outlets in the United States. An example of a de facto standard is the placement of hot water faucets on the left side of a sink and cold water faucets on the right side. Standards serve two primary purposes: to make our lives easier and to save us money. Imagine the cost of installing many different kinds of electric outlets to suit all the different types of appliance plugs that could be developed, or the inconvenience of trying both water faucets to discover which is cold and which is hot. Standards are supposed to prevent these problems and provide a commonly accepted authority for the design and function of all types of equipment.

Standards for the design and function of telecommunications equipment are no different in goal and purpose from those for any other type of technical standard. Telecommunications standards are prescribed for the design and function of equipment as simple as a telephone keypad to the design and function of the complex and sophisticated computer equipment and services provided by a mobile telecommunications network.

In tangible terms, a telecommunications standard is simply a document that establishes engineering and technical requirements for processes, procedures, and methods that have been decreed by authority or adopted by consensus. The primary goal of the telecommunications standards process is to encourage the interconnectivity of telecommunications equipment and services by establishing and promoting technical recommendations in these areas. The telecommunications industry achieves this goal by creating and maintaining voluntary specifications that can optimize equipment compatibility. Standards are typically considered to be *recommendations*; i.e., they are prescribed as purely voluntary. However, business needs usually show that it may not be very lucrative to stray from standardized designs and functions.

IS-41 (Interim Standard 41) is the technical standard that prescribes the network model, functions, protocols, and services that provide mobile telecommunications network intersystem operations.

Scope of a Standard

Standards encompass many areas of engineering and technical requirements. They can be very brief and promote an existing engineering solution to a new problem (usually by reference), or they can be quite extensive (hundreds to thousands of pages) and establish the details for the processes and methods for developing an entire system. As stated before, the primary goal of telecommunications standards is to establish and promote technical recommendations that enable the interconnectivity of equipment and services. However, standards do not dictate an *implementation* or the methods for developing an individual solution to provide equipment or service. Thus, business competition among manufacturers is preserved while enabling interoperation of their equipment and services.

Telecommunications standards can cover many aspects of the design and function of equipment. They include network models used for developing network architectures; descriptions of functions as perceived by an end user and as provided within and between networks; descriptions of protocols or the methods of communication within and between functional and network entities; testing procedures for establishing equipment compatibility and standards compliance; and any procedures that assist in achieving the goals of the standardization process.

Who Makes and Uses
Telecommunications Standards?

There are many organizations concerned with the development and use of telecommunications standards. The standards-making process requires cooperation at three basic levels: between and among industrial concerns (i.e., cellular service providers and equipment manufacturers), between industrial concerns and government concerns (e.g., service providers and regulatory agencies), and between nations.

Cooperation among all the concerned parties is not always possible, hence the existence of multiple standards (try taking your hair dryer to another country and plugging it in). A good example of multiple telecommunications standards is the transmission carrier standard T-1 used in North America (operating at 1.544 Mbits/s) and standard E-1 used in Europe (operating at 2.048 Mbits/s). To accommodate many conflicting interests, international standards-making organizations concentrate on producing what are known as *base* standards. Base standards contain variants, national options, and alternative methods for implementation-dependent needs. Adopting these variants means that an *implementation* is compatible with the standards, but there is no guarantee that equipment based on different variants can work together.

Standards Groups, Trade Groups,
and User Groups

There are more than 200 organizations that prepare international standards. In fact, these organizations have developed about 15,000 technical standards. Ninety-six percent of all standards are developed by three international organizations: the International Telecommunications Union (ITU, formerly the CCITT), the International Organization for Standardization (ISO), and the International Electrotechnical Commission (IEC). Nearly half of these standards apply to telecommunications, information technology, and related fields.*

The ITU is a treaty organization of the United Nations whose activities include standardizing telecommunications and spectrum management, regulating radio telecommunications, and managing frequency assignments that have international significance. The ITU also plays a key role in the evolution of seamless global telecommuni-

*Source: TIA *Standards and Technology Annual Report* (*STAR*), TIA, Arlington, Virginia, 1994, 1995.

cations technology. The ITU membership consists of national delegations from more then 180 countries. The ISO is a voluntary, nongovernment organization mainly providing standards for information technology. This group develops standards to facilitate international trade in goods and services. ISO membership comprises primarily national standards-making bodies including the American National Standards Institute (ANSI). More than 100 nations contribute to the ISO. The ISO and ITU work closely together in areas of common interest. The IEC is also a voluntary, nongovernment organization primarily working in the area of electrical and electronic engineering. The IEC is a sister organization to the ISO, and its membership consists of about 50 contributing nations.

Functional standards are adapted from international base standards and contain only a limited subset of permissible variants. Adaptation from international base standards to national functional standards is provoked by the following types of organizations:

- Regional or national standards groups
- Trade groups
- User groups

Examples of regional or national standards groups are ANSI and the National Institute of Standards and Technology (NIST). Examples of trade groups are the Cellular Telecommunications Industry Association (CTIA) and the Personal Communications Industry Association (PCIA, formerly Telocator). Examples of user groups are the Institute of Electrical and Electronic Engineers (IEEE) and the North American ISDN Users Forum (NIUF). These organizations usually provide well-defined requirements as input to the functional standards.

As part of the standards-making process, agreed-upon test specifications and methods are developed to ensure that equipment designed to the different variants (permitted within the functional standards) will work together. Independent test organizations can perform conformance tests and certify that telecommunications equipment and products comply with the standards. Figure 2.1 depicts a flow diagram of the overall standards-making process and the relationship among the contributors to standards development.

American National Standards and the TIA

There are many national organizations providing standards for North America. In this context, North America refers primarily to the United States and Canada. At the forefront of North American telecommunications standardization is ANSI. As a U.S. national standards-making

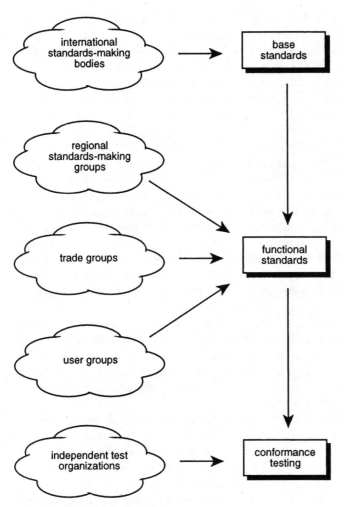

Figure 2.1 The standards-making process.

body, ANSI is responsible for accrediting other U.S. standards-making bodies. Among these are the Alliance for Telecommunications Industry Solutions (ATIS), the Electronic Industries Association (EIA), and the Telecommunications Industry Association (TIA). The ATIS was formed during the impending Bell System divestiture to develop open industry standards to replace the closed, de facto Bell System specifications. An interesting note is that although most of these Bell System specifications are still prevalent throughout the telecommunications industry (currently sponsored by Bell Communications Research or

Bellcore), they are not considered prescribed standards (although they have become de facto standards in many cases). This is due to the fact that these specifications were developed in a closed forum, considering the business concerns of only the limited number of companies making up the Bell System. Even in the postdivestiture era, these specifications remain primarily focused on the interests of the Regional Bell Operating Companies (RBOCs) and are not considered to be national standards.

The EIA was initially formed as a radio manufacturers' group and evolved to cover all areas of electronics information and communications technology. The TIA was formed in 1988 from the combination of the Information and Telecommunications Technology Group of the EIA and the U.S. Telecommunications Suppliers Association. The charter of the TIA is the formation of new land mobile telecommunications standards, although it does develop standards for technologies as diverse as fiber optics and satellite communications. The TIA is associated with the EIA and is an ANSI-accredited standards-making-body. The TIA has developed most of the currently used standards for cellular telecommunications in the United States, including IS-41.

The TIA primarily develops what are known as *interim standards,* hence the title *IS*-41. These standards are considered interim because they have a limited life—originally 5 years and now 3 years. All interim standards developed by the TIA have the potential to eventually become full ANSI national standards, if they are agreed upon by the larger membership of ANSI.

Comprising the TIA are many committees that develop wireless and other telecommunications standards. The committees concerned with developing wireless telecommunications standards are designated as *TR* committees; the TR designation is an artifact of the term *transmission* which was the original technology being standardized in the early days of the EIA. There are currently nine separate TR committees within the TIA, as listed in Table 2.1.

Standards projects begin with a contribution from an individual requesting the creation of a new standard in a specific technical area. A *contribution* is simply a written discourse that can be a request or proposition that an individual distributes at a TIA meeting. The scope and content of a contribution are very open, and contributions can contain any type of information. The contribution process is the primary method for furthering the development of standards in the TIA.

Aside from interim standards (ISs), the TIA publishes other types of specifications. Among these are Telecommunications Systems Bulletins (TSBs). TSBs are not considered standards and do not carry the weight of standards; however, they do provide information that concerns existing standards or other issues of great importance to the industry.

TABLE 2.1   TIA TR Standards Committees

TIA TR committees	TIA TR committee name
TR-8	Mobile and Personal Private Radio Standards
TR-14	Point-to-Point Communications Systems
TR-29	Facsimile Systems and Equipment (originally wire line only)
TR-30	Data Transmission Systems and Equipment
TR-32	Personal Radio Equipment
TR-34	Satellite Equipment and Systems (newest)
TR-41	User Premises Telephone Equipment Requirements (originally wire line only)
TR-45	Mobile and Personal Communications Public Standards
TR-46	Mobile and Personal Communications 1800 MHz

After an interim standard has been published by the TIA, there is a 3-year period within which one of three actions must be taken on the standard. An interim standard must be

1. Reaffirmed

2. Revised

3. Rescinded

Reaffirmation simply consists of a review that is intended to result in a decision that the technical content of an interim standard is still valid and does not require changes. A revision incorporates additional language into an interim standard that modifies its technical content or meaning. A rescission is the result of a review that determines that the technical content of an interim standard is no longer of value. Both the revision and rescission processes typically require the development of an official Standards Proposal in the same manner as required for a new interim standard. The TIA interim standard revision process is evident with IS-41. An initial version was published, followed by Revision A, Revision B, and Revision C.

TIA Committee TR-45

TIA Committee TR-45 (Mobile and Personal Communications Public Standards) is the committee that maintains the IS-41 and related mobile telecommunications standards. Committee TR-45 was formed in 1983 and has developed standards concerning interoperability, performance, and network operations for all areas of public mobile telecommunications in the 800-MHz region of radio frequencies. TR-

45 has since evolved to address mobile telecommunications in the 1800-MHz frequency range for personal communications services (PCS) as well. TR-45 is comprised of subcommittees. Each subcommittee is responsible for a different area of mobile telecommunications technology. Table 2.2 shows a listing of these subcommittees.

TIA Subcommittee TR-45.2 is the standards-formulating body that develops and maintains IS-41. Subcommittee TR-45.2 is composed of seven working groups (WGs). Each WG is responsible for a different area of cellular intersystem operations. Table 2.3 shows a listing of these working groups.

WG1 (stage 1 feature and service development) is chartered with developing standards that describe features and services from the subscriber's perspective. These standards encompass the basic description of a feature or service as well as the end-user interface to the network for using a particular feature or service. WG2 (stage 2 feature and service development) is chartered with developing standards that describe features and services from the network's perspective. These standards describe the messaging between nodes in the

TABLE 2.2   TIA TR-45 Subcommittees

TIA TR-45 subcommittees	TR-45 subcommittee name
TR-45.1	Analog Cellular Technology
TR-45.2	Cellular Intersystem Operations
TR-45.3	Digital Cellular Technology (TDMA)
TR-45.4	Microcellular/PCS Technology
TR-45.5	Wideband Spread Spectrum Digital Technology (CDMA)

TABLE 2.3   TIA TR-45.2 Working Groups

TIA TR-45.2 working groups	TR-45.2 working group responsibility
TR-45.2.1	Stage 1 Feature and Service Development
TR-45.2.2	Stage 2 Feature and Service Development
TR-45.2.3	Stage 3 Feature and Service Development
TR-45.2.4	Message Accounting
TR-45.2.5	Deactivated
TR-45.2.6	International Applications
TR-45.2.7	Interfaces to Other Telecommunications Networks

mobile telecommunications network. This standardized messaging is the basic mechanism that enables interoperation between equipment developed by different manufacturers. WG3 (stage 3 feature and service development) is chartered with developing standards that describe and define the details of the messages and parameters that support a feature or service. Procedures for handling the messages and parameters are also standardized within this working group. WG4 (message accounting) is chartered with developing standards that enable the mobile telecommunications network to disseminate and transport the details of a call that are used for billing and accounting purposes. WG6 (international applications) is chartered with developing standards that define and support radio and networking applications outside of North America. WG7 (interfaces to other telecommunications networks) is chartered with developing standards that define and describe the interfaces between the mobile telecommunications network and other types of networks. For example, interfaces between mobile telecommunications network and the regular wire line network (i.e., the public switched telephone network, or PSTN) have been developed within this working group.

The IS-41 standard is primarily developed within working groups 2 and 3. However, since the standards developed in each WG are usually based upon or affect the standards developed in other WGs and subcommittees, portions of IS-41 have been addressed throughout TIA Committee TR-45. Note that the responsibilities of working groups can and do change over time as existing standards evolve and as the need for new standards arises.

3

Mobile
Telecommunications
Network Signaling

The IS-41 specification standardizes computer communications that provide signaling within and between mobile telecommunications networks. Signaling is the primary mechanism used to transfer control information through the network. This chapter provides an overview of mobile telecommunications networks and describes the different types of signaling used by the network to provide service to mobile subscribers.

What Is a Mobile
Telecommunications Network?

The mobile telecommunications network can be viewed from two perspectives:

1. A logical view, where the network is represented by a generic functional model

2. A physical view, where the network is represented by the actual switches, specialized computers, and other equipment that comprise the nodes of the network

The logical view of the network is a method of describing network topology to simplify discourse and study. The network topology is simply depicted as functional entities that are interconnected by branches. Each functional entity represents one or more logical network functions while the branches represent a relationship between those functions. In the logical view, there is no prescribed mapping of the functions and relationships to a physical implementation.

The physical view of the network is quite different. From a physical perspective, the network is simply an organization of computer-based systems that are capable of intersystem communications. These communications are accomplished by the interconnection of circuits between the specialized computing platforms.

So, what exactly is a mobile telecommunications network? It is a conglomerate of *physical* equipment and communications facilities consisting of electronic computer-controlled switches, location registers, and other processing centers along with transmission circuits. The circuits support efficient and intelligent communications among the pieces of equipment. The connectivity among all these systems by communications transmission circuits provides the mobile telecommunications service that is accessed by subscribers. These circuits represent the network branches between the equipment that support the transmission of signals and user information.

Signals convey data that provide information or instructions to control the network. These control signals can be considered the primary mechanism that communicates the *intelligence* of the network. User information is the actual information content that is transferred through the mobile network. User information content is distinguished from signal information because it is information originating from a source end user that is ultimately delivered to a destination end user. The transmission of this user information through the mobile telecommunications network comprises the telecommunications service that is accessed by a network subscriber. The information content can be voice, data, facsimiles, or any other type of information that is conveyed by the network as a service to subscribers.

Overview of Signaling

Signaling is simply the process of sending signals or signaling information. It is the transfer of special information to control communication. Signaling consists of a protocol or a specialized set of rules that govern the communications of a system. The signaling protocol is defined by three criteria:

1. Syntax, i.e., how to construct the information

2. Semantics, i.e., what the information means

3. Procedures, i.e., what to do with the information

The protocol enables the effective use of the control information (i.e., signals) to provide meaningful communications within a network. Signaling is the mechanism used to operate, control, and manage the mobile telecommunications network. A good example of a signal is the common ringing alert signal that we are all familiar with when someone is calling a telephone. It is distinguished from the user information provided by the telephone network (i.e., voice) since it provides an indication that a party is calling, but it is not the information that is meant to be conveyed by the caller.

Signaling and signaling protocols have become very complex, especially when used to govern telecommunications and the sophisticated services provided today. These advanced signaling protocols provide for the transfer of information among network nodes that enables what is known as *intelligent networking*. Intelligent networking is a method for providing and interpreting information within a *distributed network*. A distributed network is structured such that the network resources are distributed throughout the geographic area being served by the network. The network is considered to be intelligent if the service logic and functionality can occur at the distributed nodes in the network. The mobile telecommunications network is distributed and intelligent. Because intelligent networks require such sophisticated signaling, the signaling means has evolved from electric pulses and tones to very complex messaging protocols.

Within the context of IS-41, signaling information consists of messages that contain parameters which support the function of mobility management throughout the network. This mobility management function is key to enabling subscriber mobility in wireless networks.

The transfer of *user information*—or the traffic that is conveyed from end to end between network end users—is controlled by the network signaling protocols. User information is a portion of what is commonly known as *bearer information*. Bearer information usually contains other information besides the user data. This information generally consists of message alignment, synchronizing, or error correction sequences. This added information is similar to signaling information, but is not considered signaling information content since it applies to each message individually and not to the functioning of the network as a whole.

Network Signaling and Access Signaling

Network signaling is used between network nodes to operate, manage, and control the network to support certain types of functionality (e.g., mobility, voice traffic). Network signaling is distinct from another type of signaling known as *access signaling*. Access signaling is used to manage communications between a network end user and an *access point* of the network. The distinction between network and access signaling is one of perspective and function, since these two types of signaling relate to different portions of the network. A network end user is sometimes considered part of the network—a confusing notion—but the signaling required between an end user and the access point of the network is quite different from that required between network nodes within the network.

A good analogy to help distinguish access signaling from network signaling is the national highway system. On-ramps can be considered the mechanism supporting access onto the highway network. The highways and interchanges along the way can be considered "circuits" and "switches," respectively. On-ramps have distinctly different characteristics from those of the highways themselves. Highways can lead to many different places and points. On-ramps simply provide a direct point-to-point connection between the entrance of the on-ramp and the entrance of the highway. This idea is also true of access signaling.

Network signaling is also distinguished from access signaling because of a characteristic of network signaling known as *adaptive routing*. Adaptive routing allows signaling messages to take alternate routes between points in the network in cases of failure or congestion. In other words, the traffic can adapt to new paths in the network, if for some reason the primary path has become inaccessible. Access signaling generally has no such capability (however, there are exceptions). Just as with the highway analogy, if an on-ramp is inaccessible, traffic cannot enter the network via that access point.

Note that in Figs. 3.1 and 3.2 the communication between the radio base station and the mobile telecommunications network is designated as access signaling or network signaling. Generally, this connection is based on access signaling, since the radio base station can be considered the on-ramp onto the network of mobile switching centers (MSCs). However, some mobile telecommunications networks treat the base stations as part of the network. Although adaptive routing is not always employed, the same signaling message transport protocol can be used between the base stations and the MSC as between MSCs.

IS-41 is a network signaling protocol designed to provide mobility management signaling throughout the mobile telecommunications network. IS-41 signaling is provided among MSCs, location registers, and some specialized processing centers to support subscriber mobility within a single cellular service provider network and between many different cellular service provider networks.

In-Band Signaling and Out-of-Band Signaling

There are many different types of network and access signaling protocols. However, both network and access signaling protocols can be categorized into one of the following two signaling types: in-band signaling and out-of-band signaling.

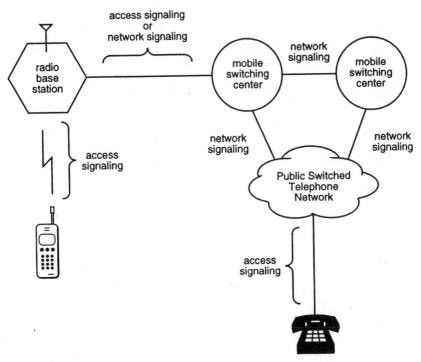

Figure 3.1 Access signaling and network signaling between network entities.

In-band signaling is a type of signaling in which the analog frequencies or digital time slots that carry the signals and signaling messages are within the bandwidth of the channel which carries the user information. In-band signaling uses a portion of the frequency band or bit stream that would otherwise be allocated for user information. An example of in-band signaling is dual-tone multifrequency (DTMF) access signaling—commonly known as Touchtone signaling—used to transmit dialed digits from a telephone to the network. The audio tones generated to inform the network of the party being called are transmitted as audio signals on the same channel as voice is transmitted during a conversation. These audio tones are generated at very precise frequencies. They consist of two specific voice frequencies that are combined to form the tone that is generated. Since they are very precise, it is difficult to exactly duplicate the tone with the human voice, minimizing the potential for inadvertent signaling errors.

Out-of-band signaling is a type of signaling in which the analog frequencies or digital time slots that carry the signals and signaling messages are outside the bandwidth of the channel which carries the

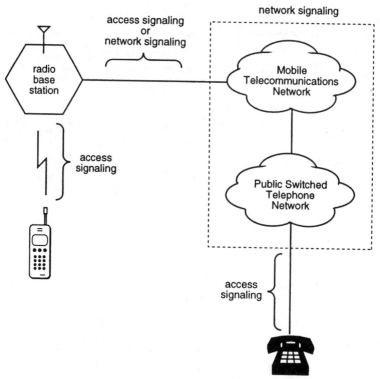

Figure 3.2 Access signaling and network signaling between network.

user information. Out-of-band signaling uses a frequency band or bit stream that is separate from the channels allocated for user information. Out-of-band signaling traffic is logically separate from the user traffic (i.e., on a different logical channel) within the same transmission line, or it may also be physically separate from the user traffic (i.e., on a completely different transmission line). An example of out-of-band signaling is ISDN, where the access signaling traffic is transmitted across the D-channel, that is separate from the B-channels, which are used to transmit user information.

Out-of-band signaling is sometimes conveyed over a single digital channel that is separate from the user information channels. This single channel can carry signaling information for many bearer, or information, channels. That is, the signaling channel is used in *common* for many information channels. Because of this characteristic, the channel is called a *common channel.* Out-of-band signaling protocols that use a common digital channel are known as *common-channel signaling protocols.* Common-channel signaling allows signaling traffic to be consolidated and sent across a separate transmission link from the user traffic. The primary advantage of sending this consolidated signaling traffic over separate transmission links is to prevent access by a user (i.e., fraudulent access). Another advantage is the high-speed signaling transmission enabled by links that can be connected over a physically separate network, distinct from the network that carries the user traffic (e.g., voice traffic). The mechanism of out-of-band common-channel signaling allows for high-performance distributed service logic across the telecommunications network as a whole, e.g., faster call setup times and the ability to efficiently support enhanced features such as 800-number service.

Signaling System No. 7

Signaling System No. 7 (SS#7) is a suite of common channel network signaling protocols defined by the ITU-T. ANSI has defined a national variant of this protocol, SS7 (without the # symbol), specifically for North American telecommunications networks. SS7 is the primary building block on which enhanced intelligent telecommunications applications are built. These applications include call control and transaction capabilities that support database access as well as a variety of intelligent network functions and mobile telecommunications services. SS7 is designed to operate over a separate network distinct from the network that carries voice or user data (i.e., it is an out-of-band common-channel signaling protocol). The scope of SS7 is extremely large since it covers all aspects of control signaling for complex digital telecommunications networks.

SS7 is based upon packet switching technology. SS7 packets (or messages) are used to convey signaling information from an originating point to a destination point through multiple switching nodes in the network. The SS7 messages contain addressing and control information that is used to select the routing of signaling information through the network, perform management functions (to provide high reliability), establish and maintain calls, and invoke transaction-based mechanisms in support of sophisticated applications. A transaction is simply a controlled exchange of information based on a query or command and the response to that query or command. The IS-41 signaling protocol provides operations for this type of transaction-based application using the SS7 protocol. The SS7 transaction operation mechanisms are used to query databases and invoke functions at remote points throughout the network. These mechanisms also support the delivery of status information and results to those database queries and invoked functions.

Since SS7 has so many advantages and provides a standard transaction-based protocol mechanism, it is ideal for performing the operations required to provide the mobility management function in the mobile telecommunications network. IS-41 is a signaling protocol that provides transaction-based operations to support subscriber mobility in the mobile telecommunications network.

Note that IS-41 signaling messages can be transported by communications protocols other than SS7 (i.e., using the CCITT X.25 packet switching protocol). However, SS7 is the more powerful and robust means for conveying the IS-41 signaling information through the mobile telecommunications network and is highly preferred over other packet switching protocols. More information on the use of the SS7 protocol is provided in Parts 2 and 3 of this book.

Overview of Intersystem Operations

The IS-41 signaling protocol provides *intersystem operations* in support of subscriber mobility management. IS-41 intersystem operations are operations that are performed between cellular systems that enable subscriber mobility in the following ways:

1. Subscribers can move between systems while a call is in progress.
2. Subscribers can originate calls while *roaming* (i.e., using their mobile stations in a system other than the home system where the cellular subscription was established).
3. Subscribers can receive calls while roaming.
4. Subscribers can activate and use supplementary call features while roaming (e.g., call forwarding).

In the context of IS-41 intersystem operations, a cellular system is defined as a single mobile switching center along with its associated location registers, radio base stations, and processing centers. The term *intersystem* refers to subscriber movement between MSCs (or, more correctly, MSC serving areas), not between radio base stations served by a single MSC. The MSCs involved in the intersystem operations can belong to a single service provider network or to different networks.

An intersystem operation consists of two elements:

1. Signaling messages (comprised of syntax and semantics)
2. Functional procedures

Intersystem signaling messages convey system and subscriber information between network nodes within a single cellular service provider network or between multiple cellular service provider networks. Functional procedures are the computer processes that are invoked at the network nodes when signaling messages are sent and received. The IS-41 intersystem operations support the following three basic mobility functions:

1. Intersystem handoff

2. Automatic roaming

3. Intersystem operations, administration, and maintenance (OA&M)

Intersystem handoff is the set of functions that enables subscribers to move between cellular systems while a call is in progress. It is a more sophisticated procedure than *intra*system handoff, which is a handoff between radio base stations subtending a single MSC. Intrasystem handoff can be completely controlled (in a proprietary manner) within a single cellular system, and hence, there is no need for intersystem operations.

Automatic roaming is the set of functions that enables subscribers to originate calls, receive calls, and use supplementary call features *seamlessly* while roaming. The term *seamless* refers to the ability of cellular systems to provide these functions *transparently* to subscribers; i.e., subscribers do not need to take any special actions to use a mobile phone as they roam from system to system. The only indication provided to subscribers that they are roaming is usually the "ROAM" indicator displayed on most mobile stations.

Intersystem OA&M is the set of functions that provide trunk maintenance between MSCs. Trunk maintenance is a set of procedures required by telecommunications switches to manage the transmission circuits between switches that are exclusively used for intersystem handoff. This management consists of blocking, unblocking, resetting, and testing of those transmission circuits. Note that trunks between the PSTN and a cellular network use other trunk maintenance procedures.

Origin of the TIA IS-41 Solution

Standardized operations are necessary for the functioning of cellular systems, since subscriber mobility is supported between different cellular service provider networks that use equipment developed by different manufacturers. TIA Committee TR-45 was established in 1983 to develop the cellular technology standards that continue to evolve. The initial version of the TIA IS-41 specification (IS-41 Revision 0), entitled *Cellular Radiotelecommunications Intersystem Operations,* was published as an interim standard by TIA Subcommittee TR-45.2 in the beginning of 1988. Contributors to this standard included cellular service provider companies and network equipment manufacturing companies. The relationship between these groups of companies is a simple one; service provider companies are customers of the equipment manufacturing companies from whom they purchase the equipment to deploy in their networks.

Without a standardized solution to intersystem operations, it would be difficult for cellular service providers to purchase equipment from different manufacturers and to directly provide subscriber mobility between cellular systems. The IS-41 specification solved this problem and continues to evolve. There have been three subsequent revisions to the initial IS-41 standard (Revisions A, B, and C), and a fourth is in development at the time of this writing. Each subsequent revision provides additional information to the previous version of the standard. Revisions are necessary for the following general reasons:

1. To add new subscriber features to the standardized set

2. To add functionality that supports new network requirements (e.g., digital radio interfaces)

3. To fix errors found during the implementation of the standard

4. To clarify text that was found to be open to many interpretations

5. To remove functionality that was found to be unnecessary

Since the initial publication of IS-41 and the deployment of networks based upon the specification, the term *IS-41 network* was coined to describe cellular networks based on the standard. This term is very generic and describes any network that uses IS-41 intersystem operations to provide mobility to subscribers. However, these networks also use other protocols, both standard and proprietary, to provide other functions that are not within the scope of IS-41. Examples include OA&M functions other than trunk maintenance and value-added features offered to subscribers that have not been addressed by IS-41 (i.e., intersystem support of proprietary features) or do not require intersystem operations.

4

Mobile Telecommunications Network Reference Models

Network reference models are used to assist engineers in standardizing network functions and interfaces between functions. The model is an integral tool that can provide a representation of the physical network nodes, logical network functions, or both. In this chapter we define reference models, explain their use in the standards process, and describe the details of the IS-41 network reference model.

Purpose and Description of a Network Reference Model

A network reference model is simply a diagram that depicts the entities of a network and the interfaces between those entities. The model encompasses the definitions of the entities and the interfaces between them and depicts a graphical representation of the mobile telecommunications system as a whole. The network entities can represent physical network nodes that contain one or more functions, or they can represent logical network functions only. The model is used to facilitate the definition and description of functions and protocols that can be standardized in the network. The model itself is usually not meant to depict a physical network implementation, rather only the basic interfaces between the minimum required functions or network nodes for the purpose of standardizing network services. A network reference model is used as the basis for a variety of network implementations, not as a description of a true physical network plan.

Physical Models versus Logical Models

A network reference model diagram consists of a representation of network entities connected by interfaces. Some models depict all network entities by a single shape (such as a square), while others depict them as different shapes with each shape representative of the kind of function provided by that network entity. An example of the latter type is the model generally used for North American SS7 networks, where circles represent end signaling points (SPs), squares with a diagonal through them represent signaling transfer points (STPs), and triangles represent service control points (SCPs). Each of these network entities provides different functions, and the depiction is meant to show external interfaces between physically distinct network nodes containing those functions. Figure 4.1 depicts the basic North American SS7 network reference model.

Many network reference models are logical models; i.e., the network entities are representative of *logical functional entities* that may be implemented as physically separate network nodes or as functions combined within the same network node. The terms *logical* and *physical* can be confusing when we begin to discuss functionality and net-

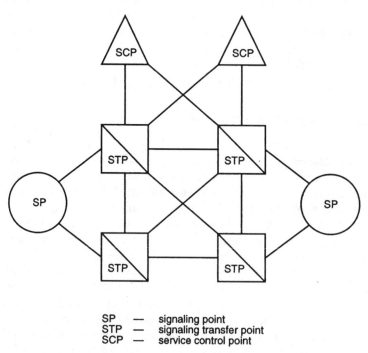

SP — signaling point
STP — signaling transfer point
SCP — service control point

Figure 4.1 Example of a physical network reference model (basic SS7 network model).

work reference models. A logical network reference model is based on logical functionality that is implementation-independent. Many physical implementations can be derived from a logical model. The logical functional entities within the model can be considered abstractions; implementations of those functional entities are left to system developers. Since a functional entity can represent one or more logical functions, the physical realization of the entity is an issue of network implementation dependency. The relationship between logical functional entities and physical network nodes can be one-to-one or many-to-one. Figure 4.2 shows the relationship of a logical view of the network to a physical view. The figure shows one possible one-to-one mapping of logical functional entities to physical network entities.

Logical models are used to represent *implementation-independent* functions and interfaces; i.e., they usually depict functionality independent of the physical implementation of that function. The interfaces between the functions can be external or internal to the functional entity. Those that are implemented as internal do not necessarily require

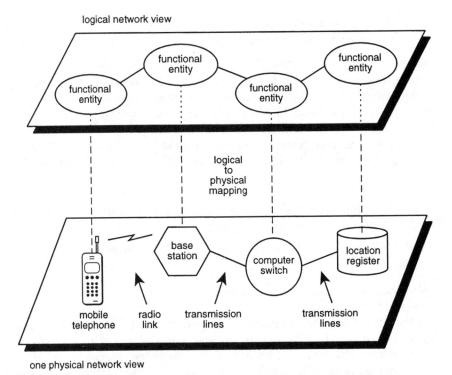

Figure 4.2 Logical and physical views of a network. Note that the logical view is mapped to only one of many possible physical views.

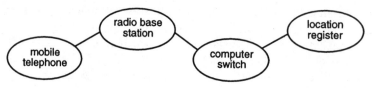

Figure 4.3   Example of a logical model.

standardization since internal interfaces usually do not provide connections between network equipment from different manufacturers.

The depiction of the functional entities in the logical view of the network (Fig. 4.3) is an example of a logical model. Based on this model, there is no reason why a network location register could not be implemented as part of the computer switch. And, in fact, the point of this depiction is to explicitly allow that network implementation, if desired. Although the model is logical, for practical reasons some functional entities are truly representative of separate physical devices, such as the mobile telephone.

Network Reference Models and the Three-Stage Specification Process

Many telecommunications services and protocols are specified via a three-stage process. The three-stage process was originally used to specify international Integrated Services Digital Network (ISDN) services, features, and capabilities, but is applicable to any telecommunications service (refer to *CCITT Recommendation I.130*). It is a generic method used to characterize services and facilitates a top-down engineering approach for designing them. The three stages are as follows:

1. Stage 1 describes the service from an end user's perspective.

2. Stage 2 describes the information flow between network entities (i.e., interfaces) to support the stage 1 service.

3. Stage 3 describes the protocol application of the stage 2 information flow.

(See Chap. 7 for a more detailed description of the three-stage specification process.)

The mobile telecommunications application services standardized by IS-41 are specified using this method, and there is a close relationship between this process and the use of a network reference model.

To provide a specific mobile telecommunications service to a subscriber, the service needs to be described in terms of what is provided to the subscriber and the actions, if any, that the subscriber needs to take. This is specified in the stage 1 description of the service. The network reference model is used as a tool to help specify the stage 2 description of the service. Signaling information, user information, or both are transferred between network entities to provide the service to a subscriber. The service can be as basic as delivering a mobile-terminated call to the subscriber, or more involved, as in the call waiting or conference calling features. The information transferred to provide a service can be derived from a database (i.e., location register), from a processing center, through switched connections, by a radio system, or by any combination of these. The information can traverse these network entities within a single cellular service provider network or between different cellular service provider networks.

The network reference model is used to define the interface between two network entities that can be standardized to support a specific service. The information flow defined across a given interface for a specific service is part of the stage 2 description for that service. The stage 2 service description is sometimes referred to as the *network perspective* of a service, since it describes the service in terms of what is provided to each network entity and the actions, if any, that those network entities need to take to provide the service. Without a network reference model, it would be difficult to establish a common point of reference for information transfer to provide a service supported by equipment developed by different manufacturers.

There are two basic philosophies for the interaction between a network reference model and the stage 2 description of a service:

1. The interface between network entities can be depicted first on the model to show the eventual need to provide a standardized stage 2 description for the interface.

2. The interface between network entities is added to the model only after the stage 2 description of the interface is specified, thereby justifying the presence of the interface on the model.

The IS-41 specification has been developed by using both of these philosophies. Early in the development of IS-41, interfaces were depicted on the model to show the eventual need to provide a standardized stage 2 description for them. As the IS-41 standard evolved, it was more appropriate to only add an interface to the model after it was justified by a newly developed stage 2 description. The latter philosophy proved to be more appropriate. It is usually not very desirable to modify an existing network reference model since existing implementations are based upon earlier versions of the model. The problem with depicting functional entities and interfaces on the model before they are standardized is that sometimes the anticipated need for them never develops. Many network reference models contain extraneous network entities and interfaces that have not been standardized. They are simply left over as artifacts of the anticipated need.

Elements of a Mobile Telecommunications Network Reference Model

Regardless of whether a mobile telecommunications network reference model is logical or physical, it consists of the following five basic network entities:

1. Radio systems
2. Switching systems
3. Location registers
4. Processing centers
5. Representations of external networks

These five network entities are principal elements of all mobile telecommunications networks. The network models also depict the interfaces, or *interface reference points,* between the network entities.

Radio systems

Radio systems consist of the following three separate subsystems:

1. Antenna systems
2. Radio transceivers
3. Radio transceiver controllers

Antenna systems convert electric signals from a radio transmitter to electromagnetic waves that comprise the radio transmission signals sent to mobile stations. Conversely, antenna systems convert the electromagnetic radio waves to electric signals that comprise the radio signals received from mobile stations. Antenna systems also manage radiated power to minimize interference.

Radio transceivers (sometimes called *base transceiver systems,* or BTSs) consist of a combination of simplex radio transmitting and receiving equipment that employs common components for both transmitting and receiving. This equipment is often referred to as simply the *radio.*

Radio transceiver controllers (sometimes called *base station controllers,* or BSCs) are equipment that control multiple radio transceivers. The controllers multiplex electric signals from many radio channels into transmission signals that are sent to the mobile net-

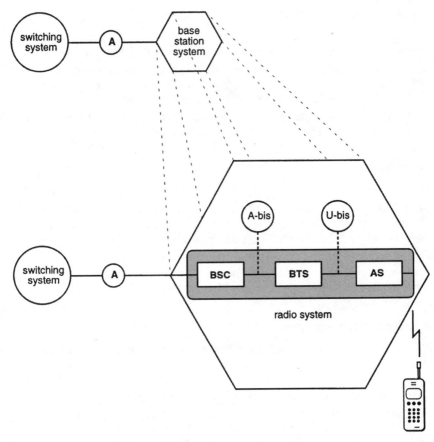

Figure 4.4 Example of a radio system reference model. The A interface is shown between the switching system and the radio system. Within the radio system, the A-bis interface is between the BSC and BTS and the U-bis interface is between the BTS and the AS. Note that communications across these interfaces fall outside the scope of IS-41.

work. The controllers also send network signals to the appropriate radio transceivers for transmission to the mobile stations. Figure 4.4 shows the relationship of the entities that comprise the radio system.

Switching systems

Switching systems provide the function of transferring transmissions from one circuit to another in the network. The switching function controls the routing of signaling and user information to specific nodes in the mobile network. The switching systems consist of the transmission facilities and computing platforms that control the switch circuits to connect calls between users.

Location registers

Location registers are data-based systems that control mobile subscriber services and contain the records and stored information related to mobile subscribers. These location registers are queried by other network functional entities to obtain the current status, location, and other information to support calls to and from mobile users. Location registers may also contain network address translation information to assist in the routing of calls to the appropriate network destination.

Processing centers

Processing centers are peripheral network computing platforms that provide services to enhance the capabilities of the network. An example of a processing center is an authentication center (AC) which uses complex algorithms to authenticate the identity of mobile subscribers. Other examples are voice-announcement systems, user messaging systems, and interworking functions that convert circuit-switched data to packet-switched data and vice versa.

Representations of external networks

External networks are integral elements in mobile telecommunications network models. They represent interconnections between the mobile network and the PSTN or other networks. Examples of these other networks are data networks, local-area networks, the ISDN, and other dissimilar mobile networks. These representations are important since mobile networks usually need to interoperate with other networks to complete calls (e.g., cellular to wire line calls).

Interface reference points

Interfaces that are depicted on network reference models are known as *interface reference points*. These reference points represent the point of connection between two adjacent physical or logical network entities. This point of connection is defined by functional and signaling characteristics and may define the operational responsibility of the interconnected network entities.

IS-41 Network Reference Model

The IS-41 network reference model is a logical model consisting of logical functional entities and interface reference points. This particular model is not meant to imply a physical implementation; however, as a practical matter, some functional entities are truly representa-

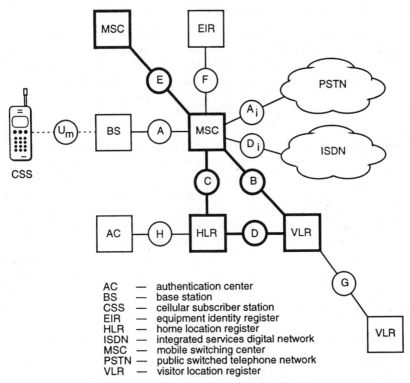

Figure 4.5 The original IS-41 network reference model (from IS-41-A and IS-41-B). Note that the CSS is depicted as mobile telephone equipment and that the PSTN and ISDN are depicted as network clouds. This is to distinguish them from the other logical functional entities that, from a practical perspective, do not imply physically separate equipment. The interfaces in **bold** represent the interfaces that are standardized in IS-41-A and IS-41-B. (*Reproduced under written permission of the Telecommunications Industry Association.*)

tive of separate physical devices. An example is the mobile station (MS), also known as the *cellular subscriber station* (CSS).

Figure 4.5 shows the first incarnation of the IS-41 network reference model. The model depicts the functional entities and interface reference points between those entities that may comprise an IS-41-based cellular network. The model shows the interfaces within and between IS-41-based mobile networks as well as interfaces between the mobile network and the PSTN and ISDN. Since the IS-41 network reference model is a logical model, it is an abstraction on which physical implementations can be based. This means that more than one functional entity can be implemented on a single physical device. When this is the case, it is not necessary for the internal interface between those functional entities to comply with the standards.

Origin of the network reference model

The first version of IS-41 (IS-41 Revision 0) did not explicitly define a network reference model. In fact, IS-41-0 considered only a single interface—that between a visited cellular system and the subscriber's home cellular system. No decomposition of each system into switches, location registers, and other functional entities was required in IS-41-0, since only basic intersystem handoff and subscriber validation functions were specified; there was no compelling need for a network reference model, given these basic standardization requirements.

The first IS-41 network reference model (see Fig. 4.5) was specified in IS-41 Revision A (IS-41-A) and was not modified until IS-41 Revision C (IS-41-C) several years later. It is interesting to note that the model was derived from the original version of the international CCITT Recommendation Q.1051, which specified the interfaces and application protocol for public land mobile networks. Only a few changes were made to the CCITT model for application to North American IS-41 networks.

Original IS-41 functional entities

Cellular subscriber station.   The cellular subscriber station is the functional entity that generally represents the mobile radio-telephone equipment. In the CCITT Q.1051 model, the CSS is known as the mobile station. The CSS was named differently to distinguish it from MS, which was used in European mobile networks. Since the CSS is a logical functional entity, it may not necessarily be a mobile telephone. In fact, it is defined in the IS-41 standard as "the interface equipment that terminates the radio path on the user side of the network." Many equipment implementations can comply with this definition; however, the most common implementation is certainly a mobile telephone. The CSS incorporates user interface functions, radio functions, and other service logic functionality.

Base station.   The base station is the functional entity that represents all the functions that terminate radio communications at the network side of the CSS interface to the cellular network. It controls radio resources and manages network information that is required to provide telecommunications services to the CSS. The BS essentially consists of the radio transceivers and radio transceiver controller functions serving one or more cells. A cell is the geographic area defined by the telecommunications coverage of the radio equipment located at a given cell site. The cell site is the physical location of a cell's radio equipment (i.e., the BS) and supporting systems. The BS incorporates radio functions and radio resource control functions.

Mobile switching center. The mobile switching center is the functional entity that represents an automatic switching system. This switching system constitutes the interfaces for user traffic between the cellular network and other public switched networks, or other MSCs in the same or other networks. The MSC provides the basic switching functions and coordinates the establishment of calls to and from the cellular subscribers. The MSC is directly responsible for transmission facilities management, mobility management, and call processing functions. The MSC is in direct contact with one or more BSs on one side and with external networks and functions on the other side. The MSC incorporates switching functions, mobile application functions, and other service logic functions.

The term *MSC* refers to the logical functionality of the wireless telecommunications switch including the signaling protocols, mobility management application, and functional interfaces. Sometimes MSC is confused with *mobile telephone switching office* (MTSO). The term *MTSO* is analogous to the wire line term *central office*. MTSO refers more to the physical architecture of the wireless switching office including the switching equipment, hardware interfaces, and the real estate where the switching system resides.

Home location register. The home location register (HLR) is the functional entity that represents the primary database repository of subscriber information used to provide control and intelligence in cellular and wireless networks. The term *register* denotes control and processing center functions as well as the database functions. The HLR is managed by the cellular service provider and represents the "home" database for subscribers who have subscribed to service in that home area. The HLR contains a record for each home subscriber that includes location information, subscriber status, subscribed features, and directory numbers. Supplementary services or features that are provided to a subscriber are ultimately controlled by the HLR. An HLR may serve one or more MSCs. The HLR incorporates database functions, mobile application functions, and other service logic functions.

Visitor location register. The visitor location register (VLR) is the functional entity that represents the local database, control, and processing functions that maintain temporary records associated with individual network subscribers. The VLR is managed by the cellular service provider and represents a "visitor's" database for subscribers who are being served in a defined local area. A visitor can be a mobile subscriber being served by one of many systems in the home service area, or a subscriber who is roaming in a nonhome, or visited, service area (i.e., another service provider area). The VLR contains subscriber location, status, and service information that is derived from the HLR. The local network MSC accesses the VLR to retrieve information for the handling of calls to and from visiting subscribers. The VLR may serve one or more MSCs. The VLR incorporates database functions, mobile application functions, and other service logic functions.

Authentication center. The authentication center is the functional entity that represents the authentication functions used to verify and validate a mobile station's identity. The AC manages and processes the authentication information related to a mobile station. This information consists of encryption and authentication keys as well as complex mathematical algorithms used to hinder fraudulent use of network services. The AC incorporates database functions used for the authentication keys and authentication algorithm functions.

Equipment identity register. The equipment identity register (EIR) is the functional entity that represents the database repository for mobile equipment–related data. An example is a database of the electronic serial numbers (ESNs) of mobile equipment along with the status of that equipment. Such a database could assist in preventing stolen or fraudulent equipment from being used to access network services. However, the functions of the EIR are not defined in any TIA standard.

Public switched telephone network (PSTN). The public switched telephone network (PSTN) is the functional entity that represents a network completely separate from the mobile telecommunications network. The PSTN refers to the regular wire line telephone network that provides service to the general public. The PSTN is commonly accessed by ordinary telephones, key telephone systems, private branch exchange (PBX) trunks, and data transmission equipment. The interface between the MSC and the PSTN represents the capabilities to originate calls from mobile phones to wire line phones and to terminate calls from wire line phones to mobile phones.

Integrated Services Digital Network. The Integrated Services Digital Network (ISDN) is the functional entity that represents a network completely separate from the mobile telecommunications network. The ISDN refers to the wire line network that provides enhanced digital services over special digital transmission lines to special digital terminals. The ISDN is commonly accessed by computer terminals and switching equipment employing digital adapters that interpret the ISDN digital signals to provide high-speed and high-bandwidth services. The interface between the MSC and the ISDN represents the capabilities to originate calls from mobile phones to destinations within the ISDN (e.g., voice mail systems) and to terminate calls from the ISDN to mobile phones.

Original IS-41 interface reference points

Table 4.1 shows the interface reference points specified in the original IS-41 network reference model (i.e., in IS-41-A and IS-41-B).

There are more interfaces and functional entities included in the model than are actually addressed by the IS-41 intersystem operations. IS-41-A and IS-41-B addressed intersystem operations for the B, C, D, and E interfaces only. The other interfaces were depicted on the model for the following reasons:

1. To show the need for future standardization in subsequent revisions of IS-41

2. To show the need for future standardization in other specifications related to IS-41

3. To show the relationship of other standard interfaces to the IS-41 protocol

The interfaces that have since been standardized (i.e., in IS-41-C or related standards) are the A, A_i, D_i, and H interfaces. The interfaces

TABLE 4.1 Original Interface Reference Points in IS-41-A and IS-41-B

(Interfaces in **bold** represent interfaces standardized in both IS-41-A and IS-41-B.)

IS-41-A/ IS-41-B Interface	Functional entities using the interface
A interface	BS-MSC
A_i interface	MSC-PSTN
B interface	**MSC-VLR**
C interface	**MSC-HLR**
D interface	**HLR-VLR**
D_i interface	MSC-ISDN
E interface	**MSC-MSC**
F interface	MSC-EIR
G interface	VLR-VLR
H interface	HLR-AC
S_m interface	Within CSS
U_m interface	CSS-BS

that have not been addressed are the F, G, and S_m interfaces. The functionality that can be provided by the F and G interfaces (i.e., the MSC-EIR and VLR-VLR interfaces, respectively) is not yet a necessity for North American mobile telecommunications. The S_m interface defines functionality within the CSS only, and it was determined to be beyond the scope of network standardization.

The other standard interfaces that relate to IS-41 but are not within its scope are the radio system interfaces. The interfaces related to the radio systems—the A, S_m, and U_m interfaces—are included in the model for the following reasons:

1. IS-41 is designed to support a variety of radio system standards that are separate from, but related to, the network.

2. IS-41 supports intersystem handoff functions that are affected by the radio systems.

Original incarnations of cellular switching and radio systems were implemented and deployed as bundled systems obtained from a single equipment vendor. Therefore, the A interface evolved as proprietary, and there was no need to standardize it. The U_m interface represents the radio, or air, interface between the mobile telephone and the radio base station. This interface was originally specified and standardized by TIA as *EIA/TIA IS-3*, and subsequently *ANSI/TIA/EIA-553*, as the AMPS protocol. Currently, there exist many standards for the radio interface such as narrowband AMPS (NAMPS), TDMA, and CDMA.

Additional IS-41 functional entities

The second generation—and current version—of the IS-41 network reference model is specified in IS-41 Revision C (IS-41-C). The following changes were made to the original reference model:

1. The name of the CSS was changed to mobile station.

2. The S_m interface was removed.

3. Short message service (SMS) functional entities and interfaces were added.

Figure 4.6 shows the second-generation IS-41 network reference model. The most prominent change from the previous model is the addition of functional entities supporting short message services. SMS is a set of services that supports the storage and transfer of short text messages (200 bytes or less) through the mobile network. The SMS functional entities were added to the model after the stage 2 description of SMS justified their presence. The name change from CSS to

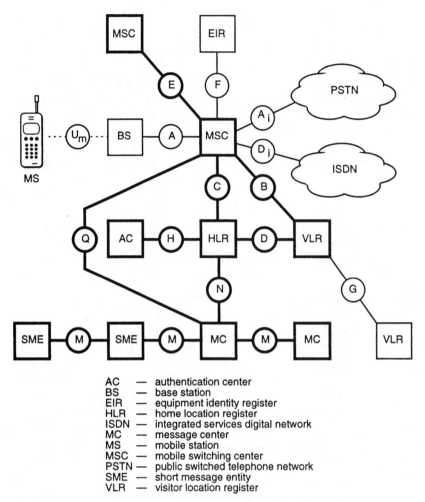

AC	— authentication center
BS	— base station
EIR	— equipment identity register
HLR	— home location register
ISDN	— integrated services digital network
MC	— message center
MS	— mobile station
MSC	— mobile switching center
PSTN	— public switched telephone network
SME	— short message entity
VLR	— visitor location register

Figure 4.6 The current IS-41 network reference model (from IS-41-C). The interfaces in **bold** represent the interfaces that are standardized in IS-41-C. (*Reproduced under written permission of the Telecommunications Industry Association.*)

MS implies no change of functionality. The need to show a distinction between this functional entity and other similar and analogous entities specified in other network standards was no longer an important consideration. The change was implemented to make the terminology of the mobile telephone more consistent with common industry usage.

Message center. The message center (MC) is a functional entity that stores and forwards short messages for SMS. The store-and-forward function provides a method of forwarding short messages to their destination recipient or storing those messages if the recipient is unavailable to receive them. The store-and-forward function can be distinguished from the real-time delivery requirements of voice calls. Short messages can be stored in a database until it is convenient for them to be sent to their specified destinations. The MC can store messages that are either sent from a mobile station (i.e., mobile-originated messages) or destined to a mobile station (i.e., mobile-terminated messages). Besides storing and delivering short messages, the MC performs signaling functions to support the other delivery functions, such as MS location and status queries and mapping of destination addresses.

Short messages can be sent to the MC from any functional entity that includes the function to support SMS message originations. Conversely, short messages can be received by any functional entity that includes the function to support SMS message terminations. These functional entities are known as *short message entities* (SMEs).

Short message entity.   The short message entity is a functional entity that can originate short messages, terminate short messages, or do both. Basically, *SME* is a generic term for any entity that can send or receive short messages via the IS-41 SMS. An SME may be associated with an IS-41 functional entity (e.g., an HLR, MC, or MSC) or with an entity external to IS-41. A mobile station also requires SME functionality for it to support the transmission of mobile-originated short messages and the reception of mobile-terminated short messages.

Additional IS-41 interface reference points

Table 4.2 shows the interface reference points specified in the IS-41-C network reference model.

The M, N, and Q interfaces were added to support the MC and SMEs. These interfaces represent signaling and bearer service communications required to provide the SMS functions.

IS-41-C and other related standards have evolved to address many of the interfaces that were included, but not yet standardized, in the original network reference model (IS-41-A and IS-41-B). The standardization of the A interface is a relatively new system requirement. The A interface standards support the interoperation of radio systems and MSCs developed by different manufacturers. Currently there are three protocols standardized by the TIA for use on the A interface: SS7-based (*TIA/EIA IS-634*), frame relay–based (also *TIA/EIA IS-634*) and ISDN-based (*TIA/EIA IS-653*).

The A_i and D_i interfaces are specified in TIA specification *TIA/EIA IS-93*. The A and D in these interface names stand for *analog* and *dig-*

TABLE 4.2   Current Interface Reference Points in IS-41-C and the Most Current
Standard Where They Are Addressed

(Interfaces in **bold** represent interfaces standardized in IS-41-C.)

Interface	Functional entities using the interface	Where addressed (most current standard)
A interface	BS-MSC	IS-634 & IS-653
A$_i$ interface	MSC-PSTN	IS-93
B interface	**MSC-VLR**	**IS-41-C**
C interface	**MSC-HLR**	**IS-41-C**
D interface	**HLR-VLR**	**IS-41-C**
D$_i$ interface	MSC-ISDN	IS-93
E interface	**MSC-MSC**	**IS-41-C**
F interface	MSC-EIR	Not standardized
G interface	VLR-VLR	Not standardized
H interface	**HLR-AC**	**IS-41-C**
M interface	**SME-SME, SME-MC, MC-MC**	**IS-41-C**
N interface	**MC-HLR**	**IS-41-C**
Q interface	**MC-MSC**	**IS-41-C**
U$_m$ interface	MS-BS	ANSI/TIA/EIA-553-A, IS-91, IS-95, IS-136*

*Note that other radio interface standards do exist, but these are the specifications most
representative of the technology supported by IS-41-C.

ital, respectively. The subscript i stands for *interface.* They are the
only interfaces in the network reference model whose letter designa-
tions (A and D) actually stand for words. They represent the interfaces between the cellular network and the analog PSTN and ISDN,
respectively. These interfaces, however, are treated as a single interface in IS-93, supporting both analog and digital network interconnection protocols. This single specified interface can be mapped onto
either the A$_i$ or D$_i$ interface of the model. The reason is that the distinction between the use of different network signaling protocols for
the PSTN and ISDN does not exist in the way the networks are
deployed today (refer to Chap. 16 for more details of interconnections
between cellular networks and the PSTN and ISDN).

The H interface is standardized in IS-41-C. It represents the interface between the HLR and the AC supporting the authentication
functions. This interface was originally standardized in TIA specification TSB51, as an addendum to the already published IS-41-B. The
information in TSB51 was subsequently refined, revised, and incorporated into IS-41-C.

Mapping the Logical Model to the Real World

The IS-41 network reference model is a *logical* model depicting logical functional entities and interfaces that may comprise a mobile telecommunications network. But how does this "logic" relate to real implementations of networks? How is the model representative of the many network implementation variations that can and do exist?

Mobile telecommunications network standards enable any cellular service provider, large or small, to conveniently and efficiently offer a standard set of mobile telecommunications services to subscribers. From a technical perspective, differences in the number, quality, and options of the standard services, as well as the deployment of additional nonstandard services, are what distinguish one service provider from another. The IS-41 standard is designed to allow many diverse implementations and has been developed, and continues to be enhanced, with these implementation considerations in mind.

A wide variety of differing IS-41 network implementations exist today that can all claim compliance to the IS-41 standard. While the model is logical and intangible, the actual network is physical and very tangible.

An individual IS-41 network usually comprises many MSCs, each deployed as a separate physical switching platform. The number of MSCs supported within a single cellular service provider network is a function of the traffic capacity and geographic service area supported. Although the IS-41 standard allows multiple MSCs to be served by a single distinct VLR, there is usually a one-to-one relationship between MSCs and VLRs. Each MSC usually employs the VLR functionality internally; i.e., the VLR is not normally deployed as a physically separate computing platform distinct from the MSC. This is due to the very close relationship between the MSC and VLR functions as defined by IS-41.

HLRs are deployed both as separate network platforms and as internal to MSC platforms. An HLR usually provides service to many MSCs in a single network. If an HLR is deployed internal to an MSC, other MSCs in the network communicate with that MSC as if it were a stand-alone HLR. A network can support multiple distinct HLR platforms, serving specified geographic areas or based upon some other division of subscribers. The HLR, though, is still considered to be a single logical entity in a given cellular service provider network; i.e., a single subscriber profile record cannot exist in more than one *active* HLR. The term *active* refers to the concept of physical network node replication. HLRs may be deployed as an *active/standby* replicated pair, since a failure will interrupt service to subscribers (see Fig. 4.7).

Many functional entities can be deployed as redundant systems. Redundancy, in the form of node replication, allows for very high ser-

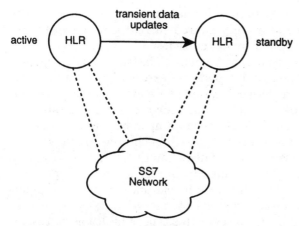

Figure 4.7   Active and standby HLR configuration. A special data link is used between the HLRs so that transient activity affecting the active database can be "mirrored" on the standby database. Each HLR of the replicated pair supports multiple links to the SS7 network. The SS7 network manages the rerouting of traffic to the standby (newly active) HLR when the previously active HLR fails.

vice availability. Replicated platforms are usually separated geographically. This avoids service interruptions based on catastrophic failures at a given site. The most common catastrophic examples used by engineers are hurricanes, earthquakes, and fires. There are two configurations for deploying replicated network nodes:

1. Load-sharing pair

2. Active/standby pair

A load-sharing pair requires both of the replicated platforms to be active and serve the network equally, with no distinction as to which platform sends, receives, or processes information. In case of failure of one of the platforms, the other can handle all the network traffic. An active/standby pair requires one platform to be active and the other to be in a standby mode. Only the active platform handles network traffic and processing. In case of failure of the active platform, the standby platform takes over and becomes the new active platform. The active/standby configuration is usually more appropriate for many mobile telecommunications network nodes (say, MSCs, HLRs, VLRs, ACs, and MCs). This is due to the nature of the real time transaction processing required by the mobile telecommunications network nodes. Multiple messages related to one or more transactions for a single subscriber need to be handled by the same platform for data consistency. In general, the transient information communicated by the transactions cannot be efficiently shared in real time between two active platforms to keep them synchronized.

In the active/standby scenario, the active platform usually updates database changes on the standby platform across a data link in *near* real time. When a failure does occur, some data are inevitably lost, but this loss is minimized by continually updating the standby database.

Some network implementations are based upon both replicated platforms being ready at the same time, with each able to support the entire traffic load in case the other fails. Note that this type of redundancy is distinct from fault-tolerant capabilities *within* a single computing platform, such as a switch. Fault tolerance enables a computing platform to operate uninterrupted or to recover quickly in the case of faults within the platform. However, this type of fault tolerance does not protect against catastrophic acts of God at a single location.

Authentication centers are usually implemented as part of the HLR. This is because there is a very close relationship between the HLR and the AC. However, ACs are also deployed as separate external platforms in the network.

Although short message services are just being introduced into the network, message centers are generally considered to be separate physical platforms, distinct from HLRs and MSCs. Short message entities are usually considered to be software applications. An SME is normally implemented within an SMS-capable mobile station. The physical MC can support SME applications as well. An SME can also be a separate physical device that can even be external to the IS-41 network, communicating with the MC via IS-41 or any other data communications protocol.

Several considerations for network implementation options are given in the IS-41 standard. There are two considerations that provide the greatest freedom to network developers:

1. IS-41 specifies the required intersystem operations only—mechanisms internal to each functional entity are left open to developers.

2. The specified IS-41 intersystem operations explicitly defined between functional entities do not preclude the participation of other entities in the operations.

These considerations are very important and powerful. They emphasize the need for standardized intersystem operations between the defined functional entities, while allowing network engineers to implement optimal solutions to their specific internal requirements.

The first consideration limits the scope of the IS-41 standard. The internal mechanisms developed for a given functional entity affect many aspects of service and performance, of which IS-41 signaling is only one part. The second consideration allows network designers and equipment manufacturers to employ a variety of schemes to implement the standard functions in IS-41, as well as a multitude of other nonstandard features and services.

Relationships between Functional Entities

The association between any two functional entities communicating with each other via IS-41 can be defined in the following logical ways:

1. A one-to-many relationship
2. A one–to–zero-or-more relationship
3. A many-to-many relationship

Figure 4.8 shows the method for depicting the relationships between functional entities. In these relationships, *many* means *one or more,* while *one* means *exactly one.* These relationships also only apply to the association between two functional entities at a single time. For example, an MS has a relationship with a single MSC at

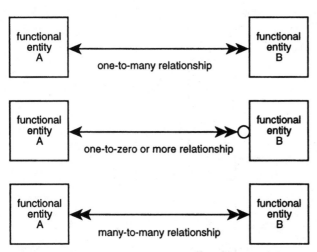

Figure 4.8 The top figure shows a one-to-many relationship from functional entity A to functional entity B and a many-to-one relationship from B to A. The center figure shows a one–to–zero-or-more relationship from functional entity A to functional entity B and a zero-or-more–to–one relationship from B to A. The bottom figure shows a many-to-many relationship from functional entity A to functional entity B and vice versa.

any time (from an IS-41 network perspective), even though it can have a relationship with many different MSCs at different times.

Defining the relationships between two functional entities in this way facilitates the understanding of the scope and purpose of the IS-41 intersystem operations specified between those entities. While it can be argued that defining these relationships limits the possible implementations of IS-41-based networks, the definitions are actually intended to limit ambiguity and interpretations of the IS-41 standard, while allowing freedom of implementation. In fact, it can even be argued that each functional entity has a many-to-many relationship (either direct or indirect) with any other functional entity! However, depicting a more confusing perspective of the reference model is not likely to be helpful. For reasons of simplicity, the many-to-many relationships were left out of the entity relationship diagram of IS-41.1-C (see Figs. 4.9 and 4.10).

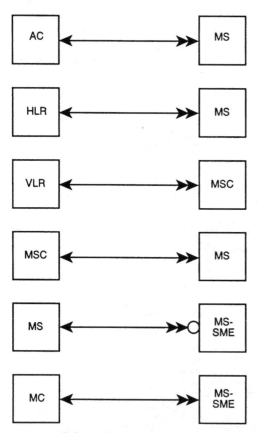

Figure 4.9 Relationships between the IS-41 functional entities as specified in IS-41.1-C.

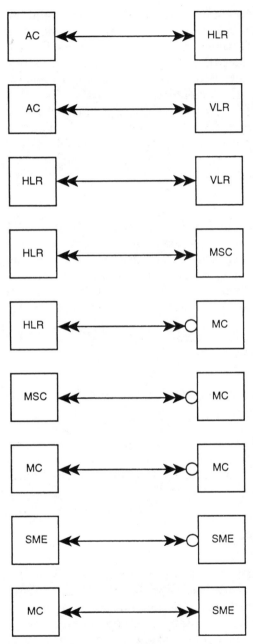

Figure 4.10   Relationships between the IS-41 functional entities *not* specified in IS-41.1-C.

Note that in Fig. 4.9 the MS-SME shows a many-to-one relationship to the MC. In the IS-41-C standard, this relationship is depicted as many-to-many. The IS-41-C standard only supports messaging and procedures from an MS-based SME to a single home MC. Although some implementations may permit an MS-based SME to communicate with multiple MCs, these implementations are outside the IS-41-C specification. Perhaps the entity relationship diagram in IS-41.1-C is meant to anticipate future SMS enhancements, or the diagram depicts an oversight that is not representative of the actual IS-41-C procedures.

Other Network Interfaces

Besides the functional entities and interfaces specified in the IS-41 network reference model, and the functions they represent, there are many other interfaces and functional entities that need to be implemented and deployed for the mobile network to be operational. The following list provides some of the basic functions that are not directly addressed by functional entities and interfaces specified in the IS-41 network reference model:

- Billing and accounting functions
- Customer service functions
- Operations, administration, and maintenance functions
- Provisioning functions

There are also many other supplementary service functions that may require additional functional entities for standardization. Examples are wireless packet data interworking functions and automated interactive voice recognition systems.

Multiple Interpretations of the
Network Reference Model

As a general consequence of the protocol standardization process, multiple interpretations of standards are inevitable. In fact, many contributors to a given standard may insist on a degree of ambiguity so that they are free to interpret the standard in an *individual* way. This concept tends to be in complete opposition to the purpose of standardization; however, it does happen.

Many efforts were made to avoid ambiguity in the IS-41-C specification. This is especially true of the network reference model. The emphasis on *logical* functional entities rather than physical nodes in the network allows for many diverse implementations, but can be confusing to those who are used to applying more physical models. From a pure academic-standards perspective, the model represents functions only; from a pragmatic perspective, mobile stations are telephone handsets, base stations are radios, MSCs are computer-controlled switches, HLRs and VLRs are data-based location registers, MCs are store-and-forward processing centers, and SMEs can be anything that originates or terminates a short message. Registers and processing centers can be combined with each other or with switches. The idea is to allow freedom of implementation of the IS-41 protocol while providing standard unambiguous descriptions. While there is no ambiguity in that goal, multiple interpretations can and do exist, and the resolution of those interpretations is continuing work for the standards committees.

5

Introduction to
Mobile Functionality

Before the details of the IS-41 protocol are described and explained, it is important to have an overall view of all the basic functions of the mobile network, even those not addressed by the IS-41 standard. This will allow a better understanding of the scope of IS-41 and place the mobile telecommunications network into an IS-41-based perspective. In this chapter we describe, in general, all the categories of mobile network functionality and identify which of these categories and general functions are standardized by IS-41.

General Scope of Mobile Functionality

Mobile telecommunications networks are similar in many ways to other communications networks. They incorporate network nodes and communications interfaces between those nodes. Generally, the communications supported by any type of network can be defined by the following characteristics:

1. *Physical and electromagnetic parameters.* The physical and electromagnetic parameters define the *means* of propagation of signals across an interface between network nodes.

2. *Channel structures.* The channel structures define the *transmission paths* that the signals take, with each path being separated by some physical means (e.g., frequency, time).

3. *A protocol.* The syntax, semantics, and procedures enable the *functions* that can be supported by communications across the network interfaces.

Although these characteristics are very closely related—because of the nature of systems based on layered protocols—IS-41 is primarily concerned with the intersystem operations of a mobile telecommunications network. These intersystem operations comprise the protocol necessary to enable functions that provide intelligent communications in the mobile network. The standard also, to some extent, addresses the other characteristics of the network; however, these other characteristics (i.e., physical and electromagnetic parameters and channel structures) are standardized in other ANSI and TIA specifications.

Mobile telecommunications networks (including IS-41-based networks) differ from other types of communications networks through the *functions* that they support. These functions are enabled by the signaling protocol. There are five generic categories of functions that enable subscriber mobility in the mobile telecommunications network:

1. Mobility management

2. Radio system management

3. Call processing and service management

4. Operations, administration, and maintenance (OA&M)

5. Terrestrial transmission facilities management

The five categories of functions named are common terminology in the industry. They are used here for convenience and understanding of the mobile functions as they relate to the entire mobile telecommunications network system. Many individual functions and subfunctions can be considered to be in more than one category, and good arguments can be made for placing some functions in one category or another. Although it is convenient to categorize the mobile functions in this way, there is not always a clear one-to-one mapping between each individual mobile network function and one of these five broad categories of mobile network functionality.

Figure 5.1 depicts a model of the five categories of functions. Terrestrial transmission facilities management supports the means for conveying communications in the network. Radio system management, call processing and service management, and mobility management provide the basic functions to enable mobile service to subscribers. OA&M provides control and management of all network functions as well as the equipment deployed in the network.

Many of the functions in these categories are not supported by the IS-41 intersystem operations; some of the functions are performed within a single switching system, and others are performed across nonstandard interfaces. Some of the functions are defined by other standards and specifications. In fact, many functions in these cate-

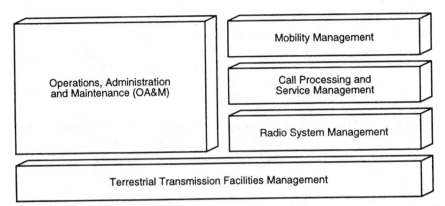

Figure 5.1 Model showing the general relationships between the five categories of mobile functionality.

gories are not standardized or specified anywhere yet! The IS-41 standard provides the intersystem operations primarily for the *mobility management* category of functions performed between systems; however, there are many IS-41 operations that also support functions considered to be in the other categories and within a single system only.

This may seem confusing at first. The reason is that many mobile network functions can be divided into subfunctions, each in a different functional category. For instance, the group of functions needed to perform intersystem handoff for a mobile subscriber can be considered to be in the radio system management category. The mobility management functions support certain radio-specific parameters that are passed between switching systems to determine and control the radio access. The radio system management functions control radio resources and manage radio channels to enable the handoff between the cells.

Mobility Management

Mobility management is a broad category of functions that enables cellular subscribers to be mobile. Mobility can be defined in a very broad sense. For instance, does a portable FM radio provide mobility? In a sense it does, since a user of the radio can receive service while she or he moves from place to place. But in cellular systems, the concept of mobility has a much narrower definition. Cellular networks are based on radio transmissions, similar to FM radio, and if a subscriber moves only within the coverage area of a single cell, he or she can obtain the same degree of mobility as the user of the portable radio. Cellular systems, however, support two-way communications and *frequency reuse* (see Chap. 1), which allow the system to support more subscribers and to provide individual radio service within and between individual cells. From both a network and subscriber perspective, mobility in cellular systems refers to the ability of the network to track a subscriber's status and location and to continue to provide service in the following situations:

1. While the subscriber is being served in a cell associated with a mobile switching center (MSC) in the home service area

2. While the subscriber is being served in a cell associated with a visited MSC

3. While the subscriber is moving between contiguous cells served by the same MSC (home or visited)

4. While the subscriber is moving between contiguous cells served by different MSCs (home or visited)

These situations introduce the mobility concepts known as *automatic roaming* and *handoff*. These capabilities are the cornerstone of mobility management in cellular telecommunications. *Automatic roaming* is the ability of a subscriber to obtain mobile telecommunications service in areas other than the home service provider area automatically, i.e., without additional actions being taken by the subscriber or a party calling the subscriber.

In this sense, mobile telecommunications *service* refers to the ability of a subscriber to make calls, receive calls, and apply any additional subscribed features to calls, such as call waiting, within the home serving system or in a visited serving system.

Handoff is a combination of mobile telephone and network functions that provides the capability for a subscriber to receive continuous service while a call is in progress, and without interruption, when the subscriber is moving between contiguous cells.

Assuming that the cellular geographic coverage area has no gaps, a subscriber can theoretically receive *seamless* and *ubiquitous* service anywhere within the coverage area, even if the coverage area is the entire land area of the earth! Unfortunately, the cellular systems today are still lacking, such that even within a defined coverage area, service may be interrupted due to radio transmission problems and system capacity limits.

Seamless roaming

The idea of a seamless and ubiquitous cellular network is a goal that many service providers try to attain—but it is very elusive. *Seamlessness* (a good industry buzzword) refers to the ability of a subscriber to obtain mobile telecommunications services while roaming in the exact same way as it is provided in the home service area (i.e., no apparent *seams* between network service areas). For instance, the sequence of dialed digits that a subscriber enters when making a call would always have the same effect, such as dialing 411 to obtain directory assistance or 911 for emergency services in North America. Even today, these services are not seamless. Sometimes dialing 411 for directory assistance or 911 for emergency service does not work everywhere. Many cellular networks still do not provide equal access to long-distance service (i.e., choice of long-distance carrier) while originating a call, while some do.

Ubiquitous mobile service

Ubiquity refers to the ability of a subscriber to obtain mobile telecommunications services *everywhere*. This does not necessarily imply seamlessness, but it does imply that a subscriber could make and receive calls anywhere, even if the *call treatment* is different. Call treatment refers to the methods that the network uses to handle a call in a certain way. Ubiquity is also a characteristic of radio coverage. It implies that there are no gaps or holes in a radio coverage area.

The terms *seamless* and *ubiquitous* are often used in conjunction to imply consistent service capabilities, accessible from anywhere and everywhere.

The MSC and VLR as combined and separate systems

Throughout this book we consider the serving system as a single entity encompassing the MSC and VLR functional entities. This simplifies the descriptions of IS-41 application processes and also is representative of a large percentage of the IS-41 implementations currently in service.

However, IS-41's definition of the MSC-to-VLR interface makes the stand-alone visitor location register a viable network entity. For example, IS-41 does not assume that the MSC is "just a switch" without a database or feature control intelligence; if this were the model for IS-41, the MSC-to-VLR interface would be burdened with signaling, limiting the potential for separate functional entities. On the other hand, the MSC is not given full control of the operation of the serving system; authentication, e.g., assigns specific processing duties to the VLR that make the idea of sharing a VLR among multiple MSCs particularly attractive. So, while we describe automatic roaming processes in terms of a single *MSC/VLR* entity, the reader should keep in mind that the potential for separation exists.

Subscriber identification

Subscriber identification is a set of processes to uniquely identify a mobile subscriber. In IS-41 networks this identification also pertains to the subscriber's *subscription* and *service profile record* maintained at the home location register. That is, the identification of a subscriber essentially refers to the identification of a service profile as well.

A unique subscriber ID is used to register and qualify a subscriber for service; however, the unique subscriber ID is also used to enable *all network functions pertaining to an individual subscriber.* The full unique subscriber ID in IS-41 is a concatenation of two unique parameters: the mobile identification number (MIN) and the electronic serial number (ESN). Note that in many IS-41 operations, only the MIN is used to identify a subscriber. The MIN-ESN combination is used primarily for the registration and authentication functions.

Mobile identification number.   The MIN is the single most important parameter used in IS-41 networks. It is a unique 10-digit decimal number programmed into the MS. This unique value is assigned to the mobile subscriber by the cellular service provider. This MIN is transmitted over the air interface (no matter which radio standard is used) during registration to inform the network of the identity of the mobile station (MS) accessing the network. The MIN is also used as a key field to access the mobile subscriber service profile record stored in the HLR (this is why we say that the subscriber ID also identifies a subscription). The service profile record contains the current location and status of the subscriber as well as other information pertaining to the subscription, such as subscribed features and credit status. Note that the format of the 10-digit MIN parameter used in the IS-41 network is a 40-bit value, with each digit encoded as 4 bits. The same 10-digit MIN value transmitted over the air interface (that is, AMPS, NAMPS, TDMA, and CDMA) is specially encoded as two fields: the 10-bit *min1* and the 24-bit *min2*. The translation between these encoding formats can be performed at either the base station or the serving MSC. The value transmitted over the air is fewer bits because of the importance of conserving bandwidth over the air, since this bandwidth is the single most limited resource in cellular systems. Figure 5.2 shows the two formats for the MIN.

The MIN was originally designed to contain a North American Numbering Plan (NANP) 10-digit number. This is because the MIN was meant to be the subscriber's directory number (i.e., the dialable telephone number) as well as the subscriber ID for network signaling. This is due to early cellular systems being designed for North America only, incorporating wire line system concepts.

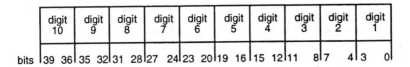

IS-41 mobile identification number (MIN)

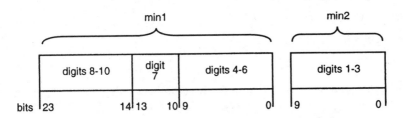

air interface mobile identification number (MIN)

Figure 5.2   The two formats of the MIN.

From an IS-41 protocol perspective, the MIN is purely the unique identifier of a mobile subscriber and subscription. In most implementations today, the MIN is usually the dialable directory number of the MS as well. However, this is not a requirement, and the MIN is not specified as the subscriber's directory number in IS-41-C.

Electronic serial number.   The ESN is a unique 32-bit serial number permanently stored in the MS equipment by the manufacturer. It identifies an MS, rather than a subscriber or subscription. The internal circuitry that provides the ESN is usually isolated from fraudulent contact and tampering. Many mobile stations are manufactured such that attempts to alter the ESN render the MS inoperative.

The ESN consists of a manufacturer's code and a serial number. The Federal Communications Commission (FCC) has mandated the presence of the ESN in the MS, so that radio transmissions can be identified, if need be. Since the MIN and ESN are used in conjunction to uniquely identify a subscriber for automatic roaming functions, the physical MS terminal is always tied to a cellular subscription. This is a problem for subscribers who need to replace the MS, since the subscription has to be modified for the new MS with a new ESN.

MSC identification

From an IS-41 perspective, the current location of a subscriber is recorded in the HLR as the mobile switching center identification (MSC ID) of the system currently serving the subscriber. The MSC ID

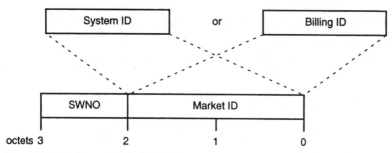

Figure 5.3 The format of the MSC ID.

is a unique three-octet number that identifies each individual MSC in the network and the group of cells associated with that MSC. The structure of the MSC ID is shown in Fig. 5.3.

The MSC ID is a concatenation of two parameters, the two-octet market identification (market ID) and the one-octet switch number (SWNO). The SWNO identifies an individual MSC within a defined market area. SWNO values are assigned for each MSC by the service provider. The market ID is split into two types of ranges of values:

- System ID (SID)
- Billing ID (BID)

The SWNO parameter is allocated by the service provider to uniquely identify each MSC within a given market identified by an SID or a BID.

SIDs. SIDs are 15-bit values assigned by the FCC that are allocated to each cellular service provider.

For cellular networks, SIDs are uniquely allocated for each *cellular geographic service area* (CGSA) covered by a licensed operator. CGSAs represent the 733 U.S. government-defined geographic areas that are individually licensed to the two 800-MHz cellular service providers in that market area (i.e., the A side and B side). A CGSA is further designated as either a *metropolitan statistical area* (MSA) or a *rural service area* (RSA). An MSA denotes one of the 305 largest urban population cellular markets. An RSA denotes one of the 428 less-populated areas defined for the country. SID values of 0 to 2175 have been allocated for North American cellular service providers.

For personal communications services networks, SIDs are allocated for each *trading area* covered by a licensed operator. Trading areas represent the 544 Rand McNally–defined geographic areas that are individually licensed to the six 1900-MHz PCS providers in that market area (i.e., the A to F sides). A trading area is further designated as either a *major trading area* (MTA) or a *basic trading area* (BTA). An MTA denotes one of 51 large PCS markets. A BTA denotes one of 493 small PCS markets. SID values of 4096 to 7679 have been allocated for North American PCS providers.

A parameter known as the *home SID* is programmed into each valid mobile station. This home SID represents the system ID of the home service area of the subscriber. When a subscriber is roaming, the home SID stored in the MS will not match the SID broadcast by the serving system cell site. This causes the "ROAM" light on the MS to activate, indicating that the subscriber is no longer being served by the home system. This home SID is never transmitted to the network.

BIDs. BIDs are 15- or 16-bit values assigned by CIBERNET, a corporate organization that is a wholly owned subsidiary of the Cellular Telecommunications Industry Association (CTIA). The BID values can be allocated to each cellular and PCS provider, upon request. BIDs are assigned to any market, which can be any geographic territory, desired by the service provider.

The purpose of a BID is to separately identify a portion of a service area (i.e., a subset of the entire area represented by an SID). Some service providers choose to operate all their markets using the same SID, thus preventing the "ROAM" light from activating on the mobile station while the subscriber is in the service provider's territory. However, a distinction among markets within a service provider's territory is still desired for business reasons, and BIDs serve this purpose. Some of these business reasons are the generation of marketing statistics, roaming agreement restrictions, and to provide a distinction among recorded call detail information.

Markets identified as a BID are still identified by an SID for broadcast over the air by cell sites. The BID overrides an SID when more specific location information is required. When BIDs are used to separately identify portions of a service area, the SID must still be used for one portion of that area. A service provider can request and use multiple BIDs for a single network to provide greater resolution of information for any purpose. BID values are derived from the same pool of market ID numbers as SID values. Market ID values of 26112 to 31103 and 32768 to 40960 have been allocated for North American cellular and PCS service providers as BIDs.

Note that BIDs are never broadcast by a cell site; rather the associated SID is broadcast, and the BID is used for call detail recording and billing purposes.

Cell site identification. The identification of the cell currently serving a subscriber is managed by the serving MSC only. This more specific location information is not normally stored at the HLR; however, it can be passed to the HLR for statistical purposes. From an IS-41 perspective, the MSC ID is adequate location information for an HLR. The IS-41 intersystem operations also pass a *serving cell ID* parameter between MSCs to support the intersystem handoff function.

Mobility management functions

Mobility management provides the functions supporting the mobility of cellular subscribers. Mobility management generally encompasses the following functional subsets:

1. Automatic roaming (including the component functions of mobile station service qualification, mobile station location management, and mobile station state management)

2. Authentication

3. Intersystem handoff

Conspicuously missing from this list of functions is a process known as *registration*. The reason that registration is not explicitly included as a mobility management function here is that the term is really representative of a subset of automatic roaming functions. The registration process primarily consists of mobile station service qualification and mobile station location management. These functions enable a subscriber to "register" with the network, i.e., to indicate to the network location register functional entities the location and status of the mobile station.

The registration-related functions generally must be performed before the network can provide services to the subscriber; however, there are exceptions. (For example, access to emergency services by dialing 911 may not require registration—because in some cases a subscription may not be necessary.) Figure 5.4 shows the cellular network registration process.

The process of registering a mobile subscriber is unique to wireless networks. In wire line networks, there is no need for a telephone user to register with the network, since the user is not mobile and the user's location and status, as far as the network is concerned, are constant and static. That is, a wire line user has no need to make the network aware of the location and status of the telephone being used to make or receive calls since it is always directly connected to the network via a physical line.

Automatic roaming. The entire process of automatic roaming includes the functions performed between the mobile station and the network

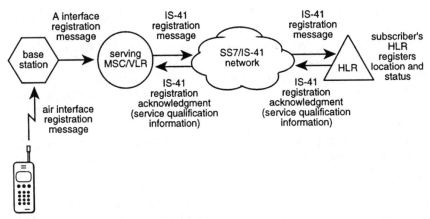

Figure 5.4 The MS sends a registration message to the serving system. The message is propagated to the home system (the HLR) to notify it of the subscriber's location and status. The HLR then responds to the serving system with service qualification information.

that allow the MS to inform the network of its current location, status, identification, and other characteristics. These functions allow a subscriber to obtain mobile telecommunications service outside of the home service provider area without requiring special subscriber actions. The mobile station location and status information is recorded in the network so that services can be provided to the subscriber (based on the MS identification) at that location.

Mobile station service qualification. MS service qualification is the set of functions that qualifies the MS (i.e., the subscriber) to use the mobile telecommunications network services. This service qualification includes functions to both validate the subscriber and manage the subscriber's service profile information.

Validation consists of examining the identification, location, and status information provided by the MS and comparing it to information stored in the subscriber's service profile. Network access can be either allowed or denied on the basis of service or financial criteria defined for that subscriber.

Managing service profile information consists of handling the specific set of features and service capabilities, including service restrictions, associated with the subscriber. An example of this information is a restriction on the types of calls a subscriber can make, such as limitations on long-distance or international calls. A subscriber's HLR service profile record is the network repository for the subscriber's service qualification information.

Mobile station location management. MS location management is the set of functions that stores, updates, and cancels the MS's (i.e., subscriber's) location. When the subscriber initially registers with the network, the current location is stored in the service profile record. As the subscriber moves from one serving area to another, the location is updated. The subscriber's location is canceled in the network when the subscriber is no longer considered active (e.g., when the MS is turned off).

When a VLR serves multiple MSCs, or when a system serves multiple markets, many levels of the MS's location resolution are possible (see Fig. 5.5):

- *The serving market ID.* If the MS is authorized for service for a particular market, the HLR will be provided a location update if the market changes, even if the serving VLR remains constant.

- *The serving MSC ID.* If the MS is authorized for service at the serving MSC's MSC ID, the HLR will be provided a location update if the MSC ID changes or if the serving VLR changes [these two are the same for a serving system (MSC/VLR) that has a single MSC ID].

- *The serving location area ID.* If the MS is authorized for service in the location area of the serving MSC, the HLR will be provided a location update if the location area changes or if the serving VLR changes.

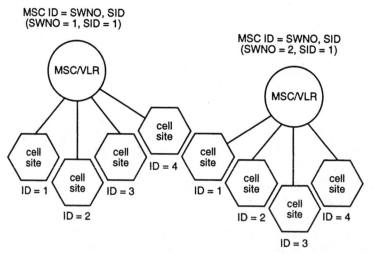

Figure 5.5 In IS-41, the location of a mobile station can be resolved to the SID only, to MSC ID (SWNO and SID), or to a location area representing a group of cells.

Mobile station state management. MS state management is the set of functions that manages the activity status of the MS (i.e., the subscriber). The subscriber's state is considered by the network to be either *active* or *inactive*. When active, the subscriber is considered available to receive calls. When inactive, calls are generally not sent to the subscriber by the network. The subscriber's state is considered inactive when the subscriber is not registered, is outside the radio coverage area, or is intentionally inaccessible based upon the modes of certain features or capabilities.

Authentication. Authentication is a mobility management process by which information is exchanged between an MS and the network for the purpose of *authenticating* (i.e., verifying with a strong degree of confidence) the identity of the MS. Authentication provides robust security against unauthorized or fraudulent access to the network above and beyond simple subscriber validation. Note that the authentication process requires an authentication-capable MS as well as the network support to perform the IS-41 authentication intersystem operations and calculations.

The authentication process is designed to be performed while a subscriber is in the home system or roaming. This means that if a subscriber with an authentication-capable MS roams to a network that is authentication-capable, the network can challenge the MS to prove authenticity.

Background and need for authentication. Cellular fraud has become a widespread problem in the wireless industry. The most predominant type of fraud in North American cellular networks today is known as *mobile cloning* fraud. Mobile cloning is accomplished by reprogramming an MS with the identification values of a valid MS (i.e., the MIN and ESN). The valid identification values are typically obtained by illegally scanning the control channels on the air interface. It is relatively easy to obtain the identification data of legitimate subscribers. These data are transmitted without any form of encryption from the MS to the base station over the air whenever registration occurs. Special radio scanners can easily capture these values during the registration process of a valid MS. Any fraudulent MS programmed with a valid subscriber identification can then make calls as if it were the original legitimate MS.

Authentication-capable systems attempt to prevent fraud and mobile cloning by supporting special algorithms that are performed in both the network and the MS. These algorithms use the mobile identification values and additional secret key values that are never transmitted over the air.

Intersystem handoff. Intersystem handoff is a mobility management process that enables the handoff of an MS from one cell to another where the two cells involved are served by two different systems (i.e., two different MSCs). Intersystem handoff transfers a call in progress from a radio channel in one cell to a radio channel in another cell in a neighboring system. In general, handoff functions are usually considered to be in the category of radio system management rather than mobility management. This is because the handoff functions involve the management of radio signal measurements, radio traffic channel allocation, and continual monitoring and control of radio resources at the base stations. However, since intersystem handoff enables the cellular subscriber to be mobile while involved in a call, it can also be considered part of mobility management as we have defined it.

Intersystem handoff provides for continuous terminal mobility management while a call is in progress. It is designed so that not only calls in progress, but also calls in progress involving special features, such as three-way calling, can be handed off between MSCs. Note that the two MSCs involved in the intersystem handoff process can belong to a single cellular service provider or to two different cellular service providers.

Radio System Management

Radio system management is a broad category of functions that manage the radio resources, connections, and radio transmission paths between the mobile station and the network, before, during, and after a call. Radio system management can be defined in a very broad sense. It encompasses the functions to establish and release radio connections, maintain those connections, and control all resources associated with the type of radio technology supported by the cellular system (i.e., analog AMPS and NAMPS, digital TDMA, and CDMA). These functions are usually specific to the interfaces between the base station systems and the MSCs where radio transmission paths are transcoded (i.e., converted from one digital coding plan to another) for terrestrial transmission in the network. Individual radio system management functions are generally beyond the scope of the IS-41 standard except for the intersystem handoff functions.

The radio system consists of the subscriber's MS and the network radio equipment—antenna systems, base transceiver systems (BTSs), and base station controllers. The MSC can also be considered part of the radio system if it contains functions that control the operations of the base station (e.g., network-controlled handoff) (see Fig. 5.6). Radio system management provides the functions that manage the radio and signaling transmission paths between the MS and the base sta-

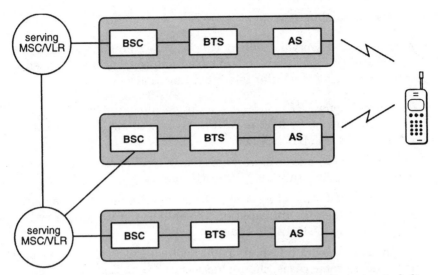

Figure 5.6 Radio system management controls aspects of the radio system including the base station controller (BSC), base transceiver system (BTS), and the antenna system (AS). The MSCs are involved for the coordination of radio resources for handoff.

tion (i.e., the U_m interface) and the radio control signaling between the base station and the MSC (i.e., the A interface).

Radio system management functions

Radio system management generally encompasses the following functional subsets:

1. Radio channel management
2. Transmission mode management
3. Power control
4. Handoff control

Note that these are only the most general functional subsets and that there are other functions performed by the radio systems in the mobile telecommunications network.

Radio channel management. Radio channel management is a set of functions that manage the set of radio channel resources within each cell site. Radio channel management generally consists of the following functions:

1. Channel allocation and initialization
2. Channel configuration

3. Traffic channel management

The term *channel* in this context can be very confusing since it has many meanings in cellular systems. The following terms are used to describe channels for cellular telecommunications:

- *Radio frequency (RF) channel*—a single allocation of contiguous spectrum (that is, 30 kHz in AMPS, 10 kHz in NAMPS, 30 kHz in TDMA, and 1.25 MHz in CDMA).

- *TDMA channel*—one or more time slots of a set of six time slots carrying information within a TDMA RF channel.

- *CDMA channel*—one of a set of information channels that is distinguished from other information channels by a unique encoding scheme within a CDMA RF channel.

- *Traffic channel*—an information channel used to carry voice or user data and, in some cases, channel-associated signaling.

- *Control channel*—an information channel that is dedicated to the transmission of non-channel-associated signaling data (e.g., mobile registrations and page responses) and supplementary service data. The control channel is generally split into a one-way channel to the MS and a one-way channel from the MS.

Channel allocation and initialization is a function that controls the assignment of radio channels to cellular subscribers to support calls and signaling. These radio channels provide the communications across the air interface between the base station and the MS.

Channel configuration is a function that manages the set of channels defined for use within a cell. Channels can be configured for different uses and for a given set of frequencies supported by a cell.

Traffic channel management is responsible for dynamically modifying the set of traffic channels to meet real-time traffic demands on the radio system.

Transmission mode management. Transmission mode management is a set of functions that manage the transmission methods and characteristics of the circuits used between the radio system and the MSC. The function determines which communication modes can operate on a particular circuit or channel. Examples of these modes are signaling only, signaling and speech, and data only. These modes are dependent on the type of air interface supported (that is, AMPS, NAMPS, TDMA, and CDMA) and the types of channels supported. The function also manages the transcoding and rate adaptation methods for the circuits supported.

Power control. Power control is a set of functions that manage the real-time transmission power of the MS and the BTSs within the cell. The transmission power is defined within a certain range of values. The power can be dynamically modified based on interference levels and signal fading to provide adequate service to subscribers.

Handoff control. Handoff control is a set of functions that allow the MS to move from one cell to another (or one radio channel to another, within or between cells) while a call is in progress. Intrasystem handoff (between two cells or radio channels that subtend the same MSC) is outside the scope of IS-41 since no coordination is required between MSCs to manage the mobility of the MS. Intersystem handoff is a handoff between two cells subtending two different MSCs. This type of handoff combines functions from both radio system management and mobility management (that is, IS-41 intersystem operations) to coordinate the movement of the MS between the cells.

Handoff control includes handoff preparation methods and measurements to determine the need for an MS to change channels or cells and support the communications switch between those channels or cells.

Call Processing

Call processing (also known as *call control*) is a broad category of functions that establish, maintain, and release calls to and from the mobile subscriber. It consists of the operations performed from the initial reception of an incoming call through the final disposition of the call. A call can be defined as a temporary communication between telecommunications end users for the purpose of exchanging information. A call includes the sequence of events that allocates and assigns resources and signaling channels required to establish a communications connection. A call can be a conventional telephone call or another type of communications connection, such as data (see Fig. 5.7).

The call processing functions establish and release terrestrial transmission paths for calls as well as invoke and manage call features that provide a specific variation on the way a call is treated. Many call processing functions are beyond the scope of the IS-41 standard; however, many aspects of call processing involve the IS-41 mobility management functions. These aspects include call delivery to a mobile subscriber and the application of call features for a mobile subscriber.

Basics of call processing

The basic capabilities to support mobile-originated calls and mobile-terminated calls involve the coordination of many functions, most of which are transparent to the parties involved in the call. These func-

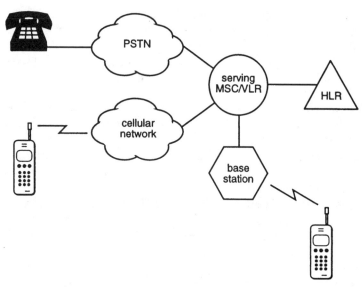

Figure 5.7 Call processing involves all network signaling, management, and connectivity to complete calls between a cellular subscriber and a wire line phone or another cellular subscriber. The phones involved may be served by different networks.

tions become much more complex when a subscriber is roaming and during an intersystem handoff. The mobile-originated call scenarios are simpler than the mobile-terminated call scenarios, since MS location and status information does not need to be obtained. Also, terminating calls to a roaming MS involves special routing to deliver the call to the system currently serving the subscriber. Note that the MS must be registered with the serving system prior to originating a call.

Dialing plan. Mobile-originated calls are any calls initiated by the MS. The dialing plan refers to the format of numbers that a mobile subscriber dials to reach a called party. These dialed digits may include special routing information, such as choice of long-distance carrier, special digits for operator access (i.e., dialing a zero), and the actual directory number address digits of the called party. The cellular dialing plan is standardized and specified in *TIA/EIA IS-52-A*.

The subscriber enters (or, euphemistically, *dials*) the number to be called. In mobile systems today, the subscriber performs *pre-origination dialing,* meaning that the numbers are dialed and sent to the system before any connection is requested. This is why mobile subscribers need to press a SEND key to transmit the dialed digits. This is in contrast to wire line systems where *post-origination dialing* is used. In wire line systems, the off-hook signal from the receiver informs the network that a call is about to be made, and the network responds with a dial tone. The called party's digits can then be dialed.

Numbering plan. The numbering plan for North America is specified by the North American Numbering Plan (NANP). The NANP defines the format of all dialable numbers in North America. Cellular service providers obtain ranges of dialable directory numbers from the North American Numbering Plan Administration (NANPA) that follow this format. These are the directory numbers that are given to subscribers when they become initially authorized for service.

The numbering plan defines the format of directory numbers (i.e., addresses) that represent a party that can be called. It differs from the dialing plan, since it only defines the address format and does not include digits that are *dialed* for special access or to provide additional information. Directory numbers for cellular systems in North America use the same numbering plan and dialing plan as wire line systems.

In wire line networks, a directory number indicates an area (i.e., a three-digit *area code*), an exchange, and a line (see Fig. 5.8). An *exchange code* (or office code) is used to indicate a specific switch that serves lines directly connected to telephones. A *station number* indicates a specific line directly connected between the switch and the telephone.

In wireless networks, a directory number indicates an area (i.e., the same three-digit area code), an exchange, and a *subscriber.* The exchange code indicates a specific MSC that serves the home subscribers. The subscriber number indicates a specific MS served by that home system. Since a mobile station's location can change, the method of routing calls geographically is contradictory to the concept of mobility. IS-41 provides signaling that allows the network to locate the MS and to route calls to the subscriber in any location.

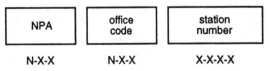

NPA	office code	station number
N-X-X	N-X-X	X-X-X-X

Figure 5.8 The basic NANP directory number format for wire line and wireless networks. *N* represents any digit within the range of 2 to 9. *X* represents any digit within the range of 0 to 9. Note that the formats are identical although the last four digits represent a different type of user.

Call processing functions

Call processing generally encompasses the following functional subsets:

1. Call establishment
2. Call release
3. Call features

Note that these are only the most general call processing functional subsets and that there are other call processing functions performed in the mobile telecommunications network.

Call establishment. Call establishment (also known as *call setup*) is a set of functions that arrange for the connection of cellular calls. There are essentially two types of cellular calls:

- Mobile-originated calls—calls that are placed from a mobile station
- Mobile-terminated calls—calls that are made to a mobile station

Mobile-originated calls are established from an MS (i.e., the calling party) to a telecommunications termination point (i.e., the called party) that can be either within or outside the mobile telecommunications network. Mobile-terminated calls are established from a telecommunications terminal (i.e., the calling party), which can be either within or outside the mobile telecommunications network, to an MS (i.e., the called party). The method for delivering a mobile-terminated call to a roaming mobile subscriber is an IS-41 feature known as *call delivery.* Note that a mobile-to-mobile call is usually treated as the combination of a mobile-originated call scenario and a mobile-terminated call scenario. Both mobile-originated and mobile-terminated calls typically make use of the public switched telephone network (PSTN) to create the connections between the calling and called parties. This is due to the extensive infrastructure and routing capabilities that exist in the PSTN. There is no need to deploy new switching capabilities for calls that can be handled by existing networks. However, connection through direct trunks between MSCs that are in close proximity is sometimes used for mobile-to-mobile calls. This architecture is typically used in high traffic areas where there are many mobile subscribers and many mobile-to-mobile calls.

The establishment of a cellular call involves two types of signaling to properly connect the call between parties: IS-41 call processing signaling and traditional call control signaling.

IS-41 call processing signaling is used to obtain the location, status, routing, and any special call treatment information about a mobile subscriber to properly complete mobile-terminated calls. It is also used to obtain call treatment and routing information for mobile-originated calls. This signaling is provided by the IS-41 intersystem operations. For intersystem handoff, IS-41 signaling is also used to control the inter-MSC trunks.

Traditional call control signaling is used to establish trunks and propagate the call, as well as information about the call, from the calling party to the called party. Three types of call control signals are used to complete calls: *supervisory signals,* to transmit the busy and idle states of lines and trunks; *address signals,* to transmit the digits of the called party's number and other information through the network; and *call progress signals,* to inform the calling party about the status of the connection that is progressing through the network (e.g., alerting, busy, routing failure).

Examples of other information provided with the address signals are the calling party's number, charge number information (also known as *automatic number identification* or ANI) used for billing, originating line or number information, network routing information, and special call treatment information. Call control signaling is generally provided by the out-of-band SS7 ISDN User Part (ISUP) protocol or an in-band multifrequency (MF) signaling protocol.

Special call treatment information includes *call features,* also known as *supplementary services.* These features can apply to mobile-originated calls (e.g., three-way calling) and mobile-terminated calls (e.g., call waiting). Depending on the feature, the signaling can be provided by IS-41 call processing signaling, traditional call control signaling, or a combination of both.

Mobile-originated calls. Mobile-originated calls are calls that a subscriber makes to any other party. The subscriber dials the digits to initiate the call processing functions with the serving system. When the serving MSC receives the dialed digits, the following procedures are performed to complete the call (see Fig. 5.9):

1. *Authentication.* The authentication procedure can be performed locally or with the home system to verify the identity of the MS. This procedure may be skipped if the subscriber dialed an emergency number such as 911.

2. *Mobile station service qualification.* The serving MSC queries the VLR to get service profile information and to validate the subscriber's ability to make a call defined by the dialed digits (e.g., an international or long-distance call). In many cases, the HLR is queried for this function for greater security.

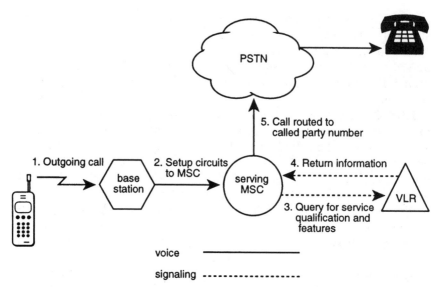

voice ————————

signaling ------------------

Figure 5.9 Basic MS call origination. The MS dials the called party's number. The serving MSC validates the subscriber and then routes the call to its destination. The HLR may also need to be accessed if the VLR does not hold the subscriber's service profile, or if the service qualification is requested. Authentication may be performed if it is supported by the MS and the serving system.

3. *Digit analysis.* The serving MSC analyzes the digits for format and routing purposes.

4. *Obtain routing information for the call.* The serving MSC routes the call to its destination based on internal routing tables using the called party's dialed digits, special instructions obtained from the VLR or HLR, or *origination triggers* that can be used to request special call origination treatment from anywhere in the network.

5. *Application of call features.* The serving MSC uses the service profile to apply appropriate subscribed mobile-originated features to the call.

Once the call is authorized, routing information is obtained, and call features are applied, the call can be established by using call control signaling techniques (for example, SS7 ISUP or MF) appropriate to the selected network.

Mobile-terminated calls. Mobile-terminated calls are calls made to a subscriber's MS. The subscriber can be in the home area or roaming. The delivery of mobile-terminated calls to a roaming subscriber is a call feature known as *call delivery* in IS-41. If the subscriber is being served by the same MSC as the one where the incoming call arrives, call termination is a relatively simple process since no location information or special signaling or routing is required. The MSC where calls are initially delivered is known as the *originating MSC*. The originating MSC may be the home MSC or a gateway MSC that may not be controlled by the home service provider.

When a subscriber is being served by the same MSC as the one where the incoming call arrived, the following procedures are performed to complete the call (see Fig. 5.10):

1. *Obtain MS registration status.* The MSC queries the HLR to determine if the MS is available to receive a call.

2. *Apply terminating call features.* The MSC queries the HLR to apply appropriate subscribed mobile-terminating features to the call.

3. *Page the MS.* The serving MSC alerts the MS and waits for it to answer.

When a subscriber is roaming and the originating MSC receives the dialed digits, the following procedures are performed to complete the call (see Fig. 5.11):

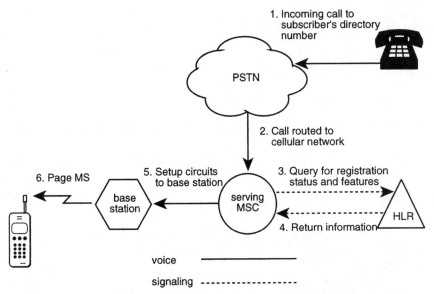

Figure 5.10 Basic call delivery to the home system. The serving/originating MSC queries the HLR for registration status and features. The HLR returns the status and feature information to the MSC which then establishes the call to the MS.

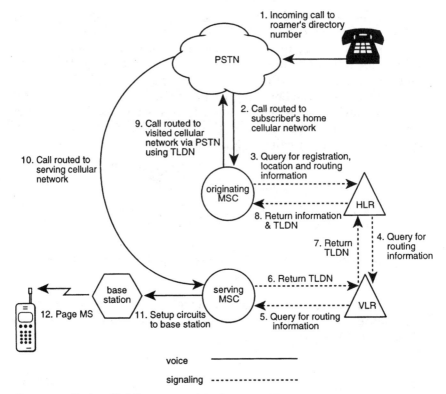

Figure 5.11 Basic call delivery to a visited system. The originating MSC queries the HLR for registration, location, and routing information to deliver the call. The HLR queries the serving system for routing information in the form of a TLDN. The serving MSC assigns a TLDN and returns the digits to the HLR which sends them to the originating MSC. The originating MSC establishes a call using the TLDN to the serving MSC which then establishes the call to the MS.

1. *Obtain MS registration status.* The MSC queries the HLR to determine if the MS is available to receive a call.

2. *Obtain serving system routing information.* The MSC queries the HLR (which subsequently queries the current serving system) to obtain routing information to deliver the call to the location where the subscriber is being served.

3. *Apply terminating call features.* The MSC queries the HLR to apply appropriate subscribed mobile-terminating features to the call.

4. *Deliver the call to the serving system.* The originating MSC establishes the call to the serving system using call control signaling techniques (for example, SS7 ISUP or MF) appropriate to the selected network.

5. *Page the MS.* The serving MSC alerts the MS and waits for it to answer.

Of course, a call to a roaming subscriber can be redirected based upon subscribed features such as call forwarding. This can be done unconditionally, or when the call cannot be completed. There are many reasons why a call may not be completed to a subscriber:

- The subscriber does not answer the call after a period of time.
- The subscriber is already engaged in a call (i.e., busy).
- The MS does not respond to paging (e.g., the subscriber has moved beyond the cellular coverage area, or the subscriber has turned off the MS and the registration has not yet been canceled).
- The subscriber's location is unknown.
- The subscriber is reported as inactive.

Temporary local directory numbers. Delivering a call to a roaming subscriber requires a *temporary local directory number* (TLDN). A TLDN is a real NANP directory number that is assigned by the serving system to roaming subscribers for a brief period (about 20 s) to support call delivery.

The use of TLDNs is necessary because calls to a roaming subscriber must be delivered somewhere other than the subscriber's home system. This is due to the fact that the subscriber's directory number is a geographic number, implying the subscriber's home system. Ranges of TLDN values are reserved by the cellular service provider from the entire set of directory numbers obtained from NANPA. These numbers are associated with a serving MSC and are never assigned to mobile subscribers.

The home system begins a series of IS-41 signaling queries to locate the subscriber. The serving MSC assigns and maps a TLDN to the subscriber's identity at the serving system. The incoming call is redirected to the serving MSC, via the PSTN, using the TLDN as the called party's number. The call can then be established to the subscriber at the serving system.

TLDNs are dynamically allocated and released when call delivery is completed.

The tromboning problem. Tromboning is a problem that occurs because of the use of TLDNs and the fact that directory numbers are geographically based. *Tromboning* refers to the shape that a call path takes during call delivery to a roaming subscriber (see Fig. 5.12). The following scenario clearly represents the problem:

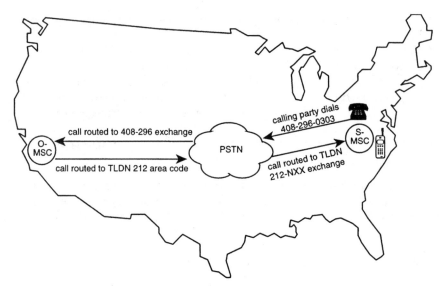

Figure 5.12 The tromboning problem. A call made to a roaming MS in New York is routed to the MS's home MSC in San Jose and then redirected back to New York via a TLDN to the serving MSC.

Kirk has a mobile subscription and lives in San Jose, California. His number is 408-296-0303. Kirk travels to New York with his MS. Babbette lives in New York and calls Kirk on his mobile phone. The call is established from New York to San Jose, since it is a 408 number. Kirk's home MSC makes the appropriate signaling queries and receives a TLDN from the serving MSC in New York, which is a 212 number. The call is then redirected from San Jose to New York where the call is completed when Kirk answers the phone.

The actual call path in this scenario traverses the United States twice, to complete the call. Call setup time is increased, and there are two long-distance legs to the call: one is charged to the calling party, and the other is charged to the roaming subscriber.

This problem exists today, even in IS-41-C. The solution to the problem requires additional signaling and changes to network implementations. Note, however, that solving this problem may not be very desirable to long-distance carriers who make money from its existence.

Call release. Call release (also known as *call disconnect*) is a set of functions that release cellular calls. Calls can be disconnected by one of the following three entities:

1. Calling party

2. Called party

3. Network

The cellular calling or called party disconnects by hanging up, e.g., pressing the END key. The network can disconnect a call for a variety of reasons, most due to a direct failure or to avoid the failure of a system involved in the call. Cellular calls can be disconnected by entities in the network regardless of whether the call is mobile-originated or mobile-terminated. In wire line networks, the disconnection is normally under control of the calling party.

Call release involves the release of resources used to establish and maintain the call so they can be available for other calls. These resources include lines, trunks, and stored program control switch resources such as timers, memory, and real-time software processes.

Call features. Call features (also known as *supplementary services*) are services provided to cellular subscribers that offer variations on the *basic* service. Basic service refers to the ability of a subscriber to simply make or receive calls. In fact, in IS-41 networks, the ability of a subscriber to receive calls while roaming is even considered to be a feature, known as *call delivery*. The IS-41 protocol provides intersystem operations to support the application of many call features for mobile subscribers while roaming and during intersystem handoff.

Call features typically enhance the way a call is treated, and they can apply to mobile-originated and mobile-terminated calls. Some features affect only the cellular subscriber, and others affect the other party, or parties, involved in the call. Examples of the more conventional call features are call forwarding, call waiting, three-way calling, and calling party number identification. An example of a mobile-terminated feature is call forwarding, since the calls originally intended for the mobile subscriber are to be redirected to another termination point. An example of a mobile-originated feature is three-way calling since it allows a mobile subscriber to originate a new call while already engaged in a call.

Operations, Administration, and Maintenance

OA&M is a broad category of functions that enable the cellular service provider to monitor and control the mobile telecommunications network. OA&M tends to be a catchall term describing all activities involved in managing and operating a telecommunications network.

Operations generally include coordinating and operating the network on a day-to-day basis. Personnel in a network operations center monitor alarms and traffic through the network.

Administration generally includes the functions to ensure the network operates at a high quality of service. It involves correct network sizing for users. The difference between operations and administration is that operations involves collecting data, whereas administration involves analyzing those data.

Maintenance generally includes troubleshooting, reporting, testing, and repair. It incorporates all functions that ensure that the network does not fail.

OA&M functions are primarily used to observe, record, and analyze operational characteristics of the network. OA&M functions are generally beyond the scope of the IS-41 standard; however, there are a few OA&M intersystem operations specified in part 4 of IS-41 (IS-41.4). These operations are used for a very small subset of the OA&M functions necessary to operate a mobile network.

The typical scope of OA&M functionality includes both unsolicited methods and inquiry and response methods for obtaining operational information about the mobile network. These statuses can be obtained on demand, on a priority basis, or on a time-interval basis. The most crucial information obtained is the following mobile network measurement and statistical data:

1. Network and service performance

2. Network and service availability

3. Network and service usage

These data are used to optimize the resources, performance, and quality of service of the network. The data are also used as a marketing tool to determine the types of network services accessed by subscribers.

Most cellular service providers have implemented a centralized *operations center* as part of the network to perform the OA&M functions (see Fig. 5.13). The operations center is customized to support

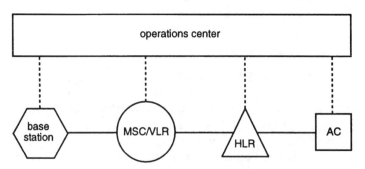

Figure 5.13 Typically, a centralized operations center controls the telecommunications network elements.

and manage the equipment and services specific to the network. Communications are supported between the operations center and the other network elements either directly or indirectly through mediating elements.

OA&M generally encompasses the following functions:

1. *Configuration management*—to control the provisioning, configuration, and modification of the network elements and the functionality within those elements

2. *Network engineering*—to manage the parameters and statistics used to evaluate the efficiency of the cellular system

3. *Network change control*—to manage changes in network elements while the network is operating

4. *Circuit management*—to control the inter-MSC trunks (i.e., trunks directly between adjacent MSCs) in the network to support intersystem handoffs

5. *Fault management*—to reduce the number of network element failures as well as minimize the effects of a particular failure

6. *Performance management*—to evaluate and report on the behavior of equipment and the effectiveness and efficiency of network systems, services, and elements

7. *Overload control*—to manage traffic overload conditions at the network elements

Note that OA&M functions are almost always implementation-dependent. The functions listed here are only the most general functions, and many other OA&M functions are performed in the mobile telecommunications network.

Terrestrial Transmission Facilities Management

Terrestrial transmission facilities management is a broad category of functions that manage the physical means of providing voice and data communications services to cellular subscribers. These functions manage the wire-based trunk and line types and support the physical rates and encoding of the trunks and lines associated with the switching system (i.e., the MSC). The management of these facilities and resources is beyond the scope of the IS-41 standard except in the case of inter-MSC trunks supporting intersystem handoff.

The typical scope of transmission facilities management functionality includes digital transmission functions for both bearer and signaling channels. This is true even for analog AMPS-based systems since the analog voice can be digitally encoded for transmission across T-1 facilities. The primary characteristics for digital transmission channels include the following:

1. Selection of speech voice coding and data interworking techniques (since transmission rates and encoding schemes are different on radio channels than on terrestrial transmission channels)

2. Selection of transmission modes (e.g., voice, data)

3. Companding techniques

These characteristics apply to particular channels and for particular call types. They are dependent on the physical wire line facilities and the types of bearer and signaling communications supported by the MSC.

Summary of IS-41's Functional Scope

This chapter described a comprehensive overview of the basic functions required to operate a mobile telecommunications network. The IS-41 standard provides signaling in the form of intersystem operations to support a subset of these functions. This section provides an outline of the functions supported by IS-41 only. These functions and the protocol details of the IS-41 mobile application part (MAP) are described in subsequent chapters.

Mobile network functions addressed by IS-41

As discussed previously, there are five generic categories of communications functions that provide, support, and enable subscriber mobility in the mobile telecommunications network. IS-41 provides many but not all of the functions in each of these categories. Table 5.1 shows the functions supported by IS-41.

Areas of IS-41 application

The IS-41-C standard specification is a large and complex document. It is sometimes not easy to grasp the power that the protocol provides. IS-41 can be applied to a variety of network architectures and services to enable the mobility of subscribers. IS-41-C was written to support the features specified in the *TIA/EIA IS-53-A* cellular features standard. IS-41 can support additional features and services; careful study of the operations and parameters provided reveals a multitude of tools to support many, many services. Keep in mind that basic cellular telephony may be the most prevalent application of IS-41, but it is not the only application.

TABLE 5.1   Functions Supported by IS-41

Functional category	Functions supported by IS-41
Mobility management	MS service qualification MS location management MS state management Authentication Intersystem handoff
Radio system management	Handoff control (intersystem handoff only)
Call processing	Mobile-originated calls Mobile-terminated calls Call release (for intersystem handoff only) Call features and control
OA&M	Circuit management (for inter-MSC trunks only)
Terrestrial transmission facilities management	None

Intersystem signaling and intrasystem signaling. IS-41 provides intersystem signaling operations, with a system defined by the standard as a single wireless telecommunications switch and related functional entities (i.e., location registers and processing centers). In deployed networks, a system is sometimes viewed as a single service provider network, consisting of many MSCs, VLRs, and at least one HLR. *Inter*system operations are usually viewed as the operations performed between one cellular service provider's functional entity and another cellular service provider's functional entity, as in the scenario of a roaming subscriber (i.e., between a *serving* MSC and the subscriber's service provider HLR). But IS-41 can be, and is, used just as appropriately for *intra*system operations. The IS-41 operations work just as well between functional entities within a single service provider network.

There are some deployed networks that use all proprietary protocols for operations within their own networks and use protocol conversion techniques to perform the IS-41 operations for intersystem signaling only. These networks can offer additional proprietary features. The major drawback of not using IS-41 operations for intrasystem operations is that there is no standard allowing the interconnection of different manufacturer equipment within that system. These systems are usually comprised of equipment manufactured by a single vendor, which limits the ability of the cellular service provider to deploy another vendor's equipment.

IS-41 capabilities. The IS-41 signaling operations transfer information and invoke commands between functional entities. Taken from this simplistic perspective, IS-41 can be applied to a variety of wireless applications. Although the original impetus for IS-41 was public 800-MHz cellular telecommunications, the standard can be used to support many capabilities:

- Personal communications services (PCS)
- Network interfaces to wireless PBXs
- Personal base stations and related applications
- Rudimentary international roaming (see Chap. 16)
- Enhanced mobile station functions (e.g., short message services)
- Voice mail systems
- Interactive voice response systems

These capabilities are only a sample of the applications of the IS-41 protocol. As more capabilities are desired by cellular service providers, they have a greater desire for equipment manufactured by different vendors, and the need for standardized intersystem operations becomes greater.

IS-41 Revision C Explained

6

The Road to IS-41 Revision C

Since the mid-1980s, the TIA has been developing the IS-41 standard for cellular intersystem operations. The standard was always intended to evolve with the growing demands of the wireless marketplace. Initially, only basic capabilities to provide mobility were addressed. These capabilities eventually became the building blocks that now allow cellular service and PCS providers to offer a rich set of services to subscribers. This chapter follows the evolutionary path of IS-41 that leads to the sophisticated mechanisms that support truly advanced wireless services.

IS-41 Revision 0 (IS-41-0)

In the beginning, there was Interim Standard-41, named simply because 41 was the next available number for a new TIA standard. In 1984, the development of the first version of a cellular internetworking standard had begun. IS-41-0 was eventually published in February 1988, and it addressed the following basic wireless intersystem functions:

1. Basic intersystem handoff
2. Basic service qualification
3. Basic OA&M (circuit management)

Basic intersystem handoff included the core functions still used today. For example, handoff measurements, handoff forward, handoff back, and inter-MSC trunk release were all supported. Multiple intersystem handoffs were allowed as a propagation of the basic handoff-forward procedures.

Basic service qualification simply addressed the operations to validate a mobile subscriber. Validation was a function to address *financial accountability* of a roaming subscriber. This validation could only be performed when the roaming subscriber was visiting a neighboring system connected by direct point-to-point signaling links to the subscriber's home system. These operations were specified to occur between a visited MSC and a home MSC across a generic data network. There were no HLRs, VLRs, or other functional entities specified to manage more sophisticated functionality.

Basic OA&M included the same functions specified today in IS-41-C (i.e., inter-MSC circuit maintenance).

IS-41-0 did not specify a network reference model. In fact, there was no need since there were only two functional entities involved, connected by a data network to transport the IS-41 application signaling. The communications specified included the CCITT recommendations for the Transaction Capability Application Part (TCAP) as an application service element. This application protocol was specified for use over X.25, as the primary choice, or North American SS7 (1984 version), as the second choice. The X.25 protocol was used in all the original field trials for IS-41.

IS-41 Revision A

IS-41-0 was a good start, but it obviously did not satisfy the true wireless needs of either what subscribers wanted or what the cellular service providers wanted to offer. Within 3 years of the publication of IS-41-0, IS-41-A was published (in January 1991). IS-41-A was the first significant step to providing elaborate wireless capabilities and services while improving on the basic functions specified in IS-41-0.

The first network reference model appeared in IS-41-A. This model introduced the concept of providing standard interfaces between particular functional entities with defined responsibilities. The HLR, VLR, and AC, along with interfaces to the PSTN, were the start to defining a true distributed intelligent network supporting wireless telecommunications.

IS-41-A addressed the following wireless intersystem functions:

1. Intersystem handoff
2. Enhanced service qualification
3. Location management
4. Mobile station state management

5. Call delivery

6. Cellular feature support

Intersystem handoff was essentially the same as in IS-41-0 except for some minor modifications to the handoff procedures, and the use of ANSI TCAP as the transaction protocol. Enhanced service qualification included subscriber validation from any serving system that has a roaming agreement with the home system and subscriber profile management. These functions were born from the introduction of an HLR.

The concept of an HLR also supported the MS location management and MS state management procedures. Call delivery to a roaming subscriber via a temporary local directory number (TLDN) was introduced. This same mechanism is used as the basis for call delivery today.

IS-41-A supported the first group of cellular features to enhance subscriber service. The following six features were specified in the initial version of *TIA/EIA IS-53* (Revision 0):

1. Call forwarding—unconditional

2. Call forwarding—busy

3. Call forwarding—no answer

4. Call waiting

5. Three-way calling

6. Feature control

One of the most important changes specified in IS-41-A was the use of ANSI SS7. ANSI TCAP was specified instead of CCITT TCAP, as well as the other ANSI SS7 standards (1988 version). X.25 continued to be specified for network transport, but it was no longer favored over SS7.

IS-41 Revision B

IS-41-B was published in December 1991, only a short time after IS-41-A. Although there were many significant changes in IS-41-B, they were clearly anticipated before IS-41-A was even published. Changes were made primarily to intersystem handoff and network routing mechanisms.

IS-41-B addressed the following wireless intersystem functions:

1. Path minimization

2. Flash feature support after handoff

3. Support for TDMA handoff parameters

4. Use of SS7 global title translation

Although these functions are significant, they are quite subtle. Networks implementing IS-41-A could provide many enhanced services without these functions. Path minimization and intersystem handoff to a third MSC are fairly rare occurrences. Although not as efficient and flexible as global title translation, SS7 point code routing works just fine (see Chap. 17). Because of these reasons, IS-41-B was not implemented for quite a while, and in many cases not at all. Networks that support IS-41-A may have no need for many of the enhancements that IS-41-B specified. And many networks today are skipping IS-41-B implementations and developing IS-41-C.

Not long after IS-41-B was published, implementation and deployment issues began to arise with IS-41-A. Most of the IS-41-A issues also applied to IS-41-B, since the primary functions remained unchanged. Ambiguities were also discovered in the standards. This is not surprising since these standards addressed some very complex issues and the industry was growing very rapidly. Thus, the development of Telecommunications Systems Bulletins (TSBs) began. These publications do not carry the weight of interim standards; however, they are provided as addenda to the published standards and to solve particular problems specific to one or more of the standards.

Within 3 years of the publication of IS-41-B, several TSBs were published to add enhancements or correct problems with both IS-41-A and IS-41-B. The following TSBs directly pertain to IS-41 functionality:

- *TIA/EIA TSB41*—IS-41-B technical notes
- *TIA/EIA TSB51*—authentication and signaling message encryption
- *TIA/EIA TSB55*—IS-41-A/B forward compatibility rules
- *TIA/EIA TSB64*—support for CDMA intersystem handoff
- *TIA/EIA TSB65*—resolutions to border cell problems

These TSBs provide complex functionality that could be published as interim standards in their own right.

TSB41—IS-41-B technical notes

TSB41 resolved several ambiguities and errors in IS-41 that resulted in incompatible implementations. The TSB documents 38 problems found in IS-41-B that are listed as *decision points*. It also provides replacement text for IS-41-B. Most of the decision points and the replacement text address both technical and subtle portions of IS-41-B; yet they are necessary to avoid incompatibilities that translate into poor quality of service for subscribers.

TSB51—authentication and signaling message encryption

TSB51 provided completely new functionality to support authentication, signaling message encryption, and voice privacy. The TSB specifies intersystem operations to support authentication-capable MSs. The functions supported are global challenge, unique challenge, and base station challenge incorporating the CAVE algorithms, as well as updating of shared secret data (SSD) and the call history count.

TSB55—IS-41-A/B forward compatibility rules

TSB55 provided modifications and additions to IS-41 Revision A network implementations to support forward compatibility to IS-41 Revision B and C. The TSB also applies to Revision B forward compatibility to Revision C implementations, although the title of the document implies the rules apply to Revision A only. An example of these rules is that Revision A systems should not reject protocol messages from Revision B or C systems simply because additional parameters that are not recognized are present in the message.

TSB64—support for CDMA intersystem handoff

TSB64 provided intersystem handoff support for CDMA mobile stations. New parameters and intersystem operations were added as well as provisions for mobile-assisted handoff (MAHO).

TSB65—resolutions to border cell problems

TSB65 provided solutions to *border cell problems* which occur when an MS is at or near the border between two cells served by different MSCs. When this situation arises, it is possible that incoming calls to the MS can be delivered to the wrong MSC, and the MS does not receive pages.

Motivation for IS-41 Revision C

IS-41-C represents the next major leap forward in North American wireless network technology. Published in February 1996, IS-41-C was, and is, the culmination of hundreds of person-years of effort. It is a major rework of the entire standard comprising approximately 1500 pages of text and diagrams, which is about 6 times bigger than IS-41-B! The motivation to develop IS-41-C came from many factors:

- The desire to provide many more features and services to mobile subscribers

- The desire to resolve compatibility problems found in earlier revisions of IS-41

- The desire to integrate the TSBs into a single standard specification

- The desire to incorporate state-of-the-art technologies such as intelligent networking principles and philosophies

IS-41-C adds a new part, *IS-41.6,* that provides detailed signaling procedures to assist with the implementation of the intersystem operations. It also adds support for 19 additional features (beyond the five supported by IS-41-A) and short message services (SMS). It is backward-compatible to IS-41-A and IS-41-B, and provisions were added for better forward compatibility with subsequent versions.

The TSBs were enhanced and rolled into the standard in the following ways:

- TSB41 technical notes were added to the appropriate text.

- TSB51 authentication was added and enhanced.

- TSB64 intersystem handoff for CDMA was added with modifications to CDMA-specific parameters and procedures for handoff and automatic roaming.

- TSB65 border cell resolutions were added with additional operations supporting precall handoff between serving and border systems.

Note that the additions to IS-41-C that included modifications to the TSBs supersede IS-41-B systems that incorporated those TSBs.

Other capabilities incorporated into IS-41-C include the following:

1. Support for *TIA/EIA IS-124 (DMH)* which provides real-time transfer of call detail records supporting billing and fraud management

2. Enhanced network addressing

3. Intelligent network trigger points

4. More service management controls at the HLR

IS-41-C has evolved into a very sophisticated specification that supports intelligent networking concepts; 24 cellular features; short message services; AMPS, NAMPS, TDMA, and CDMA radio systems; and many, many other capabilities. With this functionality, IS-41-C can support today's cellular systems as well as the next-generation PCS systems with a great deal of flexibility.

7

The IS-41-C
Specification
Structure

The collection of text, tables, and figures that comprise a revision of the IS-41 standard—particularly the 1500 or so pages in IS-41 Revision C—can be imposing at first. Without a clear understanding of its organization, it is difficult to make effective use of the standard. This is our goal in this chapter, to facilitate understanding of the IS-41-C specification structure, including how it differs from past revisions of IS-41.

Background

As the complexity of telecommunications systems has increased, so, too, have the demands on the organizations of people creating standards for these systems. Tools—including analysis techniques, specification processes, and even highly specialized languages—have been created to assist in the task. While some of these tools have proved their value over time, most are in a constant state of refinement as standards committee members try to find that optimum mix of clarity, brevity, and sufficiency.

Indeed, the act of writing a good standard involves a reasonable measure of art in addition to technical expertise. And since they are produced in standards committees, most standards are also laden with compromises. Some are for the better, others for the worse.

IS-41 has grown up in this environment and has managed to remain, for the most part, structurally consistent over the course of its revisions. Its structure has been largely influenced by two factors:

(1) the three-stage specification process and (2) the three major functional areas addressed in the standard: intersystem handoff; automatic roaming; and operations, administration, and maintenance (OA&M). We discuss these influences as an introduction to how the IS-41 revisions are organized.

Influence of the Three-Stage Specification Process on the IS-41 Structure

As we discussed in Chap. 4, the three-stage specification process is popular within the telecommunications standards community. Originally developed for use in the ISDN service standardization efforts, the basic technique is described in *ITU-T Recommendation I.130*. The three stages of the process are:

1. Describe the service from the user's perspective.
2. Derive a functional model for the service in terms of (*a*) the information flows between the functional entities involved in the service and (*b*) the actions required of these functional entities in order to provide the service to the user.
3. Design protocols and procedures that realize the functional information flows and actions derived in stage 2.

Application of the three-stage process to IS-41

The I.130 three-stage process is an example of a standards development tool that is in a constant state of refinement. And the application of this tool to IS-41 is no exception: some parts of the I.130 process have been kept as is, while others have been modified to better suit IS-41. The following sections describe some of the key differences and similarities between the I.130 and IS-41 processes.

Stage 1—service description.   While stages 2 and 3 are embodied in IS-41, the stage 1 cellular service descriptions associated with IS-41 are specified in a separate standard, known as *TIA/EIA IS-53*. Separating the user perspective in stage 1 from the network aspects in stages 2 and 3 is a step in the right direction. This separation is one component of the three-stage process that is almost universally adopted by telecommunications standards committees throughout the world. The effect is to divorce, insofar as possible, service definition from implementation constraints. User needs are given priority over the limitations of the network. The result is a more market-driven development process for new service standards than would be the case if the service definition were constrained by current network limitations.

In the case of IS-41-C, the service descriptions are in *TIA/EIA IS-53 Revision A (IS-53-A)*, which was completed almost a year before IS-41-C. The stable—or "frozen"—set of service specifications in IS-53-A was a practical necessity for completing a standard the size of IS-41-C; the job was difficult enough without having to deal with a last-minute service change that would have a ripple effect throughout the standard.

Creating a separate stage 1 standard also enables the parallel development of the network and air interface standards. There remains the need for coordination between the various standards groups, but each should be working to fulfill the common requirements defined in stage 1.

The IS-53-A stage 1 service definitions are structured similarly to those of the ITU-T approach specified in Annex A of *ITU-T Recommendation I.210*. Each service is defined by using the following basic format:

- Normal procedures with successful outcome

- Exception procedures or unsuccessful outcome

- Alternate procedures

- Interactions with other cellular services

Note that, as in I.210, IS-53-A provides specifications of service interactions; e.g., between unconditional call forwarding and do-not-disturb services. This is a critical set of requirements, particularly in a multivendor environment. It is at stage 1 that the rules are established, not only for how a service should operate on its own, but also for how it should interact with other services.

A key difference between the IS-53-A and I.210 stage 1 methods is that IS-53-A does not employ the Specification and Description Language (SDL) diagram technique recommended in I.130 and described in Annex D of I.210. Apparently, SDL diagrams were deemed useful but not absolutely necessary for IS-53-A purposes.

Stage 2—functional model and information flows. The application of ITU-T stage 2 techniques to the IS-41 environment has been more limited. The ITU-T approach for ISDN, as introduced in I.130 and defined in detail in *ITU-T Recommendation Q.65,* involves a high level of abstraction and formalism encompassing five steps. We illustrate these formal steps in a very informal manner; we use them to model a low-technology service, trash collection.

Step 1. Derive a functional model for the service in terms of abstract functional entities (FEs) and relationships between functional entities. In general, each service model would involve a new set of

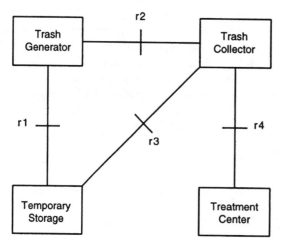

Relationships:

  r1 - Between trash generator and temporary storage.

  r2 - Between trash generator and trash collector.

  r3 - Between temporary storage and trash collector.

  r4 - Between trash collector and treatment center.

Figure 7.1 In step 1 of stage 2, the functional entities and relationships are defined.

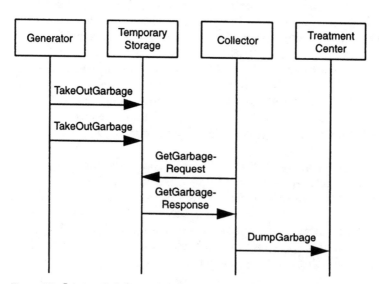

Figure 7.2 In step 2, information flow diagrams are defined, illustrating several cases of operation of the service.

FEs and relationships. Our trash collection service involves four FEs: the generator of the trash, temporary storage for the trash, the collector of the trash, and the trash treatment center (see Fig. 7.1).

Step 2. Define the abstract information flows between FEs required for proper service operation. Create information flow diagrams—often referred to as *Ping-Pong diagrams*—to illustrate these information flows for typical cases of operation of the service (see Fig. 7.2).

Step 3. Create diagrams to completely describe the actions of each FE associated with each information flow. These graphic descriptions are called Specification and Description Language diagrams. An SDL diagram for the temporary storage FE is provided in Fig. 7.3.

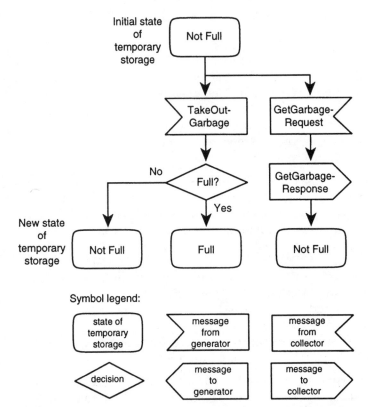

Figure 7.3 A portion of an SDL diagram for the temporary storage functional entity. Another part of the diagram would describe the actions taken when the storage is full and a message is received.

TABLE 7.1   Scenarios for the trash collection service

Scenario	Generator	Temporary storage	Collector	Treatment center
1	Household	Trash can	Sanitation engineer	Dump
2	Business	Trash bin	Sanitation engineer	Incinerator

Step 4. Identify all the actions required of each FE. For example, some of the actions associated with the collector functional entity in our example would be as follows:

- Schedule periodic trash collection.
- Transfer trash from multiple temporary storage locations to the collector.
- Transfer collected trash to the treatment center.
- Notify the generator if a problem with the temporary storage is detected.

Step 5. Allocate the abstract FEs to physical equipment, with each set of allocations referred to as a *scenario.* Table 7.1 illustrates a couple of scenarios for our trash collection example.

IS-41 takes a much more simplified approach. Although it is not formally described in any TIA document, we can consider the IS-41 stage 2 process to involve only two steps:

Step 1. Use the IS-41 network reference model (described in Chap. 4) as the functional basis for providing the service. Define new FEs only in cases where (*a*) the existing IS-41 FEs do not already encompass the required service-providing functions or (*b*) it is not desirable to add the required new functions to the existing IS-41 FEs. When a new FE is required, the IS-41 network reference model must be updated. The addition of authentication functions to IS-41 is an example of this step. In this case, a new FE, called the *authentication center* (AC), was added to the IS-41 network reference model rather than add the required authentication functions to the HLR FE's responsibilities; conversely, authentication functions were added to the existing VLR FE rather than create yet another new FE.

Step 2. Describe the required information flows between FEs in terms of existing or proposed IS-41 messages and parameters, as opposed to the Q.65 approach whereby an intermediate, abstract representation of the required information flows is defined and then mapped to actual messages and parameters in stage 3. Furthermore, IS-41 stage 2 information flow descriptions are in the form of ping-

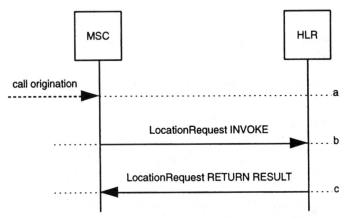

Figure 7.4 An example of the IS-41 approach to stage 2. The ping-pong diagrams describe signaling scenarios in terms of actual IS-41 messages, like LocationRequest, rather than in abstract terms. A text description of each step accompanies the diagram. (*a*) A call origination is received by the originating MSC. (*b*) The originating MSC sends a LocationRequest Invoke message to the HLR associated with the called mobile station. (*c*) The HLR determines the required call routing and returns this information to the originating MSC in the LocationRequest Return Result message.

pong diagrams with accompanying text to describe what is happening in each step (see Fig. 7.4). The more formal SDL diagram technique is not employed.

As we can see, the IS-41 approach to stage 2 is similar to that of Q.65 in some aspects but different in others—not necessarily better or worse, but definitely different. Given its track record, we must assume that the approach is good enough for the job.

Stage 3—protocols and procedures. The I.130 goal of stage 3—the design of the messages, parameters, and procedures required for the service in question—is common to the IS-41 approach. In fact, I.130 does not go into any detail regarding message formats and other presentation issues. Thus, IS-41 can be considered to be an application of the I.130 stage 3 process.

Influence of the Three Major Functional Areas on the IS-41 Structure

IS-41 encompasses three main categories of functions:

1. Intersystem handoff
2. Automatic roaming

3. Intersystem operations, administration, and maintenance

These are three distinct functional areas, and the distinctions are reflected in the structure of IS-41:

1. Intersystem handoff functions are defined in part 2 of IS-41 (IS-41.2).
2. Automatic roaming functions are defined in part 3 of IS-41 (IS-41.3).
3. Intersystem OA&M functions are defined in part 4 of IS-41 (IS-41.4).

However, all these functions share a common need: an intersystem signaling system over which IS-41 functional messages can flow. Therefore, it made sense to place the description of the IS-41 signaling systems in a separate part of the standard. This fifth part, called IS-41.5, also contains the stage 3 message and parameter definitions associated with the IS-41 functions. Again, this was a logical choice: All the implementation pieces are in a single place. However, the procedural aspects of stage 3—what to do with the messages and parameters—remained divided between IS-41.2, IS-41.3, and IS-41.4 until IS-41-C. A sixth part, IS-41.6, was then introduced, containing the detailed procedures for intersystem handoff and automatic roaming; OA&M was considered sufficiently unique to warrant leaving its procedures in IS-41.4.

Finally, IS-41.1 provides a very limited introduction to the standard. It also serves as a convenient place to present the IS-41 network reference model (see Chap. 4), the framework on which the subsequent parts of the standard are based.

The next section provides more details regarding the structure of each version of IS-41.

Structure of the Versions of IS-41

IS-41 Revisions 0, A, and B share a common document structure of five parts. Each of these five parts, in turn, has a structure. The part structures changed very little as the standard evolved from IS-41-0 to IS-41-B, although the content changed significantly.

As we discussed in the previous section, IS-41-C represents a significant structural departure from the prior IS-41 versions. IS-41-C incorporates a sixth part, IS-41.6, in the standard, and all the preexisting parts are modified in one way or another.

Table 7.2 provides a side-by-side comparison of the structure of IS-41 Revisions 0, A, and B to that of IS-41 Revision C.

TABLE 7.2   Comparing the IS-41-C Document Structure to Prior Revisions of IS-41

Structure of IS-41 Revisions 0, A, and B	Structure of IS-41 Revision C
Part 1: IS-41.1, Functional Overview	Part 1: IS-41.1-C, Functional Overview
Introduction	Introduction
References	References
Definitions	Definitions and documentation conventions
Symbols and abbreviations	Symbols and abbreviations
Network reference model (Note 1)	Network reference model
Cellular intersystem services	Cellular intersystem services
General background and assumptions	General background and assumptions
Restrictions	Restrictions
Part 2: IS-41.2, Intersystem Handoff	Part 2: IS-41.2-C, Handoff Information Flows (Note 2)
Introduction	Introduction
Configuration	(Note 3)
	References
	Terminology
	Intersystem handoff operations (Note 4)
Intersystem handoff message flow diagrams	Basic intersystem handoff scenarios (Note 5)
Intersystem handoff procedures	(Note 6)
	Annex A (Note 7)
Part 3: IS-41.3, Automatic Roaming	Part 3: IS-41.3-C, Automatic Roaming Information Flows (Note 8)
Introduction	Introduction
Configuration	(Note 3)
	References
	Terminology
	Automatic roaming operations (Note 9)
Automatic roaming message flow diagrams	Basic automatic roaming scenarios (Note 10)
	Voice feature scenarios (Note 11)
	Short message service scenarios (Note 12)
Automatic roaming procedures	(Note 13)
	Annex A (Note 14)
Part 4: IS-41.4, OA&M (includes stage 2 and stage 3 information)	Part 4: IS-41.4-C, OA&M Information Flows and Procedures
Introduction	Introduction
Configuration	(Note 3)
	References
	Terminology
OA&M message flow diagrams	Intersystem OA&M operations
OA&M message procedures for handoff	OA&M message procedures for handoff
OA&M procedures for automatic roaming	(Note 15)
Inter-MSC trunk testing	Inter-MSC trunk testing
Issues for further study	Issues for further study

TABLE 7.2   Comparing the IS-41-C Document Structure to Prior Revisions of IS-41 (*Continued*)

Structure of IS-41 Revisions 0, A, and B	Structure of IS-41 Revision C
Part 5: IS-41.5, Data Communications	Part 5: IS-41.5-C, Signaling Messages and Parameters
Introduction	Introduction
	References
	Terminology
Architectural concepts and terminology	IS-41 protocol architecture
Application layer	(Note 16)
Presentation layer	(Note 17)
Session layer	(Note 17)
Transport layer	(Note 17)
Carriage services	Data transfer services (Note 18)
MAP message definitions and functions	Application services
MAP compatibility guidelines and rules	MAP compatibility guidelines and rules (Note 19)
Appendix A and Appendix B (Note 20)	
	Part 6: IS-41.6-C, Signaling Procedures (Note 21)
	Introduction
	Technology and concepts
	Basic call processing
	Intersystem procedures
	Voice feature procedures
	Common voice feature procedures
	Operation timer values
	Annexes A through E (Note 22)

1. The section covering the network reference model is not included in IS-41-0.
2. IS-41.2-C includes stage 2 information only whereas prior revisions also included stage 3 procedures.
3. The configuration information is covered in the IS-41.1-C assumptions and restrictions.
4. This section defines the information flows for individual IS-41-C intersystem handoff operations.
5. This section defines the information flows for common IS-41 handoff scenarios involving multiple IS-41 intersystem handoff operations.
6. The stage 3 descriptions of intersystem handoff procedures are contained in IS-41.6-C.
7. IS-41.2-C also includes an Annex which describes the handoff functionality added to IS-41-C in support of IS-124. See Appendix B for a description of IS-124.
8. IS-41.3-C includes stage 2 information only whereas prior revisions also included stage 3 procedures.
9. This section defines the information flows for individual IS-41-C automatic roaming operations.
10. This section defines the information flows for common IS-41 automatic roaming scenarios involving multiple IS-41 automatic roaming operations.
11. This section defines the information flows for common IS-41 automatic roaming scenarios involving voice features (e.g., call forwarding).
12. This section defines the information flows for common IS-41 automatic roaming scenarios involving short message services.
13. The stage 3 descriptions of automatic roaming procedures are contained in IS-41.6-C.
14. IS-41.3-C also includes an annex which describes the assumptions associated with authentication, voice privacy, and signaling message encryption.

15. The stage 3 descriptions of the OA&M procedures for automatic roaming are contained in IS-41.6-C.

16. This section provides an overview of the application layer structure. In IS-41.5-C, this information is provided in the Application Services section.

17. The presentation, session, and transport layers are "null" layers in IS-41; i.e., no protocol is specified for these layers.

18. Both X.25-based and SS7-based data transfer (or carriage) services are defined in this section.

19. This section is found in IS-41-B and IS-41-C only.

20. IS-41-0 also includes two appendices with additional explanations of the MAP messages and parameters.

21. IS-41.6-C is a new part added in Revision C which includes the stage 3 procedures related to intersystem handoff and automatic roaming. Most of the procedures for OA&M remain in IS-41.4-C; the exception is the OA&M procedures for automatic roaming which are included in IS-41.6-C.

22. IS-41-C also includes six annexes with additional informative material related to the procedures.

8

The IS-41 Protocol Architecture

For the typical mobile telecommunications network equipment vendor, IS-41 represents a collection of protocols for data communication—protocols the vendor must implement to support the services associated with IS-41. In this chapter we identify the individual protocols included in IS-41, the functions of each, and how they collectively fit together. In the IS-41 standard, this structure is called the *IS-41 protocol architecture.*

Data Transfer versus Information Transfer

The fundamental partition within the IS-41 protocol architecture lies between *application services* and *data transfer services.*

Application services make possible the transfer of information between application processes in the mobile telecommunications network. The nature of this application information—its semantics and syntax, together with its movement between application processes— functionally defines the network. For example, the basic call delivery function—allowing a subscriber to receive calls when outside of the serving area of her or his home system—would not be possible if the subscriber's identity and current location (i.e., application information) were not transferred from the visited system to the subscriber's home system.

On the other hand, from the perspective of the protocol entities providing the data transfer services, application information is nothing more than a collection of octets which must be moved from one node in the network to another. The octets could just as well correspond to

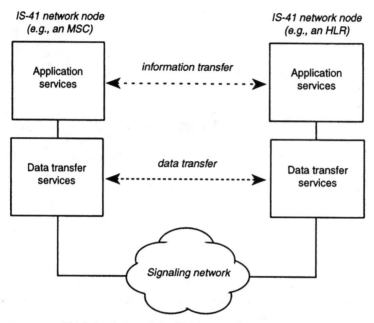

Figure 8.1   A high-level view of the IS-41 protocol architecture.

a grocery list as a subscriber's location information; the "blue collar" job of data transfer would still have to be done.

Figure 8.1 illustrates this high-level view of the IS-41 protocol architecture as it applies to two IS-41 network nodes (in the figure, an MSC and an HLR) communicating via a signaling network.

Relationship with the OSI Reference Model

Since this is a book about data communications protocols, we would be remiss if we did not try to relate the IS-41 protocol structure to that of the Open Systems Interconnection seven-layer reference model for data communications (OSI model), defined in the ITU X.200 series of recommendations. We do this at various points throughout the discussion in this chapter; at this stage it is enough to explain how the IS-41 division into application services and data transfer services fits into the OSI scheme of things.

According to IS-41-C, application services encompass the application, presentation, session, and transport layers, while data transfer services cover the network, data link, and physical layers of the OSI model (see Fig. 8.2).

At this stage, OSI experts may be thinking, "They are mistaken here. The transport layer should be categorized under data transfer

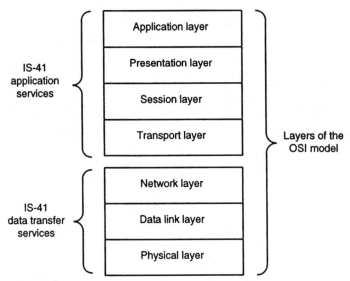

Figure 8.2 Relationship of the IS-41 protocol architecture to the layering of the OSI reference model.

services—that is what the transport layer provides: end-to-end data transfer services." This is a valid point, worthy of some explanation. The IS-41 protocol model is based on the Signaling System No. 7 (SS7) protocol structure, rather than the OSI model. In particular, the IS-41 model relies on the functions of SS7 which support transaction-oriented services. These functions are called SS7 *transaction capabilities* (TCs) and are used, e.g., to provide the query-and-response class of procedure associated with retrieving information from a remote database (see Fig. 8.3). The rationale for using the SS7 model was described in IS-41 Revision 0 (IS-41-0):

> Transaction Capabilities for the Intersystem protocol are functions which control information transfer between two or more signaling nodes via a signaling network. Since the cellular radio intersystem protocol is transaction oriented, the decision has been made to adopt the TC procedures similar to those defined for CCITT Signaling System No. 7 (SS7).*

SS7 transaction capabilities are defined in the 1988 version of ITU Recommendation Q.771—the version that was current when IS-41-0 was being developed—to encompass application layer protocols and

---

*IS-41 Revision 0 referenced the CCITT version of SS7 while subsequent versions of IS-41, including Revision C, are based on the ANSI version of SS7; however, both versions of SS7 specify the same basic protocol model.

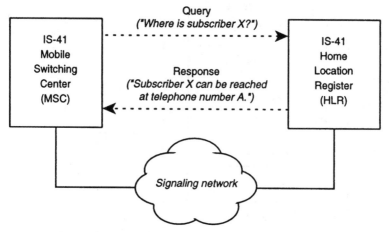

Figure 8.3   A typical query-and-response style of transaction.

services, called Transaction Capabilities Application Part (TCAP), plus the supporting presentation, session, and transport layers, collectively called the Intermediate Service Part (ISP).* IS-41 application services (see Fig. 8.4) comprise SS7 TC plus another set of application layer protocols and services called the IS-41 mobile application part, or MAP.† Thus, due to its inclusion in SS7 TC, the transport layer is considered to be part of IS-41 application services.

Of course, the real value to IS-41 of using the SS7 protocol structure is not that it places the transport layer in one category versus another; rather, it is that SS7 provides an appropriate platform on which to model (and build) the functions required for mobile network signaling. In fact, in all the revisions of IS-41, the transport layer is specified as the *null layer*—no transport layer protocol has ever been defined for IS-41.

IS-41 Data Transfer Services

Data transfer services, referred to as *carriage services* prior to IS-41-C, consist of the network, data link, and physical layer services defined by the OSI model (see Fig. 8.5).

Since IS-41 application services are supported by SS7 protocols, the choice of data transfer protocols would seem obvious—use SS7! Of course, system design decisions are rarely that straightforward.

---

*Q.771 refers to the Intermediate Service Part (ISP); however, Application Service Part (ASP) is the term used in the analogous ANSI standard, T1.114 (1988), on which revisions A, B, and C of IS-41 are based.
†See IS-41 Application Services for a more detailed description.

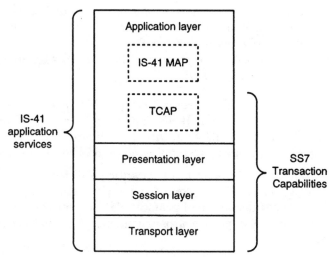

Figure 8.4   IS-41 application services comprise the Mobile
Application Part (MAP) plus SS7 Transaction Capabilities.

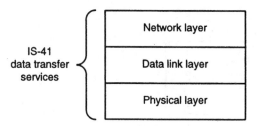

Figure 8.5   IS-41 data transfer service layers.

Other factors had to be considered by the developers of the original
IS-41 standard, IS-41-0:

- The conflicting goals of maximum flexibility (i.e., for feature and
  capacity growth) and minimum complexity* in the implementation
  of IS-41
- The desire to deploy IS-41 in the field as quickly as possible
- The practical need to reflect existing implementation practices in
  the standard

---

*TR45.2, the subcommittee with responsibility for IS-41, is a technical standards
group; discussion of financial cost issues associated with IS-41 implementation is considered taboo. In reality, of course, complexity usually implies cost, so the word *complexity* has come to be used as a suitable euphemism for *cost*.

The result of the deliberations on these issues was a "multifooted" protocol structure, whereby application services may be supported by a number of data transfer schemes (see Fig. 8.6).

Two alternative protocol sets were recommended in IS-41-0 and remain part of the standard to this day:

1. *An X.25-based protocol set.* This protocol set provides a low-complexity, readily available solution. It was particularly well suited for the initial application of IS-41 to direct dedicated facilities between pairs of communicating systems (see Fig. 8.7).

2. *An SS7-based protocol set.* This protocol set anticipates the growth in the number of subscriber calling features and processing capacity requirements. It provides an ideal protocol platform for the task of connecting multiple systems via a backbone signaling network (see Fig. 8.8).

We provide an overview of these protocol sets in the following two subsections on pages 118 and 119.

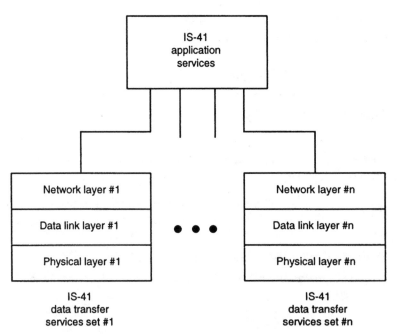

Figure 8.6 IS-41 application services can be supported by a number of data transfer service sets.

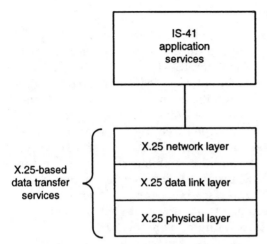

Figure 8.7 X.25 provides one set of data transfer services to IS-41 application services.

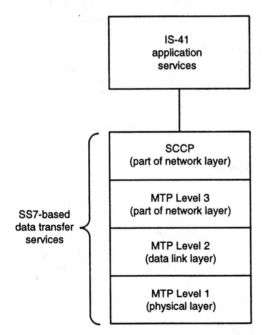

Figure 8.8 SS7 provides another set of data transfer services for use by IS-41 application services, where SCCP is the SS7 Signaling Connection Control Part and MTP is the SS7 Message Transfer Part.

X.25-based data transfer services

Although SS7-based services were defined in IS-41-0 and IS-41-A, both standards specified X.25-based data transfer services as "the minimum acceptable configuration" for IS-41. As a result, X.25 was the primary method of data transfer between nodes in the early stages of IS-41 deployment. X.25 finds significant use even today, although SS7-based data transfer is the clear direction for IS-41's future.

X.25 specifies protocols for each of the three IS-41 data transfer service layers—the network, data link, and physical layers. However, X.25 was designed as a data terminal equipment to data circuit-terminating equipment (DTE-DCE) interface. This is *access signaling,* as discussed in Chap. 3; i.e., the interface is asymmetric, with one side (the DCE) acting as the access point into a packet-switched network, through which the DTE can transfer data to another DTE.

However, for IS-41's purposes, X.25 is used over dedicated signaling facilities between pairs of IS-41 nodes. In X.25 terminology, each IS-41 node is considered a DTE; therefore, the IS-41 interface is DTE-DTE. Figure 8.9 illustrates the typical DTE-DCE use of X.25 versus IS-41's use for DTE-DTE communications.

To address this requirement, IS-41 references an X.25-compatible version of the network layer protocol, specified in ISO 8208. ISO 8208 includes the requirements that X.25 places on DTEs, but also specifies how DTEs should operate in the DTE-DTE configuration.

Similarly, the IS-41 X.25-based data link layer is aligned with ISO 7776, which is the DTE counterpart of the DCE procedures specified in X.25.

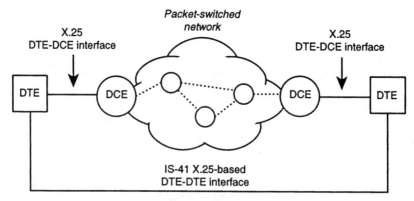

Figure 8.9 X.25 is generally considered an access protocol to a packet-switched network, operating between DTE and DCE. However, IS-41 requires a DTE-DTE interface.

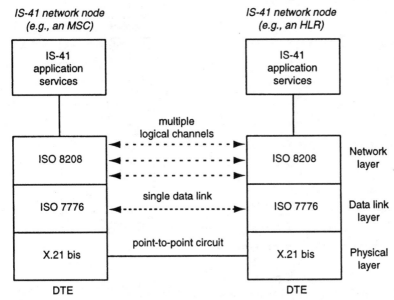

IS-41 network node
(e.g., an MSC)

IS-41 network node
(e.g., an HLR)

Figure 8.10   The IS-41 X.25-based data transfer protocol structure. The physical layer provides a single circuit, over which a single data link is maintained. ISO 8208, like X.25, provides application services with multiple logical channels, or *virtual circuits*.

Finally, the IS-41 X.25-based physical layer is as specified in X.25; i.e., the protocol is in accordance with X.21-bis, which defines the physical layer interface between a DTE and an ITU V-series modem.

Figure 8.10 illustrates the resulting protocol structure for an implementation supporting X.25-based data transfer services.

SS7-based data transfer services

In IS-41-B and IS-41-C, both SS7 and X.25 are identified as "technologically capable of supporting the requirements" of IS-41 data transfer; however, X.25 is no longer referred to as the minimum acceptable configuration. In fact, SS7 has become the IS-41 data transfer service of choice, particularly for large carriers.

IS-41 SS7-based data transfer services comprise the SS7 Message Transfer Part (MTP) and Signaling Connection Control Part (SCCP), specified in the 1988 versions* of ANSI T1.111 and ANSI T1.112,

---

*Note that IS-41 Revisions A, B, and C reference the Issue 1 (1988) versions of the ANSI MTP and SCCP standards. It is possible that a more current version of SS7 will be specified in the future.

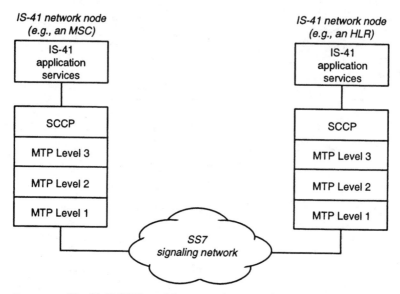

Figure 8.11   The IS-41 SS7-based data transfer protocol structure.

respectively. According to the SS7 protocol structure, the MTP provides physical, data link, and a portion of the network layer service; the SCCP provides the balance of the network layer service (see Fig. 8.11).*

Message Transfer Part.   With the exception of IS-41-0, the SS7 MTP specified in IS-41 is as defined in ANSI T1.111; i.e., no restrictions or limitations on the use of ANSI MTP are indicated in IS-41. IS-41-0 included a lot of MTP information that was deleted in subsequent revisions in favor of a simple reference to T1.111.

Signaling Connection Control Part.   Unlike the MTP, the SS7 SCCP specified in IS-41 is a subset of the ANSI standard, T1.112. Each IS-41 revision identifies the particular limitations on the use of ANSI SCCP that apply. However, one characteristic is common across all revisions: Only SCCP class 0 connectionless service is used. In other words, each SCCP message is a "datagram," containing source and destination address information. The connection-oriented services defined in the ANSI SCCP standard—involving connection establishment and connection termination phases—are not used in the IS-41 SCCP layer.

---

*Although we do not go into detail regarding the functions of the MTP and SCCP layers in this book, Chap. 17 discusses some IS-41 implementation-related issues. For a more thorough treatment, see *Signaling System #7* by Travis Russell.

IS-41 Application Services

IS-41 specifies a single set of application services—and supporting protocols—comprising the SS7 transaction capabilities (TCs) defined in ANSI T1.114, along with the IS-41-specific application services and protocols, called the mobile application part (MAP), as shown in Fig. 8.12.

The TC transport, session, and presentation layers are null layers in IS-41, just as they are described in ANSI T1.114. They are included in the IS-41 protocol model primarily to align with the ANSI definition of transaction capabilities. In fact, a typical IS-41 implementation would have the TCAP entity using the IS-41 data transfer services directly (see Fig. 8.13). Thus, it is appropriate to concentrate our discussion of IS-41 application services on the MAP and TCAP entities and their constituent elements (see Fig. 8.14).

IS-41 use of ANSI TCAP

As illustrated in Fig. 8.14, the IS-41 mobile application part is supported by the ANSI Transaction Capabilities Application Part. TCAP, in turn, is composed of two sublayers:

1. The *component sublayer,* which provides the communications tools MAP uses to realize IS-41 operations

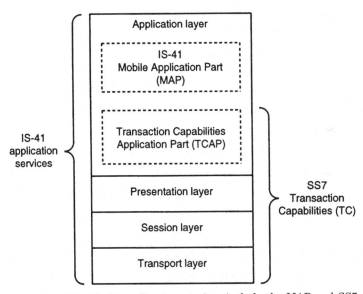

Figure 8.12 The IS-41 application services include the MAP and SS7 Transaction Capabilities. TC, in turn, comprises TCAP along with the supporting presentation, session, and transport layers.

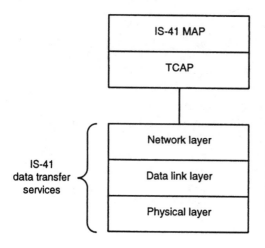

Figure 8.13 In a typical IS-41 implementation, TCAP would access data transfer services directly, rather than go through presentation, session, and transport layer interfaces.

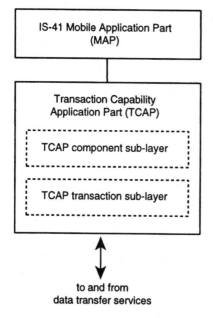

Figure 8.14 The structure of the IS-41 MAP and TCAP entities.

2. The *transaction sublayer,* which provides the communications tools MAP uses to associate multiple operations as parts of a single, logical transaction between two functional entities

The concept of an *operation* is key to IS-41, as evidenced by its formal title, "Cellular Radio Telecommunications Intersystem

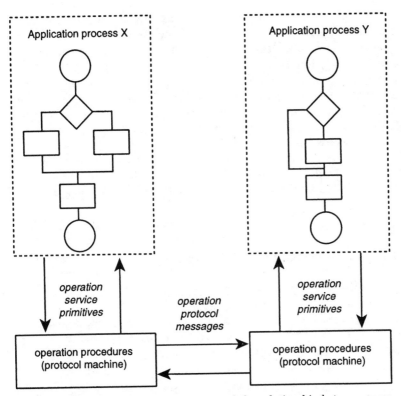

Figure 8.15 The elements of an operation and the relationship between operations and application processes.

Operations." We consider an operation as encompassing a collection of application-layer communications mechanisms (see Fig. 8.15):

1. The operation procedures that specify the rules governing the information content and exchange between operation users. A set of related operation procedures is sometimes referred to as a *protocol machine.*

2. The operation service primitives that an application process uses to access the operation's communications capabilities.

3. The operation protocol messages exchanged between peer protocol machines.

An application process in one functional entity uses operations to access another peer entity's application processes. For example, an application process in an MSC uses the LocationRequest operation to access a particular application process in an HLR; this would involve sending a LocationRequest Invoke message from the MSC protocol machine to the HLR protocol machine. Likewise, if the HLR application process is so defined, the HLR may send the results of the application process execution to the invoking entity using its protocol machine.

The ANSI TCAP component sublayer defines each of these operation elements: TCAP's component-handling procedures map the component-handling *service primitives* onto *components* (i.e., messages). TCAP provides six component types as the basis for operation definition, only four of which are used in IS-41:

1. *Invoke (Last)*. This component is used to request initiation of the remote application process. Its format and IS-41 encoding are shown in Table 8.1.

TABLE 8.1   Format of the TCAP Invoke (Last) Component

Component element name	Description
Component type identifier	This field identifies the component as an Invoke.
Component length	This is the length (in octets) of the Invoke component from the next element to the end of the component.
Component ID identifier	This indicates that the component ID follows.
Component ID length	This is the length (in octets) of the component ID that follows.
Component ID	In the Invoke component, this is the invoke ID, i.e., the identifier assigned by the TCAP user which allows correlation of invokes and responses.* The invoke ID is one octet in length.
Operation code identifier	This indicates that the operation code follows. It is coded in IS-41 so that IS-41 operation codes can be defined outside of the ANSI TCAP standard, T1.114, that is, as private TCAP.
Operation code length	This is the length (in octets) of the operation code that follows.
Operation code	The operation code is split into an operation family followed by an operation specifier; each is one octet in length. A different operation specifier is assigned to each IS-41 operation (e.g., LocationRequest).
Parameter set identifier	This indicates that a parameter set follows.
Parameter set length	This is the length (in octets) of the parameter set that follows.
Parameter set	This is the set of parameters associated with the Invoke component of the IS-41 operation (e.g., MIN, ESN).

*In general, TCAP allows the component ID to include an invoke ID and a linked ID; however, the linked ID is not used in IS-41 Revisions 0 through C.

2. *Return Result (Last).* This component is used to return the results of executing the application process. Its format is shown in Table 8.2.

3. *Return Error.* This component reports the unsuccessful completion of an invoked application process. Its format is shown in Table 8.3.

4. *Reject.* This component reports the receipt and rejection of an incorrect component or transaction. Its format is shown in Table 8.4.

5. *Invoke (Not Last).* This component is not used in IS-41.

6. *Return Result (Not Last).* This component is not used in IS-41.

Likewise, the ANSI TCAP transaction sublayer provides seven "package types" to define operation associations, only five of which are used in IS-41. All TCAP package types used in IS-41 share a common format,* shown in Table 8.5.

1. *Query with permission.* This package initiates a TCAP transaction and informs the destination node that it may end the transaction.

2. *Query without permission.* This package initiates a TCAP transaction and informs the destination node that it may not end the transaction.

TABLE 8.2   Format of the TCAP Return Result (Last) Component

Component element name	Description
Component type identifier	This field identifies the component as a Return Result.
Component length	This is the length (in octets) of the Return Result component from the next element to the end of the component.
Component ID identifier	This indicates that the component ID follows.
Component ID length	This is the length (in octets) of the component ID that follows.
Component ID	In the Return Result component, this is the correlation ID, i.e., the received operation invoke ID to which this Return Result component applies. The correlation ID is one octet in length.
Parameter set identifier	This indicates that a parameter set follows.
Parameter set length	This is the length (in octets) of the parameter set that follows.
Parameter set	This is the set of parameters associated with the Return Result component of the IS-41 operation.

---

*The Abort package type has a different format, but is not used in IS-41. For a description, see *Signaling System # 7*, by Travis Russell.

TABLE 8.3 Format of the TCAP Return Error Component

Component element name	Description
Component type identifier	This field identifies the component as a Return Error.
Component length	This is the length (in octets) of the Return Error component from the next element to the end of the component.
Component ID identifier	This indicates that the component ID follows.
Component ID length	This is the length (in octets) of the component ID that follows.
Component ID	In the Return Error component, this is the correlation ID, i.e., the received operation invoke ID to which this Return Error component applies. The correlation ID is one octet in length.
Error code identifier	This indicates that an error code follows. It is coded in IS-41 so that IS-41 error codes can be defined outside of the ANSI TCAP standard, T1.114.
Error code length	This is the length (in octets) of the error code that follows.
Error code	This indicates the reason why the associated operation did not complete successfully. A number of error codes are defined in IS-41.
Parameter set identifier	This indicates that a parameter set follows.
Parameter set length	This is the length (in octets) of the parameter set that follows.
Parameter set	In IS-41, a single parameter may be contained in the parameter set; it is named FaultyParameter.

3. *Conversation with permission.* This package continues a TCAP transaction and informs the destination node that it may end the transaction.

4. *Conversation without permission.* This package continues a TCAP transaction and informs the destination node that it may not end the transaction.

5. *Response.* This package ends the TCAP transaction.

6. *Unidirectional.* This package is not used in IS-41.

7. *Abort.* This package is not used in IS-41.

By defining standard component and package formats—and managing the procedural aspects of component and package exchange—TCAP provides consistent and versatile operation-based communication facilities to the TCAP user. In the IS-41 case, the TCAP user is the mobile application part.

TABLE 8.4   Format of the TCAP Reject Component

Component element name	Description
Component type identifier	This field identifies the component as a Reject.
Component length	This is the length (in octets) of the Reject component from the next element to the end of the component.
Component ID identifier	This indicates that the component ID follows.
Component ID length	This is the length (in octets) of the component ID that follows.
Component ID	In the Reject component, this is the correlation ID; i.e., the received operation invoke ID or correlation ID to which this Reject component applies. The correlation ID is one octet in length.
Problem code identifier	This indicates that a problem code follows. The problem code encoding specified in the ANSI TCAP standard, T1.114, is used in IS-41.
Problem code length	This is the length (in octets) of the problem code that follows.
Problem code	This indicates the reason why the component or transaction portion was rejected. The problem code is split into a problem type followed by a problem specifier; each is one octet in length.
Parameter set identifier	This indicates that a parameter set follows.
Parameter set length	This is the length (in octets) of the parameter set that follows.
Parameter set	In IS-41, a single parameter may be contained in the parameter set; it is named FaultyParameter.

IS-41 Mobile application part

The "mobile personality" of the IS-41 protocol architecture is embodied in the IS-41 mobile application part, or MAP. Other protocol architectures build on the SS7 transaction capabilities—along with suitable data transfer services—to support distributed telecommunications network applications (e.g., the 800-number database service). However, MAP is defined specifically for distributed mobile telecommunications network applications.

Up until now, we have shown MAP as purely an application-layer entity; this is the way it is illustrated in IS-41-C. However—once again assuming the OSI point of view—MAP can be considered to encompass two sets of functions (see Fig. 8.17):

1. The communications functions—also called *application service element* (ASE) functions—within a mobile network functional entity (e.g., an HLR) that are considered to be part of the OSI application layer

TABLE 8.5   Format of TCAP Packages

Package element name	Description
Package type identifier	This field identifies the package type, e.g., query with permission.
Total TCAP message length	This is the total length (in octets) of the TCAP message.
Transaction ID identifier	This indicates that the transaction IDs follow.
Transaction ID length	This is the length (in octets) of the transaction IDs that follow.
Transaction IDs	Transaction IDs (TIDs) are used to enable transaction association. One or two TIDs may be present in an IS-41 TCAP package; each is four octets in length with the originating TID (if present) followed by the responding TID (if present), although one must be present in IS-41. Table 8.6 shows the association between TIDs and package types for IS-41. Figure 8.16 shows how the TIDs are used in typical IS-41 transactions.
Component sequence identifier	This indicates that a component sequence follows.
Component sequence length	This is the length (in octets) of the component sequence that follows.
Component sequence	This is a sequence of one or more TCAP components; however, IS-41 requires (i.e., based on the figures in IS-41 illustrating the TCAP formats) that there be a single component per transaction.

TABLE 8.6   Association between TCAP Packages and Transaction IDs

Package type identifier	Originating TID	Responding TID
Query with permission	Present	Not present
Query without permission	Present	Not present
Response	Not present	Present
Conversation with permission	Present	Present
Conversation without permission	Present	Present

2. The data processing functions—also called *application processes*—within a mobile network functional entity that are considered to reside above the OSI application layer and that make use of the IS-41 ASE functions

The MAP ASE functions are the MAP-specific operations that are defined using TCAP, operations such as LocationRequest,

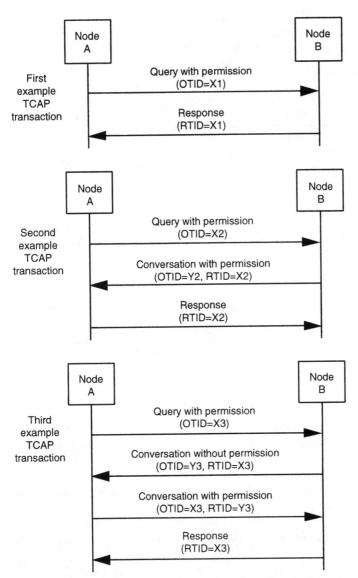

Figure 8.16 Examples of three TCAP transactions showing how the originating transaction IDs (OTIDs) and responding transaction IDs (RTIDs) are manipulated.

RoutingRequest, and QualificationDirective (which we discuss in later chapters). The components of these operations, e.g., LocationRequest Invoke and LocationRequest Return Result, are often referred to as IS-41 messages; we use this convention in this book. Together with the transaction functions offered by TCAP, the

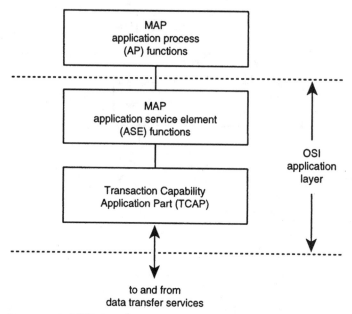

Figure 8.17 MAP functions include application process functions which lie above the OSI application layer and application service element functions which lie within the OSI application layer.

MAP ASE functions represent a broad set of communication capabilities to application processes.

The MAP application processes effectively extend the scope of the IS-41 standard beyond the OSI application layer, which deals exclusively with communications functions, and into the domain of application process definition. As a simple example, IS-41 procedures specify how an HLR shall cancel a mobile station's service qualification in one system when the HLR is notified that the MS has registered in another system. In this and many other situations, MAP steps beyond the bounds of OSI communication. Of course, this is an absolute necessity in IS-41; a judicious amount of application process specification in the standard avoids unnecessary incompatibilities between implementations. In a mobile environment, consistency is particularly important. A subscriber's serving system may change from vendor's A equipment to vendor's B, even during a call; however, from the point of view of the services provided to the subscriber, this change should not be detectable—the transition should be *seamless*. One of the ongoing challenges for the developers of IS-41 is to specify the MAP application processes to the extent that seamless service can be achieved while not constraining innovation in the individual vendor implementations.

An IS-41 MAP application service interface

An IS-41 implementation designed based on OSI layering concepts would likely include a clearly defined interface between the MAP application processes and application service element (ASE) functions. This interface separates the data processing functions of IS-41 (i.e., the processes) from the purely communications-related functions (i.e., the ASE functions), thereby off-loading communication details while providing a consistent set of communication services to a variety of application processes. However, this interface is not defined in IS-41 since it represents an internal interface found within an IS-41 implementation—clearly not an intersystem issue. Figure 8.18 illustrates a basic model for such an interface; we refer to it as the IS-41 MAP *application service interface* (ASI).

As shown in Fig. 8.18, the MAP ASI consists of a set of OSI-style service primitives of the following types:

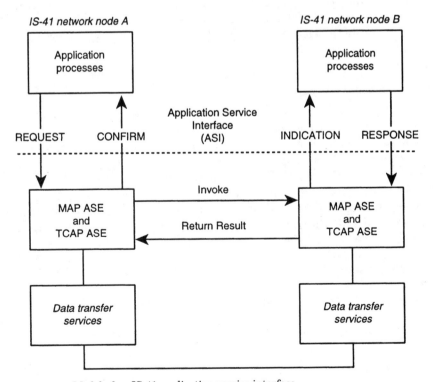

Figure 8.18   Model of an IS-41 application service interface.

1. *REQUEST primitive.* This type of primitive allows a service user A to invoke an application process in a remote peer service user B.

2. *INDICATION primitive.* This type of primitive allows the MAP ASE to indicate to service user B that a peer service user has requested an application process invocation, thereby initiating the process in B.

3. *RESPONSE primitive.* This type of primitive allows service user B to send the results of the application process execution to service user A.

4. *CONFIRM primitive.* This type of primitive allows the MAP ASE to notify service user A of the response provided by service user B.

Each MAP ASI primitive has an associated IS-41 ASE function; e.g., issuing an ASI primitive called MAP-LocationRequest-REQUEST results in the transmission of an *application protocol data unit* (APDU)—consisting of the IS-41 operation component (i.e., LocationRequest Invoke) contained in a TCAP package (i.e., query with permission)—between peer MAP ASEs. Most IS-41 ASE functions are confirmed and, therefore, are accessed by using all four of the ASI primitive types. The MobileOnChannel operation is the only exception: It uses only the REQUEST and INDICATION primitives (as described in Chap. 9).

An example of MAP operation encoding

To illustrate how MAP and TCAP elements work together to convey application information between MAP application processes, Table 8.7 provides an example of an IS-41 MAP operation encoding as it would appear in a TCAP package. The example is of the Invoke component of the RegistrationCancellation operation (described in Chap. 10). As specified in IS-41, this component would appear in a TCAP query with permission package. In this example, the Invoke component has two parameters: MobileIdentificationNumber (MIN) and ElectronicSerialNumber (ESN).

TABLE 8.7   Example of MAP RegistrationCancellation Invoke Component
Encoding

Package element name	Description	IS-41 encoding (hexadecimal)
Package type identifier	The package type is query with permission.	E2
Total TCAP message length	This length is 32 octets in this example.	20
Transaction ID identifier	This indicates that the transaction ID follows.	C7
Transaction ID length	This length is 4 octets.	04
Transaction ID	The originating transaction ID chosen for this example.	00 00 00 51
Component sequence identifier	This indicates that a component sequence follows.	E8
Component sequence length	This length is 24 octets in this example.	18
Component type identifier	This field identifies the component as an invoke.	E9
Component length	This length is 22 octets in this example.	16
Component ID identifier	This indicates that the component ID follows.	CF
Component ID length	The invoke ID length is 1 octet in this example.	01
Component ID	This is the invoke ID chosen for this example.	03
Operation code identifier	This indicates that the operation code follows.	D1
Operation code length	The operation code is 2 octets in this example.	02
Operation code	The operation code is split into the operation family (09) followed by the RegistrationCancellation operation specifier (0E).	09 0E
Parameter set identifier	This indicates that a parameter set follows.	F2
Parameter set length	This length is 13 octets in this example.	0D
Parameter identifier	This is the identifier for the first parameter, MIN.	88
Parameter length	Length (in octets) of MIN.	05
Parameter value	The 10-digit MIN (408-296-0303) encoded per IS-41.	04 28 69 30 30
Parameter identifier	This is the identifier for the second parameter, ESN.	89
Parameter length	Length (in octets) of the ESN.	04
Parameter value	The actual 32-bit ESN (all zeros in this example).	00 00 00 00

9

Basic Intersystem Handoff Functions

In this chapter, we discuss the basic IS-41-C mobile telecommunications network functions related to intersystem handoff. These functions are divided into the following five categories:*

- Handoff measurement
- Handoff forward
- Handoff back
- Path minimization
- Call release

We also examine the IS-41-C application processes that support these basic network functions. While there are other handoff-related processes in IS-41-C—e.g., for authentication, voice feature, and short message services support, we treat these separately in later chapters. In this chapter we focus on the basic functions required to hand off a call from one system to another.

IS-41-C does not include complete specifications for each of the basic intersystem handoff processes; for example, IS-41-C does not define the circumstances under which an MSC should attempt path minimization, leaving this as an implementation issue. This is a deliberate omission and an example of the standards approach taken in IS-41-C: "It is intended that the procedures defined address only

---

*The list of basic intersystem handoff functions reflects the authors' subjective categorization of certain IS-41-C functions.

the required intersystem transactions without infringing on the right of individual system operators and manufacturers to design their internal methods and procedures as they may deem best."*

In the course of describing the processes, we identify the IS-41-C mobile application part (MAP) operations (e.g., FacilitiesDirective) that are used to accomplish intersystem handoff process tasks. We summarize this information at the end of the chapter. Note that we employ the IS-41-C convention for operation component acronyms; i.e., the Invoke component acronym is in all capital letters (e.g., **FACDIR**), while the Return Result component acronym is in all lowercase letters (e.g., **facdir**). Refer to the IS-41-C standard for the descriptions of the individual IS-41-C operations and parameters. Also, the reader should consult the Glossary for the description of general terms (e.g., *serving system*) that are not explicitly defined in this chapter.

What Is Intersystem Handoff?

Handoff encompasses a set of mobile station (MS) functions and network functions that enable an MS to move from one radio channel to another radio channel while a call is in progress. There are two categories of handoff (see Fig. 9.1): intrasystem handoff and intersystem handoff.

Intrasystem handoff is a handoff between two radio channels that are controlled by the same mobile switching center (MSC). No coordination is required between MSCs to support intrasystem handoff; therefore, intrasystem handoff is not within the scope of IS-41. Intersystem handoff is a handoff between two radio channels that are controlled by two different MSCs. This type of handoff requires specialized signaling between the two MSCs to coordinate the movement of the MS between the two radio channels. The IS-41 protocol provides this signaling.

Where Are Intersystem Handoff Functions Specified in IS-41-C?

The initial version of IS-41, IS-41-0, was primarily an intersystem handoff standard: Of the 12 intersystem operations defined in this early standard, 10 were concerned with intersystem handoff and maintenance of the trunks used for intersystem handoff. While automatic roaming now accounts for the majority of IS-41 Revision C (IS-41-C) operations (see Fig. 9.2), intersystem handoff remains an important part of the IS-41 standard.

---

*From IS-41.1-C, "General Background and Assumptions."

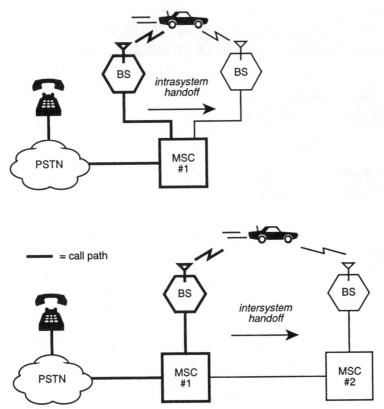

Figure 9.1 Intersystem handoff occurs between two cells controlled by two different MSCs. Intrasystem handoff occurs between two cells controlled by the same MSC.

Intersystem handoff functions are specified in three parts of IS-41-C:

1. IS-41.5-C provides the required formats—the bit-by-bit encoding—of all the IS-41-C operation components, including those used for intersystem handoff. IS-41.5-C defines both the messages (e.g., FacilitiesDirective Invoke) and the message parameters (e.g., InterMSCCircuitID, TargetCellID).

2. IS-41.6-C provides algorithmic descriptions of the procedures that are associated with sending and receiving most IS-41-C messages, including those used for intersystem handoff. As we pointed out in Chap. 8, IS-41-C does not define a service interface between the MAP application processes and the MAP application service element (ASE); therefore, the procedures in IS-41.6-C encompass both application process descriptions and ASE procedural descriptions. Furthermore, the procedures leave considerable room for implemen-

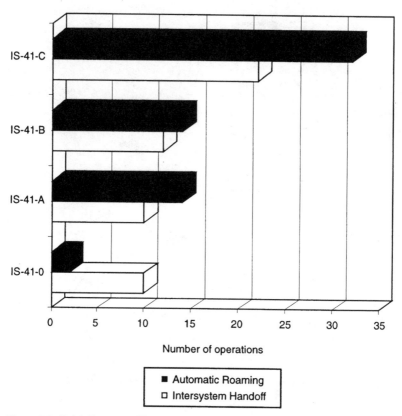

Number of operations

■ Automatic Roaming
□ Intersystem Handoff

Figure 9.2 Initially, most IS-41 operations supported intersystem handoff. A more balanced mix of handoff and automatic roaming operations existed in IS-41-A and IS-41-B. However, automatic roaming operations account for well over half the operations in IS-41-C.

tation-dependent customization; e.g., there are numerous references in IS-41.6-C to "local procedures" and "internal algorithms."

3. IS-41.2-C steps back from the protocol details contained in parts 5 and 6 and attempts to explain, by using information flow diagrams and step-by-step descriptions, how the operations are used individually and together to accomplish application process tasks. Although IS-41.2-C is a good place to begin to tackle IS-41-C intersystem handoff, IS-41.5-C and IS-41.6-C contain the definitive requirements against which an IS-41-C implementation is evaluated.*

*If inconsistencies are detected between parts, IS-41.5-C and IS-41.6-C take precedence over IS-41.2-C. Conflicting requirements between IS-41.5-C and IS-41.6-C are a more serious problem. In either case, the errors should be brought to the attention of the TIA TR45.2 subcommittee.

Issues Associated with Intersystem Handoff

Although the technical challenge of reliably executing a call handoff was (and still is) significant, the intersystem issues are relatively limited and include the following:

1. *Coordinating cell identification between neighboring MSCs.* To locate an MS for signal quality measurement and other handoff purposes, neighboring MSCs must agree on a cell identification scheme (see Fig. 9.3).

2. *Coordinating inter-MSC facility identification between neighboring MSCs.* IS-41-C handoff requires dedicated inter-MSC trunks that must be uniquely identifiable by the MSCs at both ends of the connection (see Fig. 9.3).

3. *Supporting MS characteristics after handoff.* For example, the AMPS standard allows some variability in MS characteristics (e.g., the MS power class can be 0.6, 1.6, or 4 W). The MSCs involved in the

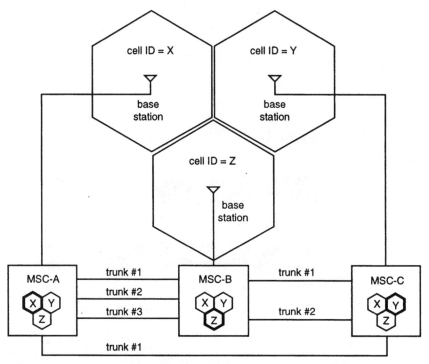

Figure 9.3 MSC-A, MSC-B, and MSC-C must have a common understanding of cell identities (X, Y, and Z) to manage the location information that they exchange. Pairs of MSCs (say, MSC-A and MSC-B) also must agree on the identification of the inter-MSC trunks that connect them.

handoff must ensure that these characteristics can be supported after the call handoff occurs.

4. *Limiting the length of the "handoff chain."* As a call is handed off from the first serving MSC (the anchor MSC) to another serving MSC and so on, a handoff chain develops, i.e., the sequence of MSCs, from the anchor to the current serving MSC, actively involved in the call at a given time. The length of the handoff chain is controlled in two different ways: by counting the number of inter-MSC facilities involved in the call and limiting that number to a value set by the individual cellular service provider and by providing a handoff-back operation, which prevents "shoelace" connections from MSC-A to MSC-B and back to MSC-A.

Early versions of IS-41 intersystem handoff (i.e., in IS-41-0 and IS-41-A) comprised a straightforward collection of functions involving a small set of messages, parameters, and procedures required to hand off analog AMPS calls. A number of factors have contributed to a rather drastic increase in the complexity of the IS-41 intersystem handoff functions in IS-41-C compared to IS-41-A (see Fig. 9.4):

1. *New air interface standards.* The development of the narrowband AMPS (NAMPS) analog air interface standard, and the time division multiple access (TDMA) and code division multiple access (CDMA) digital air interface standards, has required protocol enhancements

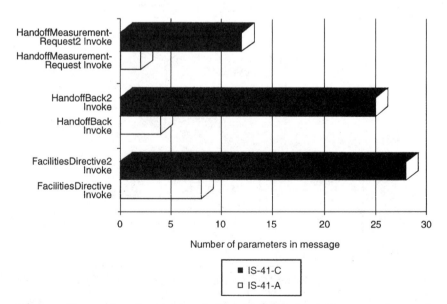

Figure 9.4 The number of parameters in the intersystem handoff messages increased dramatically between IS-41-A and IS-41-C.

to manage handoff between the different air interface types (e.g., AMPS to TDMA) in addition to handoff between the same air interface types (e.g., CDMA to CDMA). TDMA support was first incorporated in IS-41-B; CDMA and NAMPS were added in IS-41-C.

2. *Handoff path minimization techniques* (described in this chapter and initially included in IS-41-B).

3. *The addition of techniques that support handoff forward prior to the call being answered* (described in this chapter and new in IS-41-C).

4. *Feature support after handoff (included in IS-41-B and extended in IS-41-C).* MSC support after handoff of features like three-way calling and call waiting is described in later chapters.

Handoff Measurement

Handoff measurement functions may be used during the prelude to the actual call handoff, a period that includes the following major decision-making stages:

1. *Identify the need.* Is a call handoff appropriate at this time?

2. *Identify the candidates.* Which cell(s), and associated MSC(s), should be considered for call handoff purposes?

3. *Evaluate the candidates.* How suitable is each to handle the call?

4. *Select a target.* Which candidate is most suitable to handle the call?

Once these preliminaries are out of the way, the serving system can decide (1) whether to attempt call handoff and (2) what type of handoff, e.g., forward or back, is most appropriate under the circumstances.

Based on received signal quality measurements, the MS, serving system, or both can determine that there is a need to perform a handoff to another channel or cell. These three strategies are known as:

1. MS-controlled handoff

2. Network-controlled handoff

3. MS-assisted handoff (MAHO)

IS-41-C supports network-controlled handoff and MAHO only. Analog AMPS and NAMPS MSs do not provide signal measurements to support a handoff determination; thus they rely on network-controlled handoff techniques supported by IS-41-C. TDMA and CDMA MSs are able to provide measurements of received base station signal strength to the serving MSC without the need for the IS-41-C handoff measurement processes.

The radio channel received signal quality measurements taken by either the serving base station or both the serving base station and the MS (i.e., in the case of MAHO) reveal the quality of the transmission. Measurements may also be taken by adjacent base stations as the MS strays toward their coverage areas. IS-41-C's handoff measurement role is limited to requesting and conveying the received signal quality measurements between systems to assist the serving system's evaluation of handoff candidates.

Handoff measurement processes

IS-41-C handoff measurement processes include (1) the serving MSC handoff measurement process as well as (2) the candidate MSC handoff measurement process that executes in each candidate MSC.

The serving MSC handoff measurement process may be used to evaluate a candidate MSC's suitability for the role of serving MSC, that is, to accept a call handoff. Since multiple candidates may be involved, the serving MSC is capable of requesting and collecting measurement data from one or several candidate MSCs (see Fig. 9.5).

As shown in Fig. 9.5, the serving MSC uses the HandoffMeasurementRequest Invoke (**HANDMREQ**) message* to request the measurement data from the candidate MSC.

The candidate MSC handoff measurement process provides handoff suitability information to the serving MSC. This information may be either explicit or implicit. Explicit information includes measures of signal quality that the candidate MSC provides to the serving MSC. Implicit information is provided to the serving MSC simply by the candidate MSC's failure to respond to the serving MSC's request for measurement information (e.g., as candidate MSC 2 does in Fig. 9.5). In the latter case, the candidate MSC may not respond because:

1. The candidate MSC cannot support the requested radio channel characteristics; e.g., a TDMA digital channel may be required but is not available.

2. The signal quality that the candidate MSC measures does not meet its internal criteria for acceptability.

---

*Unless otherwise indicated, the HandoffMeasurementRequest2 operation may be substituted for references to the HandoffMeasurementRequest operation; e.g., either HandoffMeasurementRequest Invoke or HandoffMeasurementRequest2 Invoke messages may be used by the serving MSC. The HandoffMeasurementRequest2 operation was added to IS-41 in Revision C to support all air interface standards, including CDMA and narrowband AMPS (NAMPS) MSs; the HandoffMeasurementRequest operation supports TDMA and AMPS MSs, but not CDMA or NAMPS.

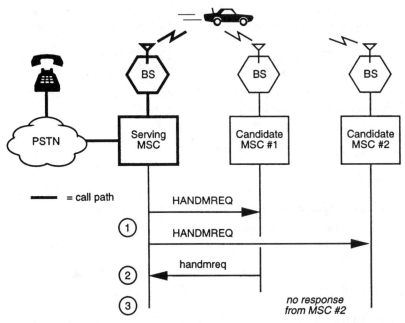

Figure 9.5 An example of the handoff measurement processes: (*1*) The serving MSC requests handoff measurements from both candidate MSC 1 and candidate MSC 2. (*2*) Candidate MSC 1 returns measurement information. (*3*) Candidate MSC 2 measures an unacceptable signal quality level, so does not respond.

3. The current traffic conditions* on the candidate MSC render it unavailable for handoff purposes at the given time; therefore, it chooses to disregard the serving MSC's request (although this information could be explicitly provided to the serving MSC via an error response).

While not sending a handoff measurement response to the serving MSC may be more efficient for the candidate MSC, it requires the serving MSC to wait the maximum time—7 s in IS-41 Revision C—before it can be sure that all candidate responses have been received.

As shown in Fig. 9.5, the candidate MSC's measurement information is conveyed to the serving MSC in the HandoffMeasurementRequest Return Result (**handmreq**) message.

*For example, radio channel resource availability influences the candidate MSC's ability to support handoff. There are many implementation-dependent strategies for allocating radio channels for handoff. For instance, some systems reserve a number of channels to support handoff from other systems, regardless of whether channels are available to new calls within that system. Other systems allocate channels for handoff only if they are available; and if not, the handoff is not accepted.

Once the serving MSC has selected the target MSC for the call handoff, it must decide on the form of handoff that is appropriate:

1. If the target MSC is already involved in the call and connected to the serving MSC via an inter-MSC circuit, a handoff back is required.

2. Otherwise, the serving MSC may attempt path minimization or choose simply to perform a handoff forward.

Handoff Forward

The IS-41-C handoff-forward functions provide the specialized form of call control signaling needed to (1) move the MS from a radio channel on the serving MSC to a compatible radio channel on the target MSC while (2) maintaining a call path between the MS and the other party to the active call by establishing a land line circuit between the serving MSC and the target MSC.

Compatibility of serving and target radio channels has a number of aspects including:

- The ability of the target to support the desired "call mode" (e.g., AMPS, CDMA, TDMA, or NAMPS)

- The ability of the target to support the characteristics of the MS (e.g., power class, use of discontinuous transmission)

- The ability of the target to support the "confidentiality modes" desired by the service subscriber (e.g., encryption of the radio channel)

Handoff forward also has a uniquely identifiable characteristic: When successfully executed, handoff forward results in an increase in the length of the handoff chain within the limit set by the MAX-HANDOFF system parameter.*

Finally, new provisions for call handoff prior to the call being answered were added to IS-41 in Revision C. Two cases are covered:

1. In the MS-originated call case, the MS is awaiting answer by the called party when handoff forward occurs. In this situation, special procedures are useful to inform the target MSC when the called party answers, e.g., for call timing purposes.

2. In the MS-terminated call case, the MS is being alerted when handoff forward occurs. In this situation, special procedures are useful to

---

*Note that the MAXHANDOFF parameter is not an IS-41-C parameter, but rather a system parameter that is programmed in each MSC by the mobile telecommunications service provider.

ensure that the MS is placed on the new target channel in the alerting state and to inform the anchor MSC when the MS answers.

IS-41-C handoff-forward processes include the serving MSC handoff-forward process and the target MSC handoff-forward process, illustrated in Fig. 9.6.

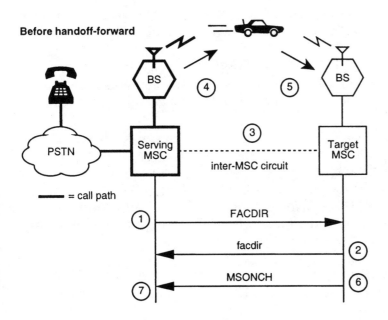

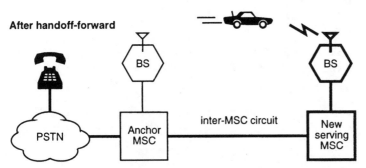

Figure 9.6 An example of the handoff-forward processes: (*1*) The serving MSC requests a handoff forward. (*2*) The target MSC accepts the handoff request. (*3*) The selected inter-MSC circuit is now ready for the handoff. (*4*) The serving MSC issues a handoff order to the MS. (*5*) The MS tunes to the new channel and is detected by the target MSC. (*6*) The target MSC notifies the serving MSC that the MS has been detected. (*7*) The serving MSC connects the call path to the inter-MSC circuit, completing the handoff.

Serving MSC handoff-forward process

The serving MSC handoff-forward process manages handoff-forward call control in the serving MSC. The key tasks for the serving MSC are as follows:

- Select and reserve the inter-MSC circuit that will provide the land line connection between the serving MSC and the target MSC.

- Request a call handoff to the target MSC. The serving MSC uses the FacilitiesDirective Invoke (**FACDIR**) message* for this purpose.

- When notified that the target MSC accepts the call handoff request via the FacilitiesDirective Return Result (**facdir**) message, direct the MS to move from the current serving radio channel to the desired target radio channel.

- When notified that the MS has moved to the target channel via the MobileOnChannel Invoke (**MSONCH**) message, connect the other party to the MS by using the inter-MSC circuit.

- If the handoff occurs while the MS is awaiting answer (i.e., in the case of an MS-originated call), wait until the called party answers and then inform the target MSC, using the InterSystemAnswer Invoke (**ISANSWER**) message (see Fig. 9.7).

- If the handoff occurs while the MS is alerting (i.e., in the case of an MS-terminated call), wait for the **ISANSWER** message from the target MSC, indicating the MS has answered the call; then acknowledge the notification by sending the InterSystemAnswer Return Result (**isanswer**) message to the target MSC and provide answer supervision toward the calling party.

Target MSC handoff-forward process

The target MSC handoff-forward process manages handoff-forward call control in the target MSC. The key tasks for the target MSC are as follows:

- Respond to the serving MSC's request for a call handoff, and—in the case of handoff acceptance—inform the serving MSC of the

---

*Unless otherwise indicated, the FacilitiesDirective2 operation may be substituted for references to the FacilitiesDirective operation; e.g., either FacilitiesDirective Invoke or FacilitiesDirective2 Invoke messages may be used by the serving MSC. The FacilitiesDirective2 operation was added to IS-41 in Revision C to support all air interface standards, including CDMA and narrowband AMPS MSs; the FacilitiesDirective operation supports TDMA and AMPS MSs, but not CDMA or NAMPS.

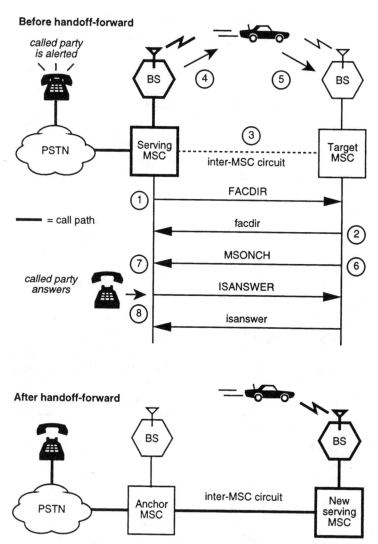

Figure 9.7 An example of the handoff-forward process while the MS is awaiting answer: (*1–7*) Same as Fig. 9.6, except serving MSC may include the HandoffState parameter in the **FACDIR** message to indicate that the handoff is occurring while the called party is alerting. (*8*) The called party answers; the serving MSC notifies the target MSC. The target MSC acknowledges the message, and the handoff is complete.

selected target radio channel; handoff acceptance is indicated by using the **facdir** message.

- Wait until the MS is detected on the target radio channel; then complete the path between the radio channel and the inter-MSC circuit.

- Notify the serving MSC that the MS has successfully moved to the new radio channel; the target MSC uses the **MSONCH** message for this purpose.

- If the handoff occurs while the MS is alerting (i.e., in the case of an MS-terminated call), wait until the MS answers; then inform the anchor MSC, using the **ISANSWER** message.

- If the handoff occurs while the called party is alerting (i.e., in the case of an MS-originated call), wait for the **ISANSWER** message from the anchor MSC, indicating the called party has answered the call; then acknowledge the notification by sending the (**isanswer**) message to the anchor MSC (see Fig. 9.7).

Handoff Back

Use of the handoff-forward functions to hand off a call back and forth between two MSCs would result in an excessively long handoff chain, composed of circular paths—creating a *tromboning* effect—between the two MSCs (see Fig. 9.8). This would deplete the inter-MSC resources available for other call handoff purposes. The IS-41-C hand-off-back functions are designed to avoid this situation.

Like handoff-forward, the IS-41-C handoff-back functions are used to move the MS from a radio channel on the serving MSC to a compatible radio channel on the target MSC; unlike with handoff forward, the target MSC is already connected to the serving MSC by an inter-MSC circuit. Therefore, once the MS is moved to the new channel, the unused inter-MSC circuit must be removed.

IS-41-C handoff-back processes include the serving MSC handoff-

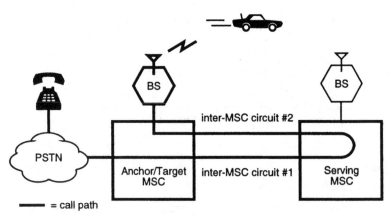

Figure 9.8 Repeated use of handoff forward between two MSCs results in a tromboning effect.

back process and the target MSC handoff-back process, illustrated in Fig. 9.9.

Serving MSC handoff-back process

The serving MSC handoff-back process manages handoff-back call control in the serving MSC. The key tasks for the serving MSC are as follows:

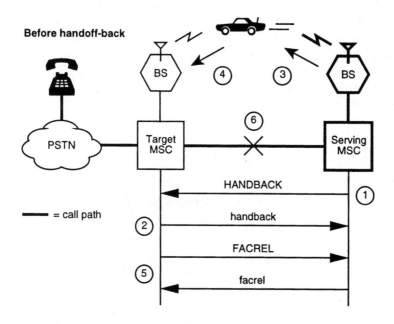

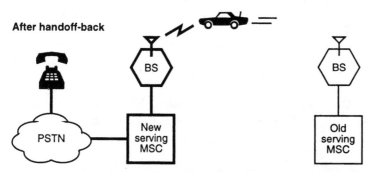

Figure 9.9   An example of the handoff-back processes: (*1*) The serving MSC requests a handoff back. (*2*) The target accepts the handoff request. (*3*) The serving MSC issues a handoff order to the MS. (*4*) The MS tunes to the new channel and is detected by the target MSC. (*5*) The target MSC requests the release of the unnecessary inter-MSC circuit and the serving MSC accepts the request. (*6*) The inter-MSC circuit is released and is now ready for other handoff purposes.

- Identify the inter-MSC circuit that already provides a land line connection between the serving MSC and the target MSC; this circuit must be released once the handoff-back process is completed.

- Request a handoff-back to the target MSC. The serving MSC uses the HandoffBack Invoke (**HANDBACK**) message* for this purpose.

- When notified that the target MSC accepts the call handoff request via the HandoffBack Return Result (**handback**) message, direct the MS to move from the current serving radio channel to the desired target radio channel.

- When notified that the MS has moved to the target channel via the FacilitiesRelease Invoke (**FACREL**) message, acknowledge the message by sending a FacilitiesRelease Return Result (**facrel**) message to the target MSC and release the inter-MSC circuit and radio channel, along with any other resources used for the call.

Target MSC handoff-back process

The target MSC handoff-back process manages handoff-back call control in the target MSC. The key tasks for the target MSC are as follows:

- Respond to the serving MSC's request for a call handoff and—in the case of handoff acceptance—inform the serving MSC of the selected target radio channel. Handoff acceptance is indicated by using the **handback** message.

- Wait until the MS is detected on the target radio channel. Then complete the path between the radio channel and the other party to the call.

- Notify the serving MSC that the MS has successfully moved to the new radio channel, and initiate the release of the inter-MSC circuit between the target and the old serving MSC; the target MSC uses the **FACREL** message for this purpose.

Path Minimization

While handoff-back functions provide a basic form of path minimization, eliminating the tromboning effect which would otherwise occur

---

*Unless otherwise indicated, the HandoffBack2 operation may be substituted for references to the HandoffBack operation; e.g., either HandoffBack Invoke or HandoffBack2 Invoke messages may be used by the serving MSC. The HandoffBack2 operation was added to IS-41 in Revision C to support all air interface standards, including CDMA and narrowband AMPS MSs; the HandoffBack operation supports TDMA and AMPS MSs, but not CDMA or NAMPS.

between the target and serving MSCs when a call is repeatedly handed back and forth between them, the IS-41-C path minimization functions refer to more complex handoff optimization techniques. These techniques were introduced in IS-41 Revision B and were updated to support the CDMA and NAMPS air interfaces in IS-41-C.

In general, MSCs in the handoff chain may use path minimization functions to eliminate unnecessary inter-MSC circuits between the anchor MSC and the target MSC. Handoff with path minimization ideally results in a single inter-MSC circuit directly between the anchor MSC and the target MSC (see Fig. 9.10), or no circuits if the anchor is the target (as shown in Fig. 9.11 in the next section). Alternatively, a tandem MSC may employ path minimization to eliminate unnecessary inter-MSC circuits between it and the target MSC. The latter option would be used only if the anchor MSC were unable to perform path minimization, e.g., if no inter-MSC circuit between the anchor and target MSCs were available, or if the handoff chain exceeded a system-specified length related to the TANDEMDEPTH system parameter.* The decision to attempt path minimization, rather than the simpler handoff forward, is an internal implementation issue outside the scope of IS-41-C.

IS-41-C path minimization processes include (1) the serving MSC path minimization process, (2) the anchor MSC path minimization process, and (3) the tandem MSC path minimization process. If the target MSC is not the anchor or tandem MSC, then the target MSC handoff-forward process is also involved, as illustrated in Fig. 9.10 and described below.

Serving MSC path minimization process

The serving MSC path minimization process manages path minimization call control in the serving MSC. The key tasks for the serving MSC, when there are no tandem MSCs in the handoff chain (this is the case in Fig. 9.10) are as follows:

- Identify the inter-MSC circuit that already provides a land line connection between the serving MSC and the anchor MSC; this circuit must be released once path minimization is completed.

- Request the anchor MSC to perform path minimization. The serv-

---

*Note that the TANDEMDEPTH parameter is not an IS-41-C parameter, but rather a system parameter that is programmed in each MSC by the mobile telecommunications service provider.

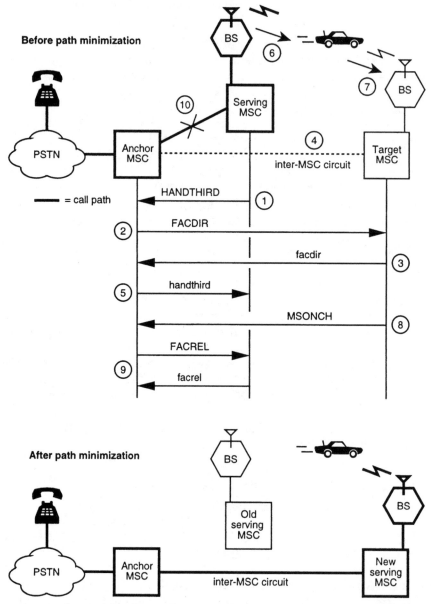

Figure 9.10 An example of the path minimization processes: (*1*) The serving MSC sends a path minimization request to the anchor MSC. (*2*) The anchor MSC attempts to determine if path minimization is possible; it sends a handoff request to the target MSC. (*3*) The target MSC accepts the handoff request. (*4*) The selected inter-MSC circuit between the anchor MSC and the target MSC is now ready for the handoff. (*5*) The anchor MSC accepts the path minimization request. (*6*) The serving MSC issues a handoff order to the MS. (*7*) The MS tunes to the new channel and is detected by the target MSC. (*8*) The target MSC notifies the anchor MSC that the MS has been detected. (*9*) The anchor MSC connects the call path to the inter-MSC circuit, completing the handoff. It then releases the unnecessary inter-MSC circuit to the serving MSC. (*10*) The inter-MSC circuit is released and is now ready for other handoff purposes.

ing MSC uses the HandoffToThird Invoke (**HANDTHIRD**) message* for this purpose.

■ When notified that the anchor MSC accepts the path minimization request via the HandoffToThird Return Result (**handthird**) message, direct the MS to move from the current serving radio channel to the desired target radio channel.

■ When notified by the anchor MSC that the MS has moved to the target channel via the **FACREL** message, acknowledge the message by sending a **facrel** message to the anchor MSC and release the inter-MSC circuit along with any other resources used for the call.

If there are one or more tandem MSCs in the handoff chain, the only change in the process is that the serving MSC must send the path minimization request toward the anchor MSC by way of the tandem MSC(s). Either the anchor MSC or possibly a tandem MSC may then provide the path minimization acceptance notification (i.e., the **handthird** message) to the serving MSC—the serving MSC is not explicitly informed of the source of the acceptance.

Anchor MSC path minimization process

The anchor MSC path minimization process manages path minimization call control in the anchor MSC. This process is initiated by the receipt of the **HANDTHIRD** message from the serving MSC (possibly via a tandem MSC). The key tasks of the process are, first, to determine if the anchor MSC is the target MSC for the handoff. If the anchor MSC is not the target (as in Fig. 9.10), then it performs the following steps:

■ Select and reserve the inter-MSC circuit that will provide the land line connection between the anchor MSC and the target MSC.

■ Request a call handoff to the target MSC using information (e.g., target cell identification) provided to the anchor MSC by the serving MSC in the **HANDTHIRD** message. The anchor MSC uses the **FACDIR** message to request the call handoff to the target MSC.

■ When notified that the target MSC accepts the call handoff request via the **facdir** message, notify the serving MSC that its path minimization request is accepted; the anchor MSC uses the **handthird** message for this purpose.

*Unless otherwise indicated, the HandoffToThird2 operation may be substituted for references to the HandoffToThird operation; e.g., either HandoffToThird2 Invoke or HandoffToThird2 Invoke messages may be used by the serving MSC. The HandoffToThird2 operation was added to IS-41 in Revision C to support all air interface standards, including CDMA and narrowband AMPS MSs; the HandoffToThird operation supports TDMA and AMPS MSs, but not CDMA or NAMPS.

- When notified that the MS has moved to the target channel via the **MSONCH** message, connect the other party to the MS, using the inter-MSC circuit established between the anchor and target MSCs.

- Initiate release of the inter-MSC circuit (or circuits, if there are one or more tandem MSCs) between the anchor MSC and the old serving MSC, using the **FACREL** message.

On the other hand, if the anchor MSC is the target MSC for the handoff, then it executes the following steps (see Fig. 9.11):

- Determine whether resources, e.g., a radio channel, are available for the call handoff; if not, reject the serving MSC's path minimization request by sending a HandoffToThird Return Error message to the serving MSC and end the process.

- Otherwise, notify the serving MSC that its path minimization request is accepted, using the **handthird** message.

- Wait until the MS is detected on the target radio channel. Then complete the path between the radio channel and the other party to the call.

- Notify the old serving MSC that the MS has successfully moved to the new radio channel, and initiate release of the inter-MSC circuit between the anchor and the old serving MSC; the anchor MSC uses the **FACREL** message for this purpose.

Tandem MSC path minimization process

The tandem MSC path minimization process manages path minimization call control in a tandem MSC. Like the anchor MSC path minimization process just described, this process is initiated when the tandem MSC receives the **HANDTHIRD** message from the serving MSC. The key tasks of the process are, first, to determine if the tandem MSC is the target MSC for the handoff. This would be the case only if there were multiple tandem MSCs in the handoff chain: The serving MSC uses handoff back if the tandem MSC—to which it is directly connected—is the target MSC. If the tandem MSC is not the target, it must determine whether it has reached the path minimization limit, based on the TANDEMDEPTH system parameter. For this, the tandem MSC performs a calculation using the following data (see Fig. 9.12):

1. The InterSwitchCount value (value X) received in the **HANDTHIRD** message sent by the serving MSC; this value represents the total number of inter-MSC circuits in the handoff chain (i.e., the number of MSCs in the chain less 1).

Before path minimization

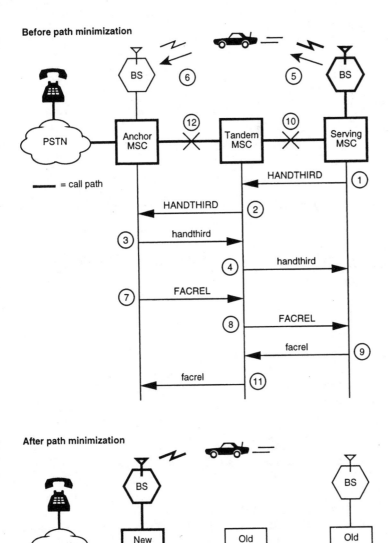

After path minimization

Figure 9.11 An example of path minimization when the anchor MSC is the target MSC: (1) The serving MSC sends a path minimization request toward the anchor MSC. (2) The tandem MSC relays the request to the anchor MSC. (3) The anchor MSC determines that it is the target for the handoff; therefore, it accepts the path minimization request. (4) The tandem MSC relays the response to the serving MSC. (5) The serving MSC issues a handoff order to the MS. (6) The MS tunes to the new channel and is detected by the anchor MSC. (7) The anchor MSC connects the call path to the MS, completing the handoff. It then releases the unnecessary inter-MSC circuit toward the serving MSC. (8) The tandem MSC requests release of the unnecessary inter-MSC circuit to the serving MSC. (9) The serving MSC accepts the release request. (10) The inter-MSC circuit between the tandem MSC and the serving MSC is released and is now ready for other handoff purposes. (11) The tandem MSC accepts the release request. (12) The inter-MSC circuit between the tandem MSC and the anchor MSC is released and is now ready for other handoff purposes.

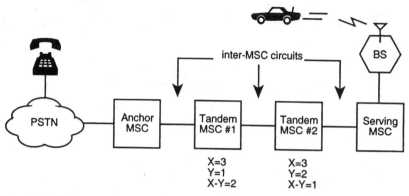

Figure 9.12 If the system parameter TANDEMDEPTH = 1, then tandem MSC 1 will not forward the **HANDTHIRD** message to the anchor MSC, since $X - Y = 2$, which is greater than TANDEMDEPTH.

2. The number of inter-MSC circuits in the segment of the handoff chain between the anchor MSC and the tandem MSC (value Y); the tandem MSC will have stored this value previously, before it performed a handoff forward.

The difference between these two numbers, $X - Y$, indicates the number of inter-MSC circuits between the serving MSC and the tandem MSC; when this number is greater than the system parameter TANDEMDEPTH, the tandem MSC is not permitted to relay the path minimization request any further down the handoff chain toward the anchor MSC. Instead, the tandem MSC must decide whether it—rather than the anchor MSC—shall attempt path minimization or simply reject the path minimization request by sending a HandoffToThird Return Error message to the serving MSC, ending the process.

However, if the tandem MSC is not the target and the path minimization limit has not been reached, then the tandem MSC performs the following steps:

- Relay the **HANDTHIRD** message back toward the anchor MSC.
- If a **handthird** message is received in response, relay the message toward the serving MSC and end the process.
- However, if a HandoffToThird Return Error message is received, the tandem MSC may either relay the message toward the serving MSC, ending the process, or attempt path minimization.

If the tandem MSC chooses to attempt path minimization—based on either of the circumstances described above—it performs the following steps:

- Select and reserve the inter-MSC circuit that will provide the land line connection between the tandem MSC and the target MSC.

- Request a call handoff to the target MSC, using information (e.g., target cell identification) provided to the tandem MSC by the serving MSC in the **HANDTHIRD** message. The tandem MSC uses the **FACDIR** message to request the call handoff to the target MSC.

- When notified that the target MSC accepts the call handoff request via the **facdir** message, notify the serving MSC that its path minimization request is accepted, using the **handthird** message.

- When notified that the MS has moved to the target channel via the **MSONCH** message, connect the other party to the MS, using the inter-MSC circuit between the tandem and target MSCs.

- Initiate release of the inter-MSC circuit(s) between the tandem MSC and the old serving MSC, sending the **FACREL** message toward the serving MSC.

Finally, if the tandem MSC is the target MSC for the handoff, then it executes the following steps:

- Determine whether resources, e.g., a radio channel, are available for the call handoff; if not, reject the serving MSC's path minimization request by sending a HandoffToThird Return Error message to the serving MSC, and end the process.

- Otherwise, notify the serving MSC that its path minimization request is accepted, using the **handthird** message.

- Wait until the MS is detected on the target radio channel. Then complete the path between the radio channel and the other party to the call.

- Notify the old serving MSC that the MS has successfully moved to the new radio channel, and initiate release of the inter-MSC circuit between the tandem MSC and the old serving MSC; the tandem MSC uses the **FACREL** message for this purpose.

Target MSC handoff-forward process and path minimization

As shown in Fig. 9.10, when the target is not the anchor MSC or a tandem MSC, then the target MSC simply executes the target MSC handoff-forward process that is described in the section entitled

Handoff Forward—the target MSC is unaware of the path minimization that is taking place.

Call Release

The IS-41-C call release functions control the release of the inter-MSC circuits and internal MSC resources at the end of a call that were used during the handoff process. Without these functions, the parties to the call could "hang up," but the handoff chain from the anchor MSC to the serving MSC would continue to exist. The call release can be initiated either by the anchor MSC, e.g., based on a call release indication received from the PSTN, or by the serving MSC, based on information it receives (or does not receive, in the case of lost radio contact) from the mobile station.

Call release processes

IS-41-C call release processes include the initiating call release process and the receiving call release process, illustrated in Fig. 9.13.

The initiating call release process sends a **FACREL** message to notify the receiving MSC of call release. When the initiator receives the response, it clears its internal resources used for the call (e.g., software processes and memory) and marks the inter-MSC circuit idle.

If the receiver is either the anchor MSC or the serving MSC, then the receiving call release process is completed simply by returning a **facrel** message to the initiating call release process and then marking the inter-MSC circuit idle. The receiver may initiate other processes, e.g., for features (see Chap. 12), to release the radio channel to the MS or the land line circuit to the PSTN. However, these processes are generally beyond the scope of IS-41-C.

In the case of a tandem MSC, both types of call release process are invoked during call release; on one side, the receiving process executes, while the initiating process executes on the other side. Also note that the initiating call release process is nested within the receiving call release process at the tandem MSC (see Fig. 9.13). This allows information (e.g., the total number of inter-MSC segments involved in the call, which is useful for recording purposes) received by the tandem MSC to be relayed toward the anchor MSC. This is a change to IS-41 introduced in IS-41-C; prior to Revision C, the operations executed in an *asynchronous* fashion, as shown in Fig. 9.14.

Before call release

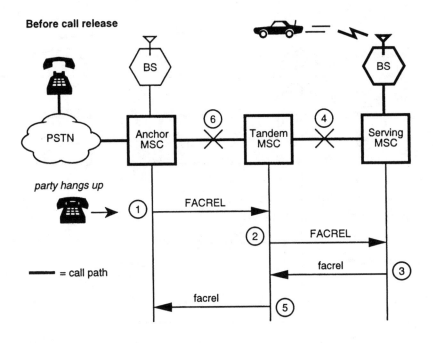

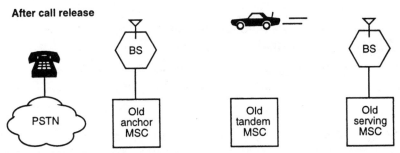

Figure 9.13 An example of the call release processes: (*1*) One of the parties to the call hangs up. The anchor MSC releases the unnecessary inter-MSC circuit toward the serving MSC. (*2*) The tandem MSC requests release of the unnecessary inter-MSC circuit to the serving MSC. (*3*) The serving MSC accepts the release request. The **facrel** message may contain billing information for use by the anchor MSC. (*4*) The inter-MSC circuit between the tandem MSC and the serving MSC is released and is now ready for other handoff purposes. (*5*) The tandem MSC accepts the release request. It forwards any billing information received in step 3. (*6*) The inter-MSC circuit between the tandem MSC and the anchor MSC is released and is now ready for other handoff purposes.

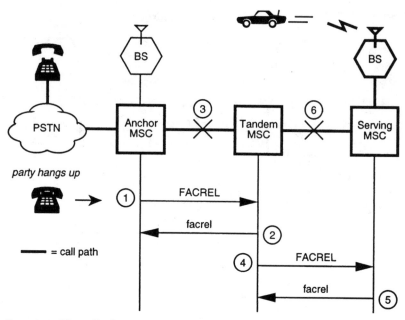

Figure 9.14 The call release processes prior to IS-41-C. Note that steps 4 and 5 are not nested within steps 1 and 2 as in IS-41-C.

Summary of IS-41-C Operations Used for Basic Intersystem Handoff

Table 9.1 summarizes the IS-41-C operations used by the intersystem handoff functions described in this chapter.

TABLE 9.1 Use of IS-41-C Operations for Intersystem Handoff

Function	IS-41-C operations used for the function
Handoff measurement	HandoffMeasurementRequest, HandoffMeasurementRequest2
Handoff forward	FacilitiesDirective, FacilitiesDirective2, InterSystemAnswer, MobileOnChannel
Handoff back	HandoffBack, HandoffBack2, FacilitiesRelease
Path minimization	HandoffToThird, HandoffToThird2, FacilitiesDirective, FacilitiesDirective2, FacilitiesRelease, MobileOnChannel
Call release	FacilitiesRelease

Chapter

10

Basic Automatic Roaming Functions

In this chapter, we discuss the basic IS-41-C mobile telecommunications network functions related to automatic roaming. These functions are divided into the following categories:*

- Mobile station (MS) service qualification

- MS location management

- MS state management

- HLR and VLR fault recovery

We also examine the IS-41-C application processes that support these basic network functions. While there are other automatic roaming-related processes in IS-41-C, e.g., for authentication, voice feature support, and short message services, we treat these separately in later chapters. This chapter focuses on the basic automatic roaming processes that allow an MS to obtain service in a visited system.

In the course of describing the processes, we identify the IS-41-C mobile application part (MAP) operations (e.g., RegistrationNotification) that are used to accomplish basic automatic roaming process tasks. We summarize this information at the end of the chapter. Note that we employ the IS-41-C convention for operation component acronyms; i.e., the Invoke component acronym is in all capital letters (e.g., **REGNOT**), while the Return Result component acronym is in all lowercase letters (e.g., **regnot**). Refer to the IS-41-C standard for the descriptions of the individual IS-41-C operations and parameters. Also, the reader should

---

*The list of basic automatic roaming functions reflects the authors' subjective categorization of certain IS-41-C functions.

consult the Glossary for the description of general terms (e.g., *serving system*) that are not explicitly defined in this chapter.

Throughout this chapter we consider the serving system as a single entity encompassing the mobile switching center (MSC) and visitor location register (VLR) functional entities. This simplifies the descriptions of IS-41-C application processes and also is representative of a large percentage of the IS-41 implementations currently in service. However, IS-41-C assigns specific processing duties to the VLR that make the concept of sharing a VLR among multiple MSCs a viable alternative to the combined implementation. Therefore, while we describe automatic roaming processes in terms of a single "MSC + VLR" entity, the reader should keep in mind that the potential for separation exists and is fully defined in IS-41-C.

What Is Automatic Roaming?

Automatic roaming encompasses a set of network functions that allow a subscriber to obtain mobile telecommunications service outside the home service provider area. These functions are automatic in the sense that they are invoked without requiring special subscriber actions. For example, the subscriber does not have to dial a code or make a call to inform the home system that the subscriber has roamed into another system—this function is provided automatically when the subscriber is detected in the new system (unlike the first generation of cellular roaming services that required manual activation or a special service call on at least a daily basis).

Registration

Detection of a subscriber in a new serving system is an example of a *registration event*. Registration events, or simply registrations, are the triggers for a number of the automatic roaming functions described in the chapter (see Fig. 10.1). The types of registrations

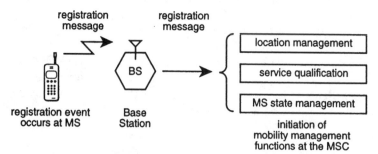

Figure 10.1   Registration can initiate other network functions. These functions must be performed before services are provided to a subscriber.

supported in a network are dependent on the protocol used for the air interface between the mobile station and the base station and on the internal algorithms implemented in the serving systems. The air interface standards for AMPS, TDMA, and CDMA support different types of registrations. The following events are common among AMPS, TDMA, and CDMA and occur most frequently:

1. Mobile station power-on

2. Timer-based (i.e., autonomous registration)

3. Transition to a new system

4. Call origination

Registration always occurs when the MS is initially turned on. This type of registration informs the network that the MS is now active and available to receive calls and other network services. Timer-based registration is more commonly referred to as *autonomous registration.* This type of registration occurs at periodic intervals while the MS is powered on. The interval between registrations ranges from around 10 min to 1 h. The periodic time value for autonomous registration is transmitted to the MS from the serving system, allowing each network to individually control the frequency of autonomous registrations. Autonomous registration helps keep the network from "losing" the current location of the subscriber while the MS is powered on and no calls are made. Registrations also occur when the MS makes a transition between systems while it is active. Registration can also occur during the process of call origination, so that the location of the MS can be updated in the network for billing of the call or the MS can again be qualified for service, if necessary.

There are other events that can cause registration to occur. For example, CDMA systems support registration when the mobile station is turned off (i.e., essentially a power-off deregistration) to inform the network that the MS is unavailable for service.

It is interesting to note that for systems not supporting a power-off deregistration procedure, a subscriber who turns off the MS can remain registered in a system for several minutes or even hours. The subscriber's registration is eventually canceled in that system when inactivity is detected after a time, when the subscriber registers in another system, or by a periodic maintenance procedure.

Where Are Automatic Roaming Functions Specified in IS-41-C?

Automatic roaming functions are specified in three parts of IS-41-C:

1. IS-41.5-C provides the required formats—the bit-by-bit encoding— of all the IS-41-C operation components, including those used for

automatic roaming. IS-41.5-C defines both the messages (e.g., RegistrationNotification Invoke) and the message parameters (e.g., QualificationInformationCode).

2. IS-41.6-C provides algorithmic descriptions of the procedures associated with sending and receiving most IS-41-C messages, including those used for automatic roaming. As we pointed out in Chap. 8, IS-41-C does not define a service interface between the MAP application processes and the MAP application service element (ASE); therefore, the procedures in IS-41.6-C encompass both application process descriptions and ASE procedural descriptions. Furthermore, the procedures leave considerable room for implementation-dependent customization; e.g., there are numerous references in IS-41.6-C to "local procedures" and "internal algorithms."

3. IS-41.3-C steps back from the protocol details contained in parts 5 and 6 and attempts to explain, using information flow diagrams and step-by-step descriptions, how the operations are used individually and together to accomplish automatic roaming application process tasks.

Issues Associated with Automatic Roaming

Automatic roaming accounts for the majority of IS-41-C operations. This can be attributed to two primary factors:

1. The complexity of the standard is being driven by new features, such as authentication and the dozens of IS-41-C calling features, while the impact of new features on intersystem handoff is limited by the definition of the anchor MSC function. The anchor MSC performs most of the feature processing, with only basic information transferred across the handoff chain.

2. Roaming is a higher-profile activity with service providers since their subscribers are usually much more likely to experience automatic roaming (e.g., to originate or receive a call in a visited system) than intersystem handoff (i.e., to move between systems during a call). Thus, there is more incentive to add value to the mobile communications experience in the area of automatic roaming.

When the subscriber roams outside the home system, a number of intersystem issues arise. From the visited system's perspective, automatic roaming issues include:

1. *Identification of the roaming MS's home system.* The typical mechanism used to relate roaming MSs to their home systems is the roaming agreement established between the visited and home system service providers.

2. *Verification of the roaming MS's identity.* At the very least, this involves obtaining validation information from the home system. Preferably, the MS is also authenticated by the home or visited system.

3. *Identification of the roaming MS's service capabilities.* This is an aspect of service qualification that allows the visited system to be aware of the MS's subscribed services.

From the home system's perspective, automatic roaming issues include

1. *Identification of the roaming MS's current location.* This is a key outcome of the location management processes.

2. *Identification of the roaming MS's current state.* Determine if the MS is available for call delivery and, if not, the method that the home system uses to process calls intended for the subscriber (e.g., deny the incoming call, forward the call to voice mail).

MS Service Qualification

The IS-41-C MS service qualification function encompasses the processes which establish a roaming MS's financial accountability and service capabilities in a serving system. Service qualification information includes one or both of the following:

1. Validation information

2. Service profile information

Validation information indicates the system—normally the home system—that is assuming financial responsibility for the roaming MS and the period for which this responsibility is assumed (e.g., one call, 1 h, 1 day). This information assures the serving system that someone will pay for the calls that the roaming MS makes; alternatively, the information may inform the serving system that validation is denied for the MS, possibly for a specified period to prevent the serving system from repeatedly re-requesting validation information from the HLR.

Service profile information indicates the specific set of features and other service capabilities, including restrictions on these capabilities, that are associated with the roaming MS; for example, features X, Y, and Z are active for the MS, and the MS is authorized to originate local calls only. The serving system uses this information to tailor the telecommunications services it provides to the MS.

An MS's HLR is the network repository for the MS's service qualification information as well as other detailed information related to the MS.

MS service qualification processes

The IS-41-C MS service qualification processes include the serving system service qualification process and the HLR service qualification process.

The serving system service qualification process is responsible for obtaining service qualification information for each visiting MS. The serving system may obtain service qualification information either by initiating a request to the HLR or by receiving a directive from the HLR.

The serving system issues service qualification requests to the HLR under various circumstances, including:

1. When the visiting MS is initially detected

2. When the previously obtained service qualification information for a visiting MS indicates a limited service area (e.g., a single location area) and the MS moves outside the service area

3. When the previously obtained service qualification information for a visiting MS indicates limited service duration and the authorized period expires

4. If for any other reason the serving system determines that it needs to retrieve service qualification information (e.g., the MS's service qualification period has expired and a call delivery request is received for the MS)

In cases 1 and 2, the serving system packages the service qualification request along with the location update information in a RegistrationNotification Invoke (**REGNOT**) message (see Fig. 10.2) to the HLR. In cases 3 and 4, the request should be via a QualificationRequest (**QUALREQ**) Invoke message* (see Fig. 10.3).

The HLR responds to the serving system request with an appropriate message, either RegistrationNotification Return Result (**regnot**) or QualificationRequest Return Result (**qualreq**). When service qualification is successful, key IS-41-C parameters in the response to the serving system include:

- AuthorizationPeriod

- OriginationIndicator

---

*IS-41-C also includes the definition and procedures for the ServiceProfileRequest operation—originally included in IS-41 Revision A—with the following caveat: "The initiating of ServiceProfileRequest procedures is not recommended due to functional overlap with QualificationRequest. This transaction is not supported and may be eliminated in the future. The QualificationRequest should be used instead." In fact, the ServiceProfileRequest operation is not included in the ANSI standard, *ANSI/TIA/EIA-41*.

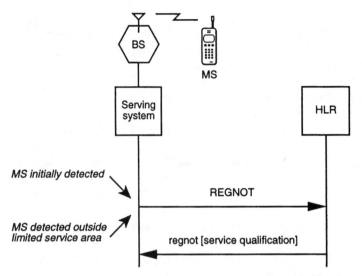

Figure 10.2 Service qualification using the RegistrationNotification operation.

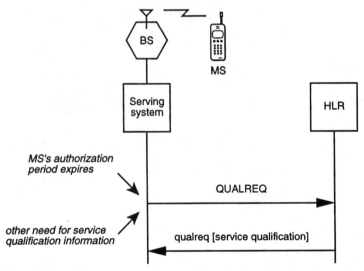

Figure 10.3 Service qualification using the QualificationRequest operation.

- TerminationRestrictionCode
- CallingFeaturesIndicator

The presence of the AuthorizationPeriod parameter in the response implies the successful validation of the MS. The AuthorizationPeriod parameter also specifies the period after which the serving system must requalify the MS; this may be specified, e.g., in terms of a certain number of hours. The OriginationIndicator parameter specifies the types of calls the MS is allowed to originate, e.g., no calls, all calls, or locals calls only. The TerminationRestrictionCode parameter indicates whether the MS is allowed to terminate calls. The CallingFeaturesIndicator parameter specifies the authorization and activity states of the MS's features.

Unsuccessful service qualification is indicated by the presence of the AuthorizationDenied parameter in the response from the HLR to the serving system. IS-41-C does not dictate how the serving system shall handle an MS that has failed service qualification; service should obviously be refused, but particular tones or announcements to the MS are left as system implementation issues. The HLR can specify the service qualification retry interval by including the DeniedAuthorizationPeriod parameter in the response along with the AuthorizationDenied parameter. This limits subsequent service qualification requests from the refused MS.

In addition to responding to serving system service qualification requests, the HLR issues service qualification directives under a number of circumstances:

1. When the roaming MS's service qualification information changes as a result of HLR administrative action (e.g., authorization has been revoked)

2. When the roaming MS's service qualification information changes as a result of subscriber action (e.g., activation of a feature)

In both cases 1 and 2, the service qualification information is conveyed to the serving system in the QualificationDirective Invoke (**QUALDIR**) message* (see Fig. 10.4); the QualificationDirective Return Result (**qualdir**) sent from the serving system to the HLR acknowledges message receipt.

MS Location Management

The IS-41-C MS location management function comprises two components: The MS location update and MS location cancellation.

---

*IS-41-C also includes the definition and procedures for the ServiceProfileDirective operation—originally included in IS-41 Revision A—with the following caveat: "The initiating of ServiceProfileDirective procedures is not recommended due to functional overlap with QualificationDirective. This transaction is not supported and may be eliminated in the future. The QualificationDirective should be used instead." In fact, the ServiceProfileDirective operation is not included in the ANSI standard, *ANSI/TIA/EIA-41*.

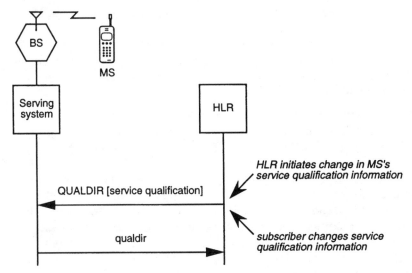

Figure 10.4 Service qualification using the QualificationDirective operation.

MS location update processes have the effect of creating or modifying an MS's temporary record in a visited system and updating the location information in an MS's record in the HLR. MS location cancellation processes have the effect of deleting an MS's temporary record in a visited system and updating the location information in an MS's record in the HLR. Location cancellation is also known as *registration cancellation* or *deregistration*.

The location update and cancellation processes often interact, e.g., to manage the movement of an MS from one serving system to another; however, this need not be the case, as we describe in this chapter.

The MS location information stored in the MS's record in the HLR is characterized by its resolution. IS-41-C supports several levels of location resolution, but, at a minimum, the MS's location is resolved to the serving VLR; the HLR will always be provided a location update if the serving VLR changes. In this case, the location of the subscriber is recorded in the HLR as the MSCID of the VLR currently serving the subscriber, usually along with the VLR's network address (e.g., the VLR's SS7 point code).

MS location update processes

The IS-41-C MS location update processes include the serving system location update process and the HLR location update process.

The serving system location update process is responsible for notifying a visiting MS's HLR of the MS's presence in the serving system.

The serving system notifies the HLR under various circumstances, including:

1. When a new visiting MS (i.e., an MS for which a temporary record does not exist in the serving system) is detected in the serving system's service area

2. When the previously obtained service qualification information for a visiting MS indicates a limited service area (e.g., a single location area) and the MS moves outside the service area

Note that the location update process is not normally triggered each time a visiting MS accesses the serving system; this would place an unnecessary signaling burden on the interface, particularly when periodic, autonomous registration is employed. While there may be other reasons for notifying the HLR on a per-system-access basis, e.g., for authentication or service qualification purposes, the HLR normally requires location update notification only when the MS changes service areas.

If the cause of the location update is the detection of a new visiting MS, the serving system creates a new temporary record; otherwise, the serving system merely modifies the preexisting record for the MS with the new location information. In both cases, the serving system packages the location update notification along with the service qualification request in a **REGNOT** message* to the HLR (see Fig. 10.2).

When the HLR receives the **REGNOT** message, it checks the MS's record for location information. If the serving system recorded in the MS record is different from the requesting system, the HLR location update process triggers the location cancellation process to cancel the MS's location in the previous serving system (see Fig. 10.5). If the MS is not authorized for service in the new location, the HLR denies the location update and includes the AuthorizationDenied parameter in the **regnot** message; otherwise, the HLR updates the MS's record with the new location information and acknowledges the location update by sending a **regnot** message to the serving system.

If the serving system receives a negative acknowledgment, it deletes the temporary record it created for the MS.

---

*Note that the RegistrationNotification operation may invoke many functions (e.g., location update, service qualification, authentication reporting, SMS functions) and sometimes these are all combined under the category of registration. However, we have chosen to describe these diverse functions separately in this book.

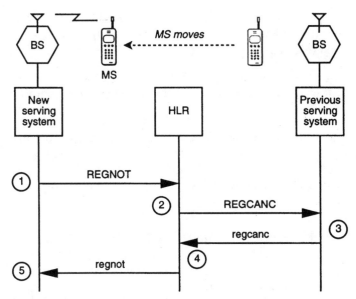

Figure 10.5 Combination of the location update and location cancellation processes: (*1*) A new MS moves into a service area; therefore, the serving system requests location update in the HLR. (*2*) The HLR has stored location information for the MS; therefore, it requests location cancellation from the previous serving system. (*3*) The previous serving system deletes the temporary MS record and sends an acknowledgment to the HLR. (*4*) The HLR acknowledges the location update. The HLR normally also provides service qualification in the **regnot** message. (*5*) The serving system provides service to the MS.

MS location cancellation processes

Location cancellation processes include the serving system location cancellation process and the HLR location cancellation process.

The serving system location cancellation process is responsible for deleting the temporary record for one or more visiting MSs in the serving system. The serving system location cancellation process may be initiated by the HLR, in the form of a directive sent from the HLR to the serving system, or locally by the serving system itself. The latter case may occur under various circumstances, all of which are serving system options not dictated by IS-41:

1. When the MS sends a power-down signal to the serving system (although not all MSs provide such an indication).

2. When the MS misses one or more autonomous registration events.

3. When the serving system determines, based on some other serving system-dependent algorithm, that a single roaming MS record shall be cleared.

4. When the serving system determines—again, based on some other serving system-dependent algorithm—that all roaming MS records shall be cleared (e.g., for serving system restoration purposes).

IS-41-C provides two mechanisms that the serving system may use to notify the HLR of the location cancellation of one or more roaming MSs:

1. The IS-41-C MSInactive Invoke (**MSINACT**) message may be sent from the serving system to notify the HLR of the location cancellation of a single roaming MS (see Fig. 10.6). However, the message must include the DeregistrationType parameter, or else the HLR will interpret the message as merely reporting the inactive state of the MS.

2. The IS-41-C BulkDeregistration Invoke (**BULKDEREG**) message may be sent from the serving system to notify the HLR of the location cancellation of all the HLR's roaming MSs currently associated with the serving system (see Fig. 10.7).

The HLR location cancellation process responds to a serving system location cancellation notification with an acknowledgment message, that is, MSInactive Return Result (**msinact**) or BulkDeregistration Return Result (**bulkdereg**). The effect of these procedures is to delete one or more temporary MS records in the visited system and to delete

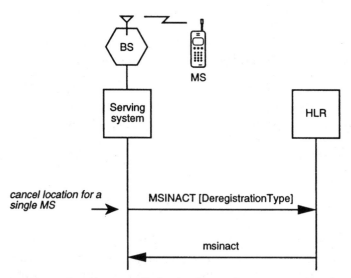

Figure 10.6 Location cancellation by the serving system using the MSInactive operation.

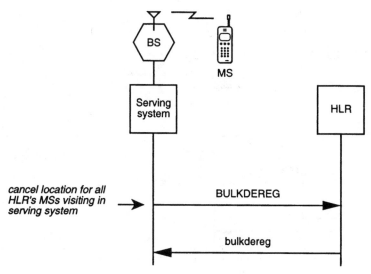

Figure 10.7 Location cancellation by the serving system using the BulkDeregistration operation.

the transient location information in one or more MS records in the HLR. The HLR still maintains service information for each of the MSs affected (e.g., subscribed features); however, this information does not include current location, which is unknown until the next successful location update for the MS.

In addition to serving system-initiated location cancellation, the HLR location cancellation process may issue a location cancellation directive to the serving system under the following circumstances:

1. When an MS registers in a new serving system, the location update process in the HLR triggers the HLR location cancellation process to clear the MS's record in the system previously serving the MS.

2. When an administrative action is executed (e.g., withdrawal of the MS's subscription).

3. When the HLR suffers a data failure, it initiates location cancellation for all its roaming MSs in all potential serving systems (assuming the HLR can identify these systems).

In cases 1 and 2, the location cancellation directive is conveyed to the serving system in the RegistrationCancellation Invoke (**REG-CANC**) message (see Fig. 10.5). This directive may be refused by the serving system in border cell situations (see the next section). Otherwise, the effect of this procedure is to delete the temporary MS record in the previous serving system.

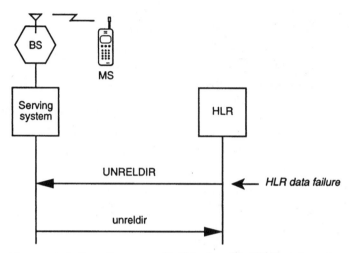

Figure 10.8 Location cancellation by the HLR using the UnreliableRoamerDataDirective operation.

For case 3, the HLR uses the UnreliableRoamerDataDirective Invoke (**UNRELDIR**) message (see Fig. 10.8). The effect of the UnreliableRoamerDataDirective procedure is to delete all temporary MS records in the visited system related to the HLR and to delete the location information in one or more MS records in the HLR. As in the case of BulkDeregistration, the HLR still maintains service information for each of the MSs affected; however this information does not include current location, which is unknown until the next successful location update for the MS. In fact, this is the rationale behind the HLR's use of UnreliableRoamerDataDirective—to direct the serving system to assist the HLR in replenishing its database with valid roaming MS location information in as expeditious a manner as possible. Without this procedure, the serving system would not necessarily inform the HLR of a visiting MS's location after an HLR data failure, limiting the HLR's ability to deliver calls to the MS.

MS location management in border cell situations

Our discussion of location management processes has, up to this point, assumed a somewhat idealized cellular radio environment. Each time the MS wishes to access the system, it scans for and selects a single, strong control channel and then uses that channel to transmit a system access message, e.g., a registration or call origination message. If the receiving system already has a valid temporary record for the MS (i.e., is the current serving system), a location update to the MS's HLR is not normally necessary; the serving system simply continues to provide service to the MS.

However, if one or more other systems are also listening for system access messages on the same control channel, we have the makings of what is known as a *multiple-access problem*. Why would multiple systems be using the same control channel? Based on the frequency reuse concept described in Chap. 1, adjacent cells would not use the same channels. However, dense urban areas or topological features of the service area, e.g., the presence of lakes, rivers, or flat, open land, may result in unintended propagation of the MS's signal well beyond the engineered boundaries of the serving cell, into the service area of a "border system."* The received system access message in the border system would be seen as an initial system access, causing a location update to the MS's HLR and a location cancellation from the HLR to the actual serving system (see Fig. 10.9). This situation is particularly problematic when the system access is a call origination since the

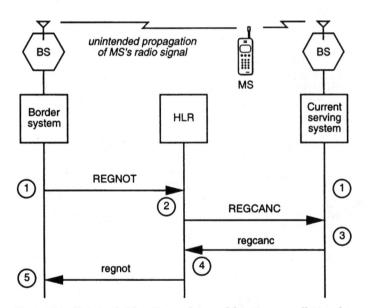

Figure 10.9 Unintended location update and location cancellation due to border cell conditions: (*1*) The MS is already registered on the serving system; however, a system access is overheard by a border system. The border system requests location update in the HLR. (*2*) The HLR has stored location information for the MS; therefore, it requests location cancellation from the current serving system. (*3*) The current serving MSC deletes the temporary MS record and sends an acknowledgment to the HLR. (*4*) The HLR acknowledges the location update. The HLR normally also provides service qualification in the **regnot** message. (*5*) The boarder system attempts to provide service to the MS.

*A border system does not necessarily serve cells immediately adjacent to the system in question, merely in close enough proximity that signals intended for one system are overheard by the other.

effect is to—at best—place the MS on a radio channel with poorer signal quality than necessary and—more likely—to drop the call altogether when the MS cannot detect the radio channel assigned by the remote system for the call.

To address this problem,* IS-41-C location management processes incorporate special procedures which allow the HLR to determine the best serving system for an MS when border cell conditions result in (1) multiple location updates for a single MS system access event or (2) one or more location updates from systems with poorer received signal quality than the current serving system. These procedures require the border systems to provide additional information, including an indication that the system access occurred in a border cell and a measure of the received signal quality, to the HLR along with the location information. This information is conveyed in the IS-41-C **REGNOT** message (see Fig. 10.10).

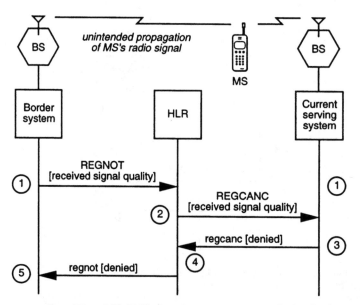

Figure 10.10 Part of IS-41-C's location management solution for border cell conditions. The steps are as described in Fig. 10.9, except that the current serving system denies the location cancellation request based on its comparison of the received signal quality provided by the border system and its own recorded measurements.

---

*Note that this is only one of several border cell problems that are addressed by IS-41-C. Other border cell issues are discussed in later chapters.

The HLR location update process makes a decision on which of several location updates is best. The HLR location cancellation process then attempts to cancel registration in the current serving system by using the IS-41-C **REGCANC** message, providing that system with the signal quality information from the best location update selected by the HLR. The serving system location cancellation process, in turn, evaluates the location cancellation directive from the HLR, using information it has stored regarding the last system access by the MS in question. If the serving system determines that it is the best system for the MS, it can deny the location cancellation directive, sending the denial indication to the HLR in the IS-41-C **regcanc** message. The HLR honors the location cancellation denial from the serving system and refuses the location update from the border system; refusal is conveyed in the IS-41-C **regnot** message.

Note that Fig. 10.10 shows the RegistrationCancellation operation embedded within the RegistrationNotification operation in a *synchronous* fashion; i.e., steps 2 and 3 occur before step 4. Prior to IS-41-C (or, more correctly, TSB65), these operations were *asynchronous,* so step 4 occurs before steps 2 and 3. This change was made to support the location cancellation denial capability.

Examples of methods that the HLR can use to determine which of several location updates is best are provided in IS-41.6-C, Annex F.

MS State Management

The IS-41-C MS state management function encompasses the processes by which the call delivery availability state of an MS is coordinated between the serving system and the HLR. The state may be either *active* or *inactive.* In the active state, the MS is available for call delivery. When in the inactive state, the MS is not available for call delivery; however, an inactive MS may be able to accept short message service (SMS) deliveries. An MS may be deemed inactive because:

1. *The MS is not registered.* When the HLR does not have a valid location for the MS, the MS is considered inactive.

2. *The MS is out of radio contact.* The MS may have missed autonomous registration events due to a loss of radio contact and may have been designated inactive by the serving system.

3. *The MS is intentionally inaccessible.* The MS may have gone into a mode (e.g., sleep mode) whereby it is intentionally inaccessible.

Keeping with its open nature, IS-41-C also allows serving systems to designate an MS inactive based on internal algorithms.

MS state management processes

MS state management processes include the serving system MS state management process and the HLR MS state management process.

Generally, the MS state is set to active by the serving system and the HLR after a successful location update for the MS. Likewise, after a successful location cancellation, the HLR sets the MS state to inactive; the state may be reset to active if the location cancellation is nested within a location update process. In this sense, the serving system MS state management process and HLR MS state management process make use of the same operations as the location update and location cancellation processes, i.e., RegistrationNotification, RegistrationCancellation, MSInactive, BulkDeregistration, and UnreliableRoamerDataDirective. Additionally, the MS state may be explicitly set to inactive in the serving system and the HLR in the following ways:

1. When the serving system sends a valid IS-41-C **REGNOT** message including a valid AvailabilityType parameter to the HLR, the serving system sets the MS's state to inactive; when the HLR successfully receives the message, it also sets the MS's state to inactive (see Fig. 10.11).

2. When the serving system sends a valid IS-41-C **MSINACT** message to the HLR, the serving system sets the MS's state to inactive; when the HLR successfully receives the message, it also sets the MS's state to inactive (see Fig. 10.12). Note that this message may also result in location cancellation for the MS if the DeregistrationType parameter is included.

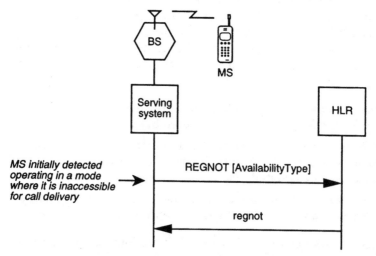

Figure 10.11 Setting the MS's state to inactive using the RegistrationNotification operation.

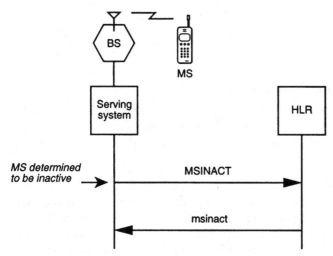

Figure 10.12 Setting the MS's state to inactive using the MSInactive operation.

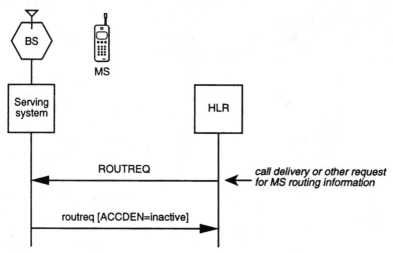

Figure 10.13 Setting the MS's state to inactive using the RoutingRequest operation.

3. When the serving system sends a valid IS-41-C RoutingRequest
   Return Result (**routreq**) message to the HLR including the
   AccessDeniedReason (ACCDEN) parameter set to inactive, the
   serving system sets the MS's state to inactive (if not already inactive); when the HLR successfully receives the message, it also sets
   the MS's state to inactive (see Fig. 10.13).

Once set, the MS's state remains inactive until a serving system sends a valid **REGNOT** message to the HLR that does not include the AvailabilityType parameter.

HLR and VLR Fault Recovery

The IS-41-C HLR and VLR fault recovery function encompasses the processes which attempt to ameliorate the effects of HLR and VLR failures in the mobile telecommunications network.

HLR fault recovery processes

The HLR fault recovery processes handle the case where the HLR experiences a failure which renders the data it uses to keep track of its roaming MSs unreliable. The UnreliableRoamerDataDirective operation is specifically designed for this purpose; the HLR sends the **UNRELDIR** message to each system (i.e., VLR) that it believes may be serving one or more of the HLR's MSs. On receipt, the serving system clears all temporary roaming records for MSs associated with the HLR and returns the **unreldir** message to the HLR as an acknowledgment (see Fig. 10.14).

IS-41-C does not specify the mechanism by which the HLR determines the visited systems that are affected. At least two strategies are possible:

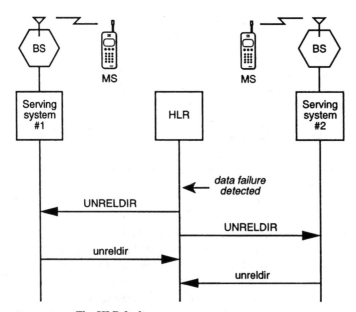

Figure 10.14   The HLR fault recovery process.

1. The HLR maintains a list of all possible serving systems in a non-volatile storage area.

2. The HLR maintains a list of active serving systems (i.e., systems that are actually serving one or more of the HLR's MSs at any particular time) in a nonvolatile storage area.

In case of failure, the HLR sends an **UNRELDIR** message to each serving system on this list.

The HLR may also use the UnreliableRoamerDataDirective operation to selectively clear a single serving system's roaming records. For example, if the HLR received repeated errors from a particular serving system, it could send an **UNRELDIR** message to the system to direct it to clear its data related to the HLR and, in effect, "start fresh."

VLR fault recovery processes

The VLR fault recovery processes handle the case where the VLR experiences a failure which renders its roaming MS data unreliable. The BulkDeregistration operation may be used for this purpose; the VLR sends the **BULKDEREG** message to each HLR that it believes may be the home system for one or more MSs visiting in the VLR's service area. On receipt, the HLR clears all location information for MSs served by the VLR and returns the **bulkdereg** message to the VLR as an acknowledgment (see Fig. 10.15).

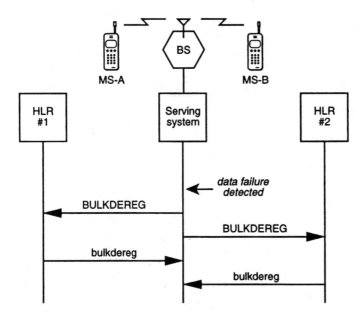

Figure 10.15 The VLR fault recovery process.

IS-41-C does not specify the mechanism by which the VLR selects the HLRs that are affected. As with HLR fault recovery, at least two strategies are possible:

1. The VLR maintains a list of all possible HLRs in a nonvolatile storage area.
2. The VLR maintains a list of HLRs associated with active visiting MSs in a nonvolatile storage area.

In case of failure, the VLR sends a **BULKDEREG** message to each HLR on this list.

Summary of IS-41-C Operations Used for Basic Automatic Roaming

Table 10.1 summarizes the IS-41-C operations used by the basic automatic roaming functions described in this chapter.

TABLE 10.1   Use of IS-41-C Operations for Basic Automatic Roaming

Function	IS-41-C operations used for the function
MS service qualification	RegistrationNotification, QualificationRequest, QualificationDirective
MS location management	RegistrationNotification, MSInactive, BulkDeregistration, RegistrationCancellation, UnreliableRoamerDataDirective
MS state management	RegistrationNotification, MSInactive, BulkDeregistration, RegistrationCancellation, UnreliableRoamerDataDirective, RoutingRequest
HLR and VLR fault recovery	BulkDeregistration, UnreliableRoamerDataDirective

Chapter

11

Authentication
Functions

In this chapter, we discuss the IS-41-C mobile telecommunications network functions related to authentication. These functions are divided into the following categories:*

- Shared secret data sharing
- Global challenge
- Unique challenge
- Shared secret data update
- Call history count update
- Authentication reporting

We also examine the IS-41-C application processes that support these network functions. In the course of describing the processes, we identify the IS-41-C mobile application part (MAP) operations (e.g., AuthenticationRequest) that are used to accomplish authentication process tasks. We summarize this information at the end of the chapter. Note that we employ the IS-41-C convention for operation component acronyms; i.e., the Invoke component acronym is in all-capital letters (e.g., **AUTHREQ**), while the Return Result component acronym is in all-lowercase letters (e.g., **authreq**). Refer to the IS-41-C standard for the descriptions of the individual IS-41-C operations and parameters. Also, consult the Glossary for a description of general terms (e.g., *serving system*) that are not explicitly defined in this chapter.

---

*The list of authentication functions reflects the authors' subjective categorization of certain IS-41-C functions. Likewise, the process descriptions are the authors' subjective interpretation of the IS-41-C procedures.

Throughout this chapter we generally consider the serving system as a single entity encompassing the mobile switching center (MSC) and visitor location register (VLR) functional entities. This simplifies the descriptions of IS-41-C application processes and also is representative of a large percentage of the IS-41 implementations currently in service. However, IS-41-C authentication assigns specific processing duties to the VLR that make the concept of sharing a VLR among multiple MSCs a viable alternative to the combined implementation. Therefore, while we describe authentication processes in terms of a single "MSC plus VLR" entity, keep in mind that the potential for separation exists and is fully defined in IS-41-C.

Note that the IS-41-C authentication procedures are intended to support mixed-version networks; e.g., the authentication center (AC) uses IS-41-C procedures while the serving system uses TSB51 procedures designed for use with IS-41-B. We do not discuss this aspect of IS-41-C; the reader is referred to the IS-41-C standard for the descriptions of these backward-compatibility mechanisms.

What Is Authentication?

Authentication is a set of functions used to prevent fraudulent access to cellular networks by phones illegally programmed with counterfeit mobile identification number (MIN) and electronic serial number (ESN) information. The functions require no subscriber intervention and provide a much more robust method of validating the true identity of a subscriber than does the basic service qualification function described in the previous chapter.

The IS-41-C authentication functions are independent of the air interface protocol used to access the network, and subscribers are never involved in the process. A successful outcome of authentication (i.e., an *authenticated MS*) occurs when it can be demonstrated that the mobile station (MS) and network possess identical results of an independent calculation performed in both the MS and the network. The AC is the primary functional entity in the network responsible for performing this calculation, although the serving system (i.e., the VLR) may also be allocated certain responsibilities, as we shall describe. The authentication calculations are based on a set of algorithms, collectively known as the *CAVE* (cellular authentication and voice encryption) *algorithm.*

The authentication process can be invoked by many events; however, it is performed most often when the following situations arise:

1. Registration
2. Call origination
3. Call termination

Note that the authentication process is not necessary for every instance of the events listed above. For example, during autonomous registration, authentication may be performed only if the subscriber has moved to a new location (i.e., the location management process has updated the subscriber's location at the HLR) provided that is part of the *authentication policy* in effect between the serving and home systems.

The authentication process and algorithm are based on the following two secret numbers:

1. Authentication key (A-key)

2. Shared secret data (SSD)

These numbers are the fundamental basis for IS-41-C authentication and are described in the following sections.

Authentication key (A-key)

The A-key is a 64-bit secret number that is the permanent *key* used by the authentication calculations in both the MS and the AC. The A-key is permanently installed into the MS and is securely stored (i.e., provisioned) at the AC in the network when a new subscription is obtained. The method for installing the A-key in the MS is implementation-dependent; however, the TIA has recommended a method for manually programming the A-key into the MS (refer to *TIA/EIA TSB50*).

Once the A-key is installed in the MS, it should not be displayable or retrievable. The MS and the AC are the only functional entities that are ever aware of the A-key; it is never transmitted over the air or passed between systems. The primary function of the A-key is as a parameter used in the calculation to create the SSD.

Shared secret data

The SSD is a 128-bit secret number that is essentially a temporary key used by the authentication calculations in both the MS and the AC. The SSD may also be shared with the serving system (specifically, the VLR functional entity within the serving system) via a number of IS-41-C messages. Unlike the A-key, the SSD is a semipermanent value. It can be modified by the network at any time, and the network can command the MS to generate a new value.

The SSD is the result of a calculation using the A-key, the ESN, and a random number shared between the MS and the network. The SSD calculation results in two separate 64-bit values, SSD_A and SSD_B. SSD_A is the value used specifically for the authentication functions. SSD_B is the value used specifically for the encryption algorithms

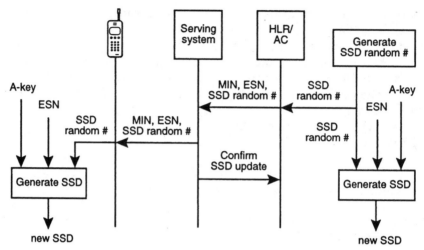

Figure 11.1 Deriving the SSD from the A-key. The HLR and AC functional entities are shown as a single unit due to their close functional relationship.

used for voice privacy and to encrypt and decrypt selected signaling messages on the radio traffic channel. Figure 11.1 illustrates the SSD generation procedure.

Where Are Authentication Functions Specified in IS-41-C?

Authentication functions are specified in three parts of IS-41-C:

1. IS-41.5-C provides the required formats—the bit-by-bit encoding—of all the IS-41-C operation components, including those used for authentication. IS-41.5-C defines both the messages (e.g., AuthenticationRequest Invoke) and the message parameters (e.g., AuthenticationResult, RandomVariable).

2. IS-41.6-C provides algorithmic descriptions of the procedures that are associated with sending and receiving most IS-41-C messages, including those used for authentication. As we pointed out in Chap. 8, IS-41-C does not define a service interface between the MAP application processes and the MAP application service element (ASE); therefore, the procedures in IS-41.6-C encompass both application process descriptions and ASE procedural descriptions. Furthermore, the procedures leave considerable room for implementation-dependent customization; e.g., there are numerous references in IS-41.6-C to *local procedures* and *internal algorithms*.

3. IS-41.3-C steps back from the protocol details contained in parts 5 and 6 and attempts to explain—using information flow diagrams

and step-by-step descriptions—how the operations are used individually and together to accomplish authentication application process tasks. Annex A in IS-41.3-C includes a description of the assumptions made in the design of IS-41-C authentication processes; these assumptions provide valuable implementation guidance.

Issues Associated with Authentication

The authentication functions are designed to validate a subscriber while being served by either the home system or a visited system. Many issues arise in the network to support authentication, including the following:

1. Which functional entity is the *authentication controller* must be determined. The IS-41-C authentication processes support authentication control from either the AC or the serving system (i.e., the VLR). The AC, as controller, can perform all authentication functions. The VLR, when the SSD key is shared, can perform only some authentication functions, as we shall describe.

2. Many systems do not currently support authentication. Although authentication has been standardized for several years, most systems do not support the capability due to implementation costs and the requirement that the mobile station support the algorithms.

3. Most mobile stations do not currently support authentication. The IS-41-C authentication processes are dependent on the MS's having the capability to perform the algorithms and support the authentication key.

4. Management of the A-key is an issue. A primary consideration for supporting authentication is A-key management, i.e., how the key is kept secure, how it is provided to the MS, and how it is maintained. Since the entire authentication process is dependent on the A-key's remaining secret, adequate management of the A-key is of paramount importance. The basic requirements of an A-key management protocol are the following:

- The A-key must be handled by a minimum number of people—none is best!
- The AC must be designed as a very secure system.
- The A-key must not be readable by any party.
- Changes to the A-key in the MS and AC must be performed in a secure manner.

The methods for fulfilling these requirements have not been addressed by IS-41 and are generally left to the network operator and its vendors to resolve.

Shared Secret Data Sharing

The IS-41-C SSD sharing function encompasses the processes by which the authentication center and the serving system (i.e., the VLR) manage the sharing of authentication responsibilities for a visiting MS. Sharing authentication responsibilities is only possible if the SSD, in addition to being shared between the AC and the MS, is shared between the AC and the serving system; this is referred to as *SSD sharing* or *shared SSD* in IS-41-C. Additionally, IS-41-C specifies that when SSD is shared with the serving system, the call history count (COUNT) information is shared as well.*

SSD sharing gives the serving system significant local control over the authentication of a visiting MS. Specifically, it can control the following network functions:

- Global challenge for all but the initial system access, when SSD sharing is not yet established
- Unique challenge, again for all but the initial system access
- The base station challenge portion of an AC-initiated SSD update (see "SSD Update")
- COUNT update

Serving system control of these network functions reduces the authentication-related signaling traffic between the serving and home systems and the associated call processing delays.

SSD sharing processes include the AC SSD sharing process, the serving system SSD sharing process, and the HLR SSD sharing process. These processes support the following tasks:

1. Notifying the AC of the serving system's capabilities for SSD sharing
2. "Turning on" and "turning off" SSD sharing
3. AC retrieval of shared COUNT information

Notifying the AC of the serving system's SSD sharing capabilities

The serving system has the responsibility to notify the AC of its capabilities for SSD sharing so that the AC can determine whether sharing is possible. This is normally done when the serving system first communicates with the home system, i.e., on initial system access by

---

*Assuming the call history count is part of the authentication "tool set" used by the home system.

a visiting MS. The serving system's SSD sharing capabilities are conveyed in the IS-41-C SystemCapabilities (SYSCAP) parameter. Two signaling scenarios are possible:

1. The MS is authentication-capable. This can be determined on the basis of the air interface protocol used to access the system (for example, IS-136 or IS-95), or information coordinated between the home and serving systems—e.g., via a roaming agreement. In this case, the first message sent from the serving system to the HLR is typically the IS-41-C AuthenticationRequest Invoke (**AUTHREQ**) message, including the SYSCAP parameter. The HLR relays this message to the AC* (see Fig. 11.2), thereby notifying the AC of the serving system's SSD sharing capabilities.

2. The MS is not authentication-capable. In this case, the first message sent from the serving system to the HLR is the IS-41-C RegistrationNotification Invoke (**REGNOT**) message, containing the SYSCAP parameter. However, IS-41-C does not define a standard way for the HLR to convey this information to the AC; therefore, when the AC is separated from the HLR (via the IS-41-C standard "H-inter-

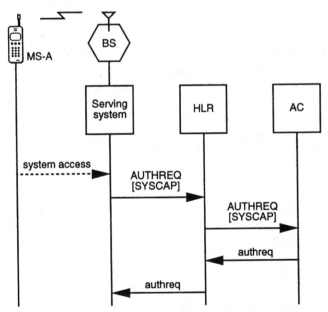

Figure 11.2 Conveying the serving system's SSD sharing capabilities to the AC.

*All communication between the serving system and the AC is via the HLR.

face"), the AC is not informed of the serving system's SSD sharing capabilities on initial system access by a non-authentication-capable MS. Of course, this is not a problem if the MS is truly not capable of authentication. However, if information in the HLR database indicates that the MS has, e.g., recently been upgraded and is now authentication-ready, the HLR must use the information in the SYSCAP parameter to determine whether a fraudulent MS is attempting to access the system. This scenario is not explicitly specified in IS-41-C.

IS-41-C also does not explicitly define the procedures to follow if the serving system's ability to handle shared authentication responsibilities changes during the period when an MS is visiting in the serving system's area, e.g., due to a failure in the subsystem that performs the authentication calculations. We can speculate on at least two approaches to this scenario:

1. If the serving system is still able to support AC-controlled authentication procedures, then the serving system can notify the AC of its modified SSD sharing capabilities at the next authentication event (e.g., call origination).

2. Alternatively, the serving system can initiate the location cancellation process for one or more visiting MSs in the system, as described in Chap. 10, that is, using the IS-41-C MSInactive Invoke message including the DeregistrationType parameter, or using the IS-41-C BulkDeregistration Invoke message. Use of MSInactive is likely the preferred approach since

   a. It allows the serving system to convey the current value of COUNT to the AC (see "Call History Count Update").

   b. It allows selective location cancellation of only those MSs for which the serving system is sharing authentication responsibilities, rather than all visiting MSs.

Turning SSD sharing on and off

Given that the serving system is capable of sharing authentication responsibilities, the AC initiates SSD sharing by sending the IS-41-C SSD and COUNT parameters to the serving system. The AC may use one of four IS-41-C messages for this purpose (see Fig. 11.3):

1. AuthenticationDirective Invoke (**AUTHDIR**)

2. AuthenticationRequest Return Result (**authreq**)*

3. AuthenticationStatusReport Return Result (**asreport**)*

---

*The information may be conveyed in the specified Return Result message only as part of the AC's response to the corresponding Invoke message (e.g., the **authreq** message is in response to the **AUTHREQ** message).

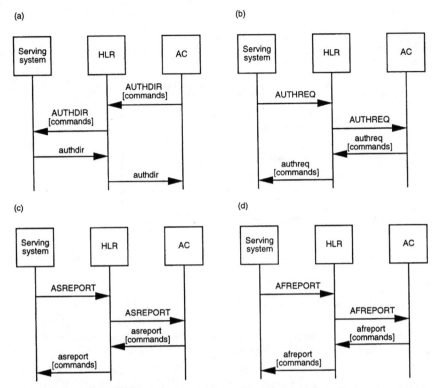

Figure 11.3 The AC may send authentication-related commands to the serving system, including the command to initiate SSD sharing, using four IS-41-C methods: (a) the **AUTHDIR** message, (b) the **authreq** message, (c) the **asreport** message, and (d) the **afreport** message.

4. AuthenticationFailureReport Return Result (**afreport**)*

Once SSD sharing is turned on, the AC may turn it off by sending the SSDNotShared (NOSSD) parameter to the serving system. Although IS-41-C specifies that the AC may use any of the four IS-41-C messages listed above to turn off SSD sharing, the **AUTHDIR** message appears to be the most practical for the following reason. When turning off SSD sharing, the AC must retrieve the current COUNT from the serving system. The response to the **AUTHDIR**—the AuthenticationDirective Return Result (**authdir**) message—includes the COUNT parameter; therefore, it provides a convenient method of COUNT retrieval under these conditions (see Fig. 11.4). AC retrieval of COUNT when using the other messages to turn off sharing (i.e., **authreq, asreport,** or **afreport**) is more problematic.

*The information may be conveyed in the specified Return Result message only as part of the AC's response to the corresponding Invoke message (e.g., the **authreq** message is in response to the **AUTHREQ** message).

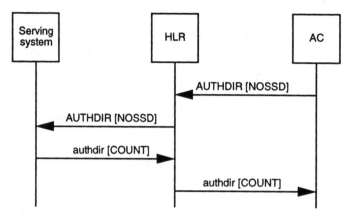

Figure 11.4 To turn off SSD sharing, the **AUTHDIR** message should be used since it allows COUNT to be retrieved from the serving system.

AC retrieval of shared COUNT information

When SSD is shared, the AC gives the serving system the current SSD and COUNT values for the MS in question. The serving system is not able to autonomously initiate a change in the MS's SSD—this is strictly an AC-controlled function (see "SSD Update"). However, the serving system may—based on its internal authentication algorithms—order the MS to increment its COUNT value without informing the AC (see "Call History Count Update"). Therefore, the final aspect of SSD sharing that has to be managed between the AC and serving system is COUNT retrieval—how the AC gets the (potentially revised) COUNT information (1) from the serving system when the AC turns off SSD sharing and (2) from the previous serving system when the MS moves to a new serving system. Four general scenarios are handled:

1. *COUNT retrieval when the AC turns off SSD sharing* was described in the previous section.

2. *The MS moves from one authentication-capable system to another authentication-capable system* (see Fig. 11.5). In this case, the initial IS-41-C message sent from the new serving system to the home system is normally an **AUTHREQ** message.* When the AC receives this message, it sends a CountRequest Invoke (**COUNTREQ**) mes-

---

*If the first message sent from the new serving system is a RegistrationNotification Invoke message and not an **AUTHREQ** message, then the procedure described for case 3 is used. However, if the MS is recognized as authentication-capable, the new serving system normally will first attempt to authenticate the MS before proceeding with the service qualification and location update processes.

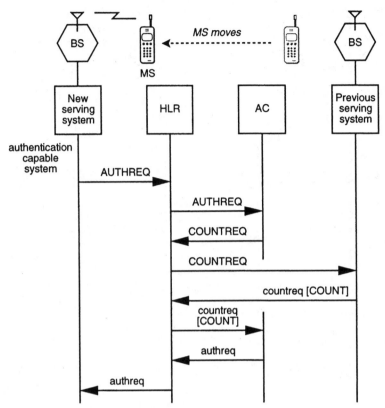

Figure 11.5 COUNT retrieval when the MS moves from one authentication-capable system to another.

sage to the previous serving system. The previous serving system replies with its stored value of COUNT in the CountRequest Return Result (**countreq**) message, and normal AuthenticationRequest processing continues.

3. *The MS moves from an authentication-capable system to a non-authentication-capable system* (see Fig. 11.6). In this case, the initial IS-41-C message sent from the new serving system to the home system is the **REGNOT** message associated with the service qualification and location update processes (see Chap. 10). The HLR does not forward this message to the AC, so the AC cannot use the CountRequest method described for case 1. Rather, the HLR initiates the location cancellation process and sends a RegistrationCancellation Invoke (**REGCANC**) message to the previous serving system. The serving system inserts the current COUNT value in the RegistrationCancellation Return Result (**regcanc**) message that is sent to the HLR. On receipt

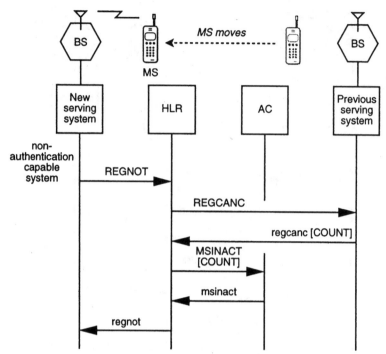

Figure 11.6 COUNT retrieval when the MS moves from an authentication-capable system to a nonauthentication-capable system.

of **regcanc**, the HLR sends an MSInactive Invoke (**MSINACT**) message to the AC, including the COUNT value received from the previous serving system. The AC acknowledges the message, completing the COUNT retrieval process.

4. *The serving system initiates location cancellation of the MS.* Chapter 10 describes two situations under which this would occur:

- In the first case, the serving system uses the **MSINACT** message to cancel location for a single MS. The HLR forwards the **MSIN-ACT** message to the AC, including the COUNT value (see Fig. 11.7). The AC acknowledges the message, completing the COUNT retrieval process.

- In the second case, the serving system uses the BulkDeregistration Invoke (**BULKDEREG**) message to cancel location for all an HLR's MSs currently visiting in the serving system. IS-41-C does not support COUNT retrieval in this case; i.e., when the serving system uses the **BULKDEREG** message to cancel location for multiple MSs, the associated COUNT information

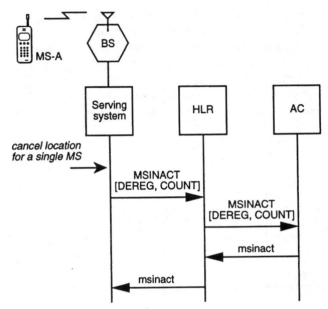

Figure 11.7 COUNT retrieval when the MS's location is canceled by the serving system using the MSInactive operation.

is lost. This is one argument for using the **BULKDEREG** message for serving system restoration purposes only.

Another case where IS-41-C does not support COUNT information retrieval from the serving system occurs when the HLR initiates the location cancellation of all its roaming MSs due to HLR data failure (see Chap. 10). The HLR uses the UnreliableRoamerDataDirective Invoke message for this purpose. The response to this message does not support transfer of COUNT information; therefore, this information is lost.

Global Challenge

The IS-41-C global challenge function encompasses the processes in which:

1. The serving system presents a numeric authentication challenge to all MSs that are using a particular radio control channel.

2. The IS-41-C AC (or the serving system if SSD is shared) verifies that the numeric authentication response from an MS attempting to access the system is correct.

This is called a *global challenge* because the challenge indicator and the random number used for the challenge are broadcast on the radio control channel and so are used by all MSs accessing that control channel. Figure 11.8 depicts the basic global challenge procedure for authentication when SSD is not shared.

Global challenge processes include the serving system global challenge process, the border system global challenge process, and the AC global challenge process.

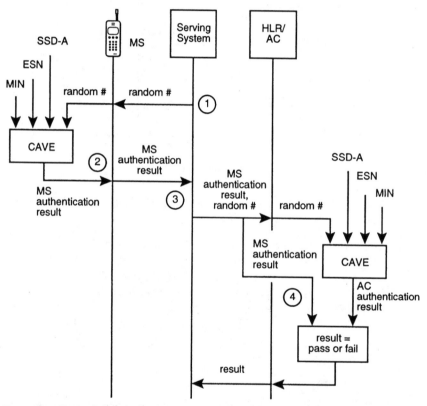

Figure 11.8 Basic global challenge authentication process when SSD is not shared: (*1*) The serving system generates a random number that is sent to the MS in the control channel's overhead message train (OMT). (*2*) The MS calculates an authentication result and transmits that result to the serving system when it accesses the system for registration, call origination, or page response purposes. (*3*) If SSD is not shared, the serving system forwards the authentication result and the random number to the AC. (*4*) The AC independently calculates an authentication result and compares it to the result received from the MS. If the results match, the MS is considered to be authentic. If they do not match, the MS may be considered fraudulent, and service may be denied. If SSD is shared, the serving system performs the calculations.

Serving system global challenge process

The details of the serving system global challenge process depend on whether SSD is shared between the serving system and the AC. The steps in the process are as follows:

1. Issue the numeric global challenge to the MSs in the serving system's service area.

2. Collect the authentication information received from each MS that attempts to access the system.

3. Prevalidate (as explained below) the MS authentication information.

Then, based on the status of SSD sharing, do either step 4 or step 5.

4. Forward the MS's global challenge response to the AC for evaluation, and complete the global challenge process as directed by the AC.

5. Evaluate the MS's global challenge response locally, and complete the global challenge process based on the results of the local evaluation.

In the following discussion of the mechanics of the global challenge process, we make a couple of assumptions:

1. The serving system and base station are capable of and configured for executing the global challenge.

2. The MS is authentication-capable, and authentication is required for the MS.

Assumption 1 is purely a deployment issue, and it is outside the scope of IS-41-C. In fact, systems can be designed to use only the unique challenge method of authentication. Assumption 2 raises the question: How does the serving system know if an MS is authentication-capable? We are aware of three general strategies:

1. Avoid the issue, at least for the initial system access; i.e., when the serving system first detects a visiting MS, assume it is authentication-capable and process system access attempts accordingly. The procedures are designed such that the AC can notify the serving system that the global challenge of subsequent system access is not required—e.g., if the MS is actually not authentication-capable—when the subscriber's profile is transferred from the home system to the serving system as part of the service qualification process. The IS-41-C AuthenticationCapability parameter provides this indicator.

2. Make the decision based on information exchanged between the home system operator and serving system operator in a roaming agreement and stored in the serving system; i.e., the *roaming agree-*

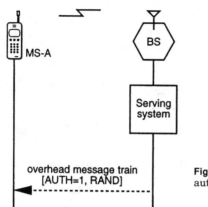

overhead message train
[AUTH=1, RAND]

Figure 11.9 Issuing the global authentication challenge.

ment can include an indication of the MS's authentication capability. As with strategy 1, the home system can override the roaming agreement indication during the service qualification process and can direct the serving system to enforce or bypass authentication of the MS.

3. Make the decision based on information provided by the MS over the air interface; e.g., if the MS attempts to access the system using the analog AMPS protocol, the serving system assumes that the MS is not authentication-capable, whereas if a TDMA or CDMA protocol is used, the serving system assumes that the MS is authentication-capable. Again, the home system can use the service qualification process to override the serving system determination (e.g., for the latest vintage of AMPS MSs that support authentication).

The serving system issues the numeric global challenge (see Fig. 11.9) by selecting a 32-bit random number (RAND) and transmitting the RAND, along with the authentication information element (AUTH) set to 1, in the overhead message train (OMT) on the radio control channel.

Prevalidating the MS information: Normal case. Under normal circumstances, an MS wishing to access the system:

1. Reads the OMT to determine whether a global challenge is in effect for access to the system (i.e., the AUTH bit is set to 1)

2. Executes the cellular authentication and voice encryption (CAVE) algorithm for the global challenge

3. Transmits the system access message (e.g., register, call origination, or page response) to the serving system including

   ■ The calculated authentication result (AUTHR)
   ■ The current value of COUNT stored in the MS

- The eight most significant bits of the 32-bit RAND, called *RANDC*

This information is collected by the serving system and prevalidated. Prevalidation involves checking that the RANDC received from the MS corresponds to the 8 most significant bits of the RAND transmitted by the serving BS. There are at least three possible reasons for differences between RAND and RANDC:

1. The MS is fraudulent and is attempting to "replay" a valid system access message that was previously recorded.

2. The MS is valid but is using a RAND that it received from one system to access a neighboring system (see Fig. 11.10); this is another border cell problem that is addressed in IS-41-C and is described below.

3. The MS did not receive the RAND transmitted by the serving base station (BS) and so chose to use the default value RAND = 0.

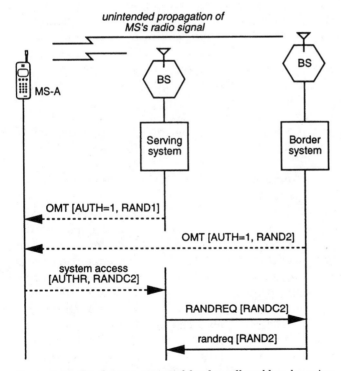

Figure 11.10 Resolving a potential border cell problem by using the RandomVariableRequest operation.

If the serving system determines—based on internal algorithms—that the mismatch may be due to a border cell condition, it initiates a request for the border system's RAND value by sending a RandomVariableRequest Invoke (**RANDREQ**) message to the border system. The **RANDREQ** message includes the RANDC that the serving system received from the MS and the serving cell identification (the latter being information coordinated between the two systems during provisioning). The border system checks its records of current and recently used RAND values in the area corresponding to the serving cell and responds with a matching RAND, if one exists, in the RandomVariableRequest Return Result (**randreq**) message (see Fig. 11.10).

If either a valid RANDC is received from the MS or a valid RAND is received from the neighbor MSC, the serving system moves on to the global challenge evaluation stage. However, if a RAND is not returned by the neighbor MSC or the serving system determines that the RANDC failure is not due to a border cell problem, then the serving system considers the access to be fraudulent. The serving system handling of the MS access in this case is based on "local recovery procedures." For example, the serving system may apply access denial treatment to the MS. If the system access attempt is a call origination or a flash request during a call, denial could be in the form of an announcement or tone, followed by call disconnect. Regardless of the MS treatment, the serving system initiates the authentication reporting process (see "Authentication Reporting"), indicating that the failure is due to a RANDC mismatch.

The serving system global challenge process also initiates authentication reporting if the MS fails to provide the AUTHR and COUNT authentication information when it is challenged to do so.

Prevalidating the MS information: Border cell case.   Under border cell conditions, it is possible for an MS to (1) detect a page request on a control channel from the serving system, (2) rescan and select a control channel from a border system, and (3) respond to the page on the control channel from a border system. If a global challenge is issued on the border system's control channel, the MS's challenge response is based on the RAND transmitted on that control channel. IS-41-C provides intersystem paging procedures (see Chap. 12) that allow successful call delivery under these circumstances. These procedures also support the transfer of the MS global challenge response from the border system to the serving system, via the InterSystemPage2 Return Result (**ispage2**) message (see later in this section for a description of the border system global challenge process).

If the serving system receives the RAND and RANDC parameters in the **ispage2** message from the border system, prevalidation of the authentication information involves checking that the RANDC corresponds to the 8 most significant bits of the RAND. If this is the case, the serving system moves on to the global challenge evaluation stage; otherwise, the serving system considers the access to be fraudulent. The serving system initiates the authentication reporting process, indicating that the failure is due to a RANDC mismatch.

Global challenge when SSD is not shared. Given no failures up to this point, the subsequent handling of the global challenge information from the MS is dependent on whether the serving system is currently sharing SSD with the AC. SSD sharing would not be established:

1. If this were the initial system access by the MS in the serving system's service area (i.e., SSD sharing would not yet have been established between the serving system and the AC)

2. If the serving system were not capable of SSD sharing

3. If the AC's authentication policy did not allow SSD sharing

In these cases, the serving system sends the MS global challenge information in an **AUTHREQ** message to the AC. The serving system* then handles the response from the AC, that is, the **authreq** message. This message can trigger the following additional processes in the serving system, depending on the message's contents:

- SSD sharing
- Unique challenge
- SSD update
- COUNT update

Additionally, the message may contain the IS-41-C DenyAccess parameter, which directs the serving system to release the system resources it has allocated for the MS access in question; this may include the release of the call in progress, if one exists.

If the MS is successfully authenticated and no additional authentication operations are requested, the serving system terminates the global challenge process and continues with normal system access processing, which may involve call origination, location update, or other serving system processes (see Fig. 11.11).

*Note that the call may have been handed off when the initial serving system—now the anchor system—receives the response from the AC.

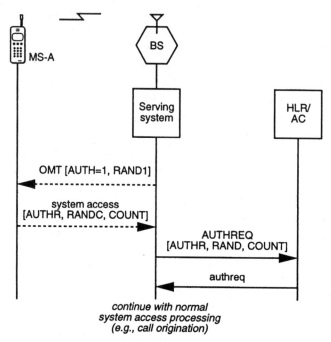

Figure 11.11   Successful global challenge when SSD is not shared.

Global challenge when SSD is shared.   If the serving system is sharing SSD, it assumes responsibility for evaluating the global challenge information it receives from the MS. Evaluation involves five tests:

1. Check that the MIN and ESN reported by the MS are valid.

2. Check that the type of air interface used by the MS to access the system is valid (e.g., is the same as the type stored in the serving system's database, if so provisioned).

3. Check that the AUTHR and COUNT information was provided by the MS.

4. Execute the CAVE algorithm, and verify that the AUTHR generated equals the AUTHR received from the MS.

5. Finally, check that the received COUNT value matches the expected value stored in the serving system's database record for the MS.

If any of these checks fail, the serving system global challenge process initiates the authentication reporting process, passing an indication of the cause of the failure. Otherwise, the serving system considers the MS to have passed the global challenge.

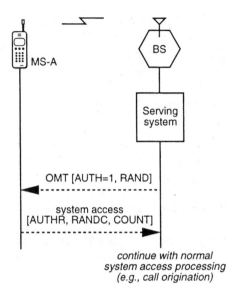

OMT [AUTH=1, RAND]

system access
[AUTHR, RANDC, COUNT]

*continue with normal
system access processing
(e.g., call origination)*

Figure 11.12 Successful global
challenge when SSD is shared.

The serving system processing beyond this point depends on the particular authentication policies in effect. If the global challenge is considered sufficient for authenticating the MS in this case, the serving system may simply terminate the global challenge process and continue with system access processing (see Fig. 11.12). Alternatively, the serving system may pass control to either the unique challenge process or the COUNT update process to perform further authentication functions.

Border system global challenge process

The border system global challenge process deals with two distinct tasks:

1. Handling of RAND requests from the serving system

2. Handling of MS access attempts under intersystem paging conditions

The first task is initiated when the border system receives a **RANDREQ** message from the serving system, including the RANDC that the serving system received from the MS and the serving cell identification (the latter is information coordinated between the two systems during provisioning). The border system checks its records of current and recently used RAND values in the area corresponding to the serving cell and responds with a matching RAND, if one exists, in the **randreq** message.

The second task is initiated when (1) the border system receives an **ISPAGE2** message for an MS and then (2) the border system receives a valid page response with global challenge information (that is, COUNT, AUTHR, and RANDC) from the specified MS. The responsibility of the process in this case is to package the received authentication information—together with the RAND value that the border system is currently using—in the IntersystemPage2 Return Result (**ispage2**) message and to send the message to the serving system.

AC global challenge process

The AC global challenge process is initiated when the AC receives an **AUTHREQ** message from the serving system that includes the IS-41-C global challenge information parameters, that is, RAND, AUTHR and COUNT. The AC global challenge process is responsible for evaluating the information in the **AUTHREQ** message.

The AC first checks that the MIN and ESN reported by the MS are valid. If this test fails, the AC sends an **authreq** message to the serving system, including the IS-41-C DenyAccess parameter, and terminates the process. This message directs the serving system to release the system resources it has allocated for the MS access in question; this may include the release of the call in progress, if one exists.

The AC then checks that the type of air interface used by the MS to access the system (as indicated by the IS-41-C TerminalType parameter included in the **AUTHREQ** message) corresponds to one of the values stored in the AC's database. If not, IS-41-C specifies that the AC may choose to deny the MS access to the network (using the DenyAccess parameter in the **authreq** message, as previously described) and terminate the global challenge process.

Otherwise, the AC executes the CAVE algorithm and verifies that the AUTHR generated by the AC equals the AUTHR received from the MS via the serving system. If the two values do not match, the AC may choose to deny the MS access to the network (using the DenyAccess parameter in the **authreq** message, as previously described) and terminate the global challenge process.

If the AC's internal algorithms are such that authentication processing continues even in the event of an AUTHR mismatch, the AC's next task is to check that the received COUNT value matches the expected value—at least to within an AC-determined range of confidence.* However, if SSD is shared with a previous serving system,

---

*It is possible that the network may lose COUNT synchronization with the MS. For example, the network may request a COUNT update, and the MS may increment its COUNT value and send an acknowledgment that is not received by the network due to poor radio transmission conditions. System failures (e.g., in the VLR or AC) may also result in the loss of COUNT information. This must be factored into the COUNT verification procedure.

then the AC must retrieve the current COUNT value from that system, as described above in "Shared Secret Data Sharing." Again, if a COUNT mismatch is detected, IS-41-C allows the AC to select access denial treatment (i.e., the AC sends an **authreq** message to the serving system, including the DenyAccess parameter) or to continue authentication processing. Note that the COUNT check is only performed during the global challenge process—this is the only time when the MS provides the COUNT information to the network. Therefore, an authentication strategy that includes COUNT would necessarily include the global challenge.

The AC processing beyond this point depends on the particular authentication policies in effect. If the MS passes the global challenge (i.e., the AUTHR and COUNT are correct), then the AC may simply terminate the global challenge process and complete the transaction with the serving system by sending it an empty **authreq** message (see Fig. 11.11). Alternatively, the AC global challenge process may pass control to any one of the following additional processes in the AC:

- SSD sharing
- Unique challenge
- SSD update
- COUNT update

Unique Challenge

The IS-41-C unique challenge function encompasses the processes in which:

1. The IS-41-C authentication controller directs the serving system to present a numeric authentication challenge to a single MS that either is requesting service from the network or is already engaged in a call.

2. Optionally, the anchor system passes the authentication challenge request to the current serving system (i.e., in the case of a handed-off call).

3. The serving system presents the numeric authentication challenge to the MS and verifies that the numeric authentication response provided by the MS is correct (Fig. 11.13 depicts the basic unique challenge procedure for authentication when SSD is not shared).

The unique challenge is so named because the challenge indicator and the random number used for the challenge are directed to a particular MS (rather than to all MSs, as in the global challenge process).

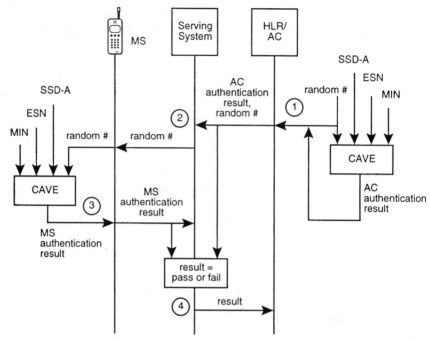

Figure 11.13 Basic unique challenge authentication process when SSD is not shared: (*1*) The AC generates a random number and uses it to calculate an authentication result. The AC sends both the random number and the authentication result to the serving system. (*2*) The serving system forwards the random number to the MS. (*3*) The MS calculates an authentication result and sends it to the serving system. (*4*) The serving system compares the result from the AC with the result from the MS. If the results match, the MS is considered to be authentic. If they do not match, the MS may be considered fraudulent and service may be denied. Either way, the serving system reports the results to the AC. If SSD is shared, the serving system may initiate the unique challenge process and will report only a failure to the AC.

In fact, the unique challenge may be used as a follow-up to the global challenge. Whether the global challenge succeeded or failed, the unique challenge serves as a double-check on the authenticity of the MS. And since the random number for the challenge is changed for each operation, the unique challenge provides a much more secure authentication test than the global challenge function.

Also, unlike the global challenge which has a single mechanism of initiation (i.e., the serving system autonomously transmits the global challenge indicator), the unique challenge can be initiated under a number of circumstances and by using a variety of IS-41-C messages. We describe these techniques in this chapter.

Note that the unique challenge is supported under call handoff conditions, whereas SSD update and COUNT update operations cannot be completed by the initial serving system (i.e., the anchor system) if the MS is involved in a call that has been handed off to another system. To support the call handoff scenario, the unique challenge processes include the serving system unique challenge process, the nonanchor serving system unique challenge process, and the AC unique challenge process.

Serving system unique challenge process

The details of the serving system unique challenge process differ depending on whether the unique challenge is controlled by the AC or by the serving system itself; the latter is possible if SSD is shared. However, note that the AC may initiate a unique challenge whether SSD is shared or not.

When controlled by the AC. Generally, the serving system begins the AC-controlled unique challenge process when it receives the IS-41-C RandomVariableUniqueChallenge (RANDU) and Authentication-ResponseUniqueChallenge (AUTHU) parameters from the AC in one of four IS-41-C messages:

1. AuthenticationDirective Invoke (**AUTHDIR**)

2. AuthenticationRequest Return Result (**authreq**)

3. AuthenticationStatusReport Return Result (**asreport**)

4. AuthenticationFailureReport Return Result (**afreport**)

Case 1 is shown in Fig. 11.14. In case 2, the **authreq** message may be a response to an **AUTHREQ** message sent from the serving system global challenge process to the AC; alternatively, the serving system unique challenge process may send the **AUTHREQ** message to the AC if the MS requires authentication and the global challenge is not in effect. In this sense, the serving system is the initiator of the unique challenge.

A complicating factor in the message exchange shown in Fig. 11.15 is intersystem handoff, i.e., when the initial serving system—the anchor system—is no longer the current serving system. IS-41-C compensates for this condition with the Authentication DirectiveForward operation. When a unique challenge is requested for an MS that is involved in a call, and the call has been handed off, the anchor system forwards the RANDU and AUTHU parameters to the current serving system—possibly via one or more tandem systems—using the AuthenticationDirectiveForward Invoke (**AUTHDIRFWD**) message.*

---

*The only exception to this procedure occurs when the unique challenge is part of an SSD update process; in this case, if the call has been handed off, the unique challenge is not attempted.

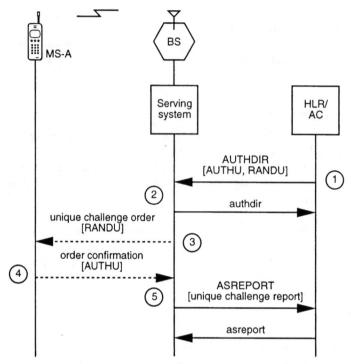

Figure 11.14   The AC-initiated unique challenge process: (*1*) The AC selects RANDU, calculates AUTHU, and sends both parameters to the serving system in an **AUTHDIR** message. (*2*) The serving system acknowledges the unique challenge request by sending an **authdir** to the AC. (*3*) The serving system sends a unique challenge order, including RANDU, to the MS in a mobile station control message. (*4*) The MS responds with its calculated AUTHU value in an order confirmation message.* (*5*) The serving system compares the AUTHU from the MS with the AUTHU from the AC and generates a unique challenge report indicating either success (i.e., the two AUTHUs match) or failure. The serving system sends the report to the AC in the **ASREPORT** message.

The response from the serving system—a unique challenge report in the AuthenticationDirective-Forward Return Result (**authdirfwd**) message—is passed to the authentication reporting process, and the unique challenge process terminates (see Fig. 11.15 and our discussion of the nonanchor serving system unique challenge process later in this chapter).

A final complicating factor exists if either the anchor system or the serving system cannot complete the unique challenge operation. For example:

---

*The Unique Challenge Order may be sent from the BS to the MS via a control channel, an analog voice channel, or a digital traffic channel. Likewise, the response from the MS to the BS may use any of these channel types.

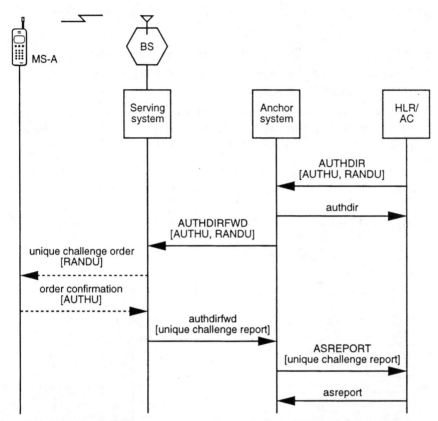

Figure 11.15 The AC-initiated unique challenge process when the MS is involved in a call that has been handed off from the anchor system to a nonanchor serving system. Message flows are similar to Fig. 11.14, with the addition of the Authentication-DirectiveForward exchange between the serving system and anchor system.

- The BS cannot send the order to the MS or receives no response from the MS due to a loss of radio contact or because the MS is turned off by the user.
- The call is handed off to a non-authentication-capable system.

In this case, the unique challenge report generated by the anchor or serving system indicates either that the challenge was not attempted or that no response was received from the MS.

Note that, in addition to the unique challenge, the **AUTHDIR**, **authreq**, **asreport**, and **afreport** messages can trigger a number of other authentication processes in the serving system, including:

- SSD update if the IS-41-C RandomVariableSSD (RANDSSD) parameter is included

- COUNT update if the IS-41-C UpdateCount (UPDCOUNT) parameter is included

- SSD sharing if the IS-41-C SSD parameter is included

- Termination of SSD sharing if the IS-41-C SSDNotShared (NOSSD) parameter is included

- Release of system resources for the current system access by the MS in question if the IS-41-C DenyAccess (DENACC) parameter is included (this is at the option of the serving system)

When more than one of these processes is requested in the same mes-

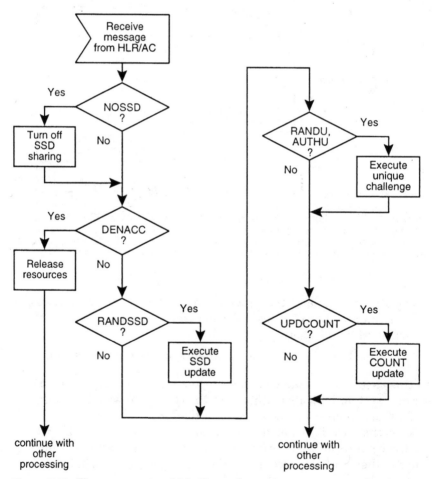

Figure 11.16 The sequence in which the serving system processes authentication commands received in a message from the AC.

sage, IS-41-C defines the required sequence of process execution (see Fig. 11.16).

When controlled by the serving system.   When SSD is shared, the serving system (i.e., the initial serving system in the case of a call hand-off) may initiate a unique challenge at any time, based on its internal authentication algorithms. In this case, the serving system selects the RANDU value for the challenge and uses the CAVE algorithm to calculate the corresponding AUTHU value. Otherwise, the operation of the serving system–initiated unique challenge process is the same as for the AC-initiated case. However, the authentication reporting process operates differently—reports are passed to the AC only in case of authentication failure, rather than on success or failure.

Nonanchor serving system unique challenge process

The nonanchor serving system (referred to simply as the serving system in the following description) has only one authentication-related task—handling the unique challenge process.

The process begins when the serving system receives an **AUTHDIRFWD** message from the anchor system, including the IS-41-C RANDU and AUTHU parameters, and proceeds as follows (see Fig. 11.15):

- The serving system directs the serving BS to send a Unique Challenge Order, including RANDU, to the MS in a mobile station control message.

- The MS responds with its calculated AUTHU value in an order confirmation message.

- The serving system compares the AUTHU from the MS with the AUTHU from the anchor system and generates a unique challenge report indicating one of the following:

   Unique challenge successful (i.e., the two AUTHUs match)

   Unique challenge not successful (i.e., the two AUTHUs do not match)

- The serving system sends the execution report to the anchor system, and the process terminates.

If the serving system cannot complete the unique challenge operation, e.g., due to loss of radio contact, the unique challenge report indicates either that the challenge was not attempted or that no response was received from the MS.

AC's unique challenge process

The AC initiates the unique challenge process based on its internal authentication algorithms. Possible triggers for the process include, but are not limited to:

- A failed global challenge
- A successful global challenge in a high-fraud area
- A call origination, call termination, or in-call flash request
- A periodic timer time-out (e.g., every few hours)

When SSD is not shared, the AC (i.e., the home service provider) must trade off the enhanced level of security provided by regular unique challenges with the added load on the signaling network between the home system and visited systems.

The AC selects the RANDU value for the challenge and uses the CAVE algorithm to calculate the corresponding AUTHU value. The AC sends these parameters to the serving system in one of four IS-41-C messages:

1. AuthenticationDirective Invoke (**AUTHDIR**)
2. AuthenticationRequest Return Result (**authreq**)
3. AuthenticationStatusReport Return Result (**asreport**)
4. AuthenticationFailureReport Return Result (**afreport**)

If the AC uses the **AUTHDIR** message, it waits for an AuthenticationDirective Return Result (**authdir**) message from the serving system; otherwise, the AC notifies the authentication reporting process that a unique challenge report is expected from the serving system and terminates the unique challenge process.

SSD Update

The IS-41-C SSD update function encompasses the processes by which the SSD in an MS is changed to a new value under the direction of the AC. Note that only the AC may initiate this operation, even when SSD is shared with the serving system.

On the network side, IS-41-C authentication procedures specify that a unique challenge is executed immediately after the SSD update to confirm that the target MS has successfully changed its SSD. On the MS side, the various authentication-capable air interface standards (e.g., IS-95, IS-136) specify that the MS initiates a *base station challenge* procedure when the MS is directed by the base station to change its SSD.* The base station challenge allows the MS to authenticate

---

*This is the only circumstance under which the base station challenge is used.

the base station issuing the SSD update; this prevents a fraudulent base station from disrupting the normal network operation by forcing the MS's SSD out of alignment with the network's SSD.

The SSD update processes include the AC SSD update process and the serving system SSD update process, in addition to the AC, serving system, and nonanchor serving system unique challenge processes previously described. The base station challenge functionality is included in the AC and serving system SSD update process descriptions.

AC SSD update process

The AC initiates the SSD update process based on its internal authentication algorithms; however, since the SSD update process requires significant network signaling, the AC does not normally change an MS's SSD without justification. For example, since the SSD and COUNT information may be transferred between systems using IS-41-C, there is the possibility that a valid authentication set (that is, MIN, ESN, SSD, and COUNT) could fall into the wrong hands. A criminal with knowledge of the CAVE algorithm and a valid authentication set would be able to simulate a legitimate MS—at least until the SSD was modified. Without knowledge of the A-key associated with the legitimate MS—presumably a far more secure piece of information than the SSD—the criminal would not be able to correctly change the SSD.

The first step in the AC SSD update process is for the AC to execute the CAVE algorithm using the MS's A-key, ESN, and a random number, called the RandomVariableSSD (RANDSSD). The result is the new, "pending" SSD.

The next step depends on whether the AC chooses to share the new SSD with the serving system. If it chooses to share SSD, the AC sends the RANDSSD and the new SSD to the serving system in one of four IS-41-C messages:

1. AuthenticationDirective Invoke (**AUTHDIR**)

2. AuthenticationRequest Return Result (**authreq**)

3. AuthenticationStatusReport Return Result (**asreport**)

4. AuthenticationFailureReport Return Result (**afreport**)

The AC then terminates the SSD update process and turns over control to the authentication reporting process, which waits for a report from the serving system. With both the RANDSSD and the new SSD, the serving system can:

- Update the SSD in the MS
- Respond to the base station challenge from the MS

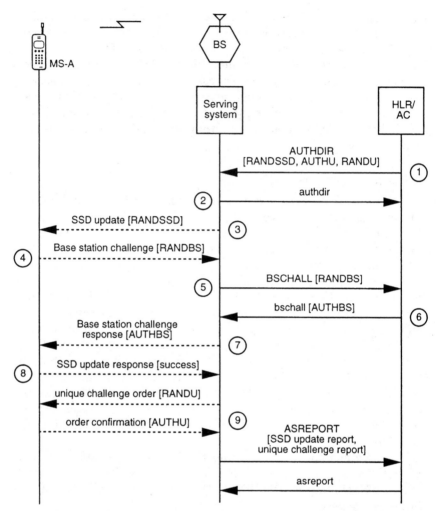

Figure 11.17   The SSD update process when the new SSD is not shared with the serving system: (*1*) The AC selects RANDSSD and calculates the new SSD. It then selects RANDU; calculates AUTHU; and sends the RANDSSD, RANDU, and AUTHU parameters to the serving system in an **AUTHDIR** message. (*2*) The serving system acknowledges the SSD update request by sending an **authdir** message to the AC. (*3*) The serving system sends an SSD Update Order, including RANDSSD, to the MS in a mobile station control message. (*4*) The MS calculates the new SSD. It then selects RANDBS and calculates AUTHBS, using the new SSD. It sends a base station challenge order to the serving system, including RANDBS. (*5*) The serving system forwards the challenge to the AC in the **BSCHALL** message. (*6*) The AC calculates AUTHBS, using the received RANDBS and the new SSD, and sends the result to the serving system in the **bschall** message. (*7*) The serving system forwards the base station challenge response from the AC to the MS. (*8*) The MS then confirms the SSD update. (*9*) The serving system executes a unique challenge using the RANDU and AUTHU values received in step 1. It sends a report, including the SSD update and unique challenge results, to the AC in the **ASREPORT** message.

- Verify the update by issuing a unique challenge to the MS

If SSD is not shared with the serving system, the AC must manage the unique challenge and base station challenge that accompany the SSD update. Therefore, the AC initiates the unique challenge process (see Fig. 11.17); i.e., it selects the RANDU value for the challenge and uses the CAVE algorithm to calculate the corresponding AUTHU value. The AC sends the RANDU, AUTHU, and RANDSSD parameters in the message to the serving system; again, any of the four messages identified above can be used (the use of the AuthenticationDirective operation is shown in Fig. 11.17). The AC notifies the authentication reporting process that an SSD update report is expected from the serving system.

However, the AC also waits for a BaseStationChallenge Invoke (**BSCHALL**) message from the serving system. If the AC receives the **BSCHALL** message, it executes the CAVE algorithm. Inputs are the random number selected by the MS and included in the **BSCHALL** message, the MS's MIN and ESN, and the new SSD. The AC returns the result of the calculation, AUTHBS, to the serving system in the BaseStationChallenge Return Result (**bschall**) message and then terminates the SSD update process, leaving the handling of the report from the serving system to the authentication reporting process.

Serving system SSD update process

The serving system begins the SSD update process when it receives the IS-41-C RandomVariableSSD (RANDSSD) parameter from the AC in one of four IS-41-C messages:

1. AuthenticationDirective Invoke (**AUTHDIR**)
2. AuthenticationRequest Return Result (**authreq**)
3. AuthenticationStatusReport Return Result (**asreport**)
4. AuthenticationFailureReport Return Result (**afreport**)

The message also contains either the new SSD, if the AC chooses to share it with the serving system, or the RANDU and AUTHU parameters, if the new SSD is not shared.

The serving system unique challenge process works as follows (see Fig. 11.17):

- If the MS is involved in a call and the call has been handed off, or if the SSD update cannot otherwise be attempted (e.g., loss of radio contact), then the serving MSC generates an SSD update report (indicating that the SSD update was not attempted), passes the report to the authentication reporting process, and terminates.

- Otherwise, the serving system directs the serving BS to send an SSD Update Order, including RANDSSD, to the MS in a mobile station control message.

- The MS responds with a Base Station Challenge Order, including the random number selected by the MS, RANDBS.

- If SSD is shared, the serving system executes the CAVE algorithm with inputs of RANDBS, the MS's MIN and ESN, and the new SSD. The result of the calculation is AUTHBS. If SSD is not shared (as shown in Fig. 11.17), the serving system sends a **BSCHALL** message to the AC. The AC responds with the AUTHBS result in the **bschall** message. The serving system returns the AUTHBS value (i.e., based on the local calculation or the AC's calculation) to the MS in an order confirmation message.

- The MS indicates a successful or an unsuccessful SSD update in an order confirmation message.

- If successful, the serving system executes the unique challenge process; otherwise, the serving system generates an SSD update report (indicating that the SSD update was unsuccessful), passes the report to the authentication reporting process, and terminates.

- The serving system generates two reports:

  1. An SSD update report, indicating that the SSD update was successful

  2. A unique challenge report, indicating whether the unique challenge was successful or unsuccessful

  It passes the reports to the authentication reporting process and terminates.

Note that the SSD update portion of the process is not executed if the MS is involved in a call that has been handed off; however, after sending the SSD update confirmation, the MS may be handed off.

Call History Count Update

The IS-41-C call history count (COUNT) update function encompasses the processes in which the IS-41-C authentication controller (i.e., the VLR if SSD is shared, the AC if SSD is not shared) directs the MS to update (i.e., increment by 1) its stored COUNT value.

The COUNT is a 6-bit parameter (i.e., a counter that increments from 0 to 63 and then back to 0) that is intended to provide additional security in case the A-key or SSD is compromised. The current value of COUNT is maintained by both the MS and the authentication controller. The respective counts should generally be the same—they may not always match exactly due to radio transmission problems or system failures in the network. If the respective counts differ by a large enough range, or frequently do not match, the AC may assume that a fraudulent condition exists and take corrective action. Note that a COUNT mismatch detection does not conclusively indicate that the particular MS accessing the system is fraudulent—only that a "clone" may exist.

The COUNT update processes include the serving system COUNT update process and the AC COUNT update process.

Serving system COUNT update process

The details of the serving system COUNT update process differ depending on whether the update is initiated by the AC or by the serving system itself; the latter is possible if SSD is shared. However, note that the AC may initiate a COUNT update whether SSD is shared or not. Figure 11.18 depicts the COUNT update procedure initiated by the AC.

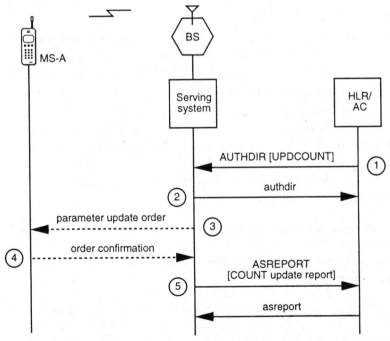

Figure 11.18 The AC-initiated COUNT update process: (*1*) The AC initiates the COUNT update by sending the UPDCOUNT parameter to the serving system, in an **AUTHDIR** message. (2) The serving system acknowledges the COUNT update request by sending an **authdir** message to the AC. (3) The serving system sends a parameter update order to the MS in a mobile station control message. (4) The MS updates its stored COUNT value and responds with an order confirmation message. (5) The serving system sends a COUNT update report indicating either success (i.e., confirmation from the MS) or failure. The serving system sends the report to the AC in the **ASREPORT** message. On receipt, the AC updates its stored COUNT value for the MS.

When controlled by the AC. The serving system begins the AC-controlled COUNT update process when it receives the IS-41-C UpdateCount (UPDCOUNT) parameter from the AC in one of four IS-41-C messages:

1. AuthenticationDirective Invoke (**AUTHDIR**)
2. AuthenticationRequest Return Result (**authreq**)
3. AuthenticationStatusReport Return Result (**asreport**)
4. AuthenticationFailureReport Return Result (**afreport**)

The serving system COUNT update process works as follows:

- If the MS is involved in a call and the call has been handed off, or if the COUNT update cannot otherwise be attempted (e.g., loss of radio contact), then the serving MSC generates a COUNT update report, indicating that the COUNT update was not attempted, passes the report to the authentication reporting process, and terminates.

- Otherwise, the serving system sends a Parameter Update Order to the MS in a mobile station control message.

- The MS increments its stored value of COUNT and responds with an order confirmation message.

- If the serving system receives the order confirmation message, it generates a COUNT update report indicating the COUNT update was successful; otherwise, it generates a COUNT update report indicating that there was no response to the COUNT update.

- The serving system passes the report to the authentication reporting process and terminates.

When controlled by the serving system. When SSD is shared, the serving system may initiate a COUNT update at any time, based on its internal authentication algorithms. Otherwise, the operation of the serving system–controlled count update process is the same as for the AC-controlled case (see Fig. 11.18). However, the authentication reporting process operates differently—reports are passed to the AC only in the case of authentication failure, rather than on both success and failure.

AC COUNT update process

The AC initiates the COUNT update process based on its internal authentication algorithms. The triggers for the process should be random so that the COUNT is not predictable by an illegitimate user.

To execute the COUNT update, the AC simply sends the IS-41-C UpdateCount parameter to the serving system in one of four IS-41-C messages:

1. AuthenticationDirective Invoke (**AUTHDIR**)
2. AuthenticationRequest Return Result (**authreq**)
3. AuthenticationStatusReport Return Result (**asreport**)
4. AuthenticationFailureReport Return Result (**afreport**)

The AC count update process then passes control to the authentication reporting process, which waits for a COUNT update report from the serving system.

Authentication Reporting

The IS-41-C authentication reporting function encompasses the processes by which the success or failure of various authentication processes is conveyed from the serving system to the AC. The processes which provide reports are:

- Global challenge
- Unique challenge
- SSD update
- COUNT update

The authentication reporting processes include the serving system authentication reporting process and the AC authentication reporting process.

Serving system authentication reporting process

We describe the mechanics of the serving system authentication reporting process in terms of three characteristics:

1. *Reporting triggers.* When is the process initiated?
2. *Reporting method.* How are reports conveyed to the AC?
3. *Report content.* What is in a report?

The first two characteristics are correlated: the reporting method depends on the reporting trigger.

Reporting triggers and methods. Each of the global challenge, unique challenge, SSD update, and COUNT update processes has associated outcomes, i.e., success or failure. While these outcomes are normally recorded, not all are reported; i.e., not every authentication process outcome is conveyed from the serving system to the AC. The five report-triggering factors are:

1. The nature of the authentication function, i.e., global challenge, unique challenge, SSD update, or COUNT update
2. The authentication version, either TSB51 or IS-41-C
3. The authentication process result, either success or failure
4. The authentication initiator, either the AC or the serving system
5. The status of SSD sharing, either shared or not shared

IS-41-C is backward-compatible with the authentication procedures defined in TSB51; however, the reporting methods defined in the two documents differ slightly. In particular, TSB51 specifies a single IS-41-C operation (called SecurityStatusReport in TSB51 and renamed AuthenticationFailureReport in IS-41-C) to report the outcome of both AC-initiated (success or failure) and serving system–initiated (failure only) authentication processes. IS-41-C specifies the AuthenticationFailureReport Invoke (**AFREPORT**) message for reporting the failure of serving system–initiated authentication processes and the AuthenticationStatusReport Invoke (**ASREPORT**) message for reporting the success or failure of AC-initiated authentication processes.

The operation initiator is generally the AC when SSD is not shared; the exception is the global challenge, which is always initiated by the serving system. When SSD is shared, either the AC or the serving system can initiate the unique challenge and COUNT update operations. Only the AC may initiate the SSD update operation.

Table 11.1 lists the IS-41-C reporting methods (i.e., reporting messages generated) under various conditions based on the five factors identified above.

Report content.   Another difference between IS-41-C and TSB51 authentication reporting lies in the content of reports. The **AFRE-PORT** message in TSB51 reports on both the success and failure of authentication processes; therefore, the TSB51 ReportType parameter supports both types of indications. The IS-41-C ReportType parameter is reduced in scope, supporting only failure indications.

IS-41-C also adds new, individual report-type parameters for each of the authentication processes that the AC can initiate: UniqueChallengeReport, SSDUpdateReport, and CountUpdateReport.

TABLE 11.1   IS-41-C Authentication Reporting Methods

Function	Initiator and version	Message sent to AC on successful outcome	Message sent to AC on failed outcome
Global challenge	IS-41-C or TSB51 serving system	No report	**AUTHREQ, AFREPORT,** or **REGNOT***
Unique challenge	IS-41-C AC	**ASREPORT**	**ASREPORT**
Unique challenge	TSB51 AC	**AFREPORT**	**AFREPORT**
Unique challenge	IS-41-C serving system	No report	**AFREPORT**
Unique challenge	TSB51 serving system	No report	**AFREPORT**
SSD update	IS-41-C AC	**ASREPORT**	**ASREPORT**
SSD update	TSB51 AC	**AFREPORT**	**AFREPORT**
COUNT update	IS-41-C AC	**ASREPORT**	**ASREPORT**
COUNT update	TSB51 AC	**AFREPORT**	**AFREPORT**
COUNT update	IS-41-C serving system	No report	**AFREPORT**
COUNT update	TSB51 serving system	No report	**AFREPORT**

*The handling of failures in the global challenge process is somewhat ambiguous in IS-41-C. For example, each of the listed messages can convey the authentication failure report when an authentication-capable MS registers and does not provide a global challenge authentication response.

AC authentication reporting process

A significant aspect of the AC authentication reporting process that is modified in IS-41-C compared to TSB51 relates to the timing of the reports the AC receives from the serving system. In TSB51, when the AC initiates an authentication operation, the timing of the report is indeterminate; i.e., there is no specification that the report shall be generated within x seconds of the receipt of the authentication-related request (see Fig. 11.19).

IS-41-C defines the Authentication Status Report Timer (ASSRT) that specifies the number of seconds (default value is 24) the AC shall wait for a report before executing recovery procedures. The ASSRT timer is started in one of two ways:

1. When the **authdir** message is received by the AC, indicating the serving system's acknowledgment of the authentication-related request

2. When the **authreq, asreport,** or **afreport** message is sent by the AC, including an authentication-related request (e.g., unique challenge)

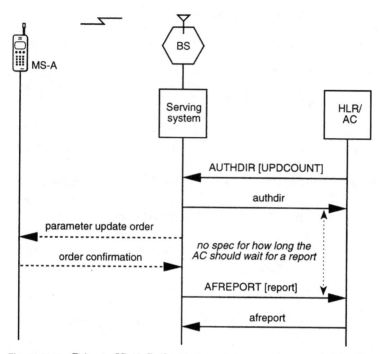

Figure 11.19 Prior to IS-41-C, the timing of the report generated by the serving system was not specified.

Summary of IS-41-C Operations Used for Authentication

Table 11.2 summarizes the IS-41-C operations used by the authentication functions described in this chapter.

TABLE 11.2   Use of IS-41-C Operations for Authentication

Function	IS-41-C operations used for the function
SSD sharing	AuthenticationDirective, AuthenticationFailureReport, AuthenticationRequest, AuthenticationStatusReport, CountRequest, MSInactive, RegistrationCancellation, RegistrationNotification
Global challenge	AuthenticationRequest, IntersystemPage2, RandomVariableRequest
Unique challenge	AuthenticationDirective, AuthenticationDirectiveForward, AuthenticationFailureReport, AuthenticationRequest, AuthenticationStatusReport
SSD update	AuthenticationDirective, AuthenticationFailureReport, AuthenticationRequest, AuthenticationStatusReport, BaseStationChallenge
Call history count update	AuthenticationDirective, AuthenticationFailureReport, AuthenticationRequest, AuthenticationStatusReport
Authentication reporting	AuthenticationFailureReport, AuthenticationStatusReport

Chapter

12

Call Processing
Functions

In this chapter, we discuss the IS-41-C mobile telecommunications network functions related to call processing. These functions are divided into the following categories:*

- Call origination functions
- Call termination functions
- Feature control functions

We describe each of the call processing functions defined in IS-41-C and provide examples of how the IS-41-C mobile application part (MAP) operations (e.g., LocationRequest) are used to accomplish call processing tasks. We summarize this information at the end of the chapter. Of course, our descriptions are not intended to be a complete definition of how each of the IS-41-C call processing functions works—for that, we defer to the standard.

Note that we use the IS-41-C convention for operation component acronyms; i.e., the Invoke component acronym is in all-capital letters (e.g., **LOCREQ**), while the Return Result component acronym is in all-lowercase letters (e.g., **locreq**). Refer to the IS-41-C standard for the descriptions of the individual IS-41-C operations and parameters. Also, consult the Glossary for a description of general terms (e.g., *serving system*) that are not explicitly defined in this chapter.

Throughout this chapter we consider the serving system as a single entity encompassing the mobile switching center (MSC) and visitor

---

*The list of call processing functions reflects the authors' subjective categorization of certain IS-41-C functions.

location register (VLR) functional entities. This simplifies the descriptions of the IS-41-C call processing functions and also is representative of a large percentage of the IS-41 implementations currently in service. However, keep in mind that the potential for separation of the MSC and VLR exists and is fully defined in IS-41-C.

What Is IS-41-C Call Processing?

Call processing is generally considered to be the broad category of functions that establish, maintain, and release calls to and from the mobile subscriber. It consists of the operations performed from the initial acceptance of an incoming call through the final disposition of the call. The call processing functions establish and release terrestrial transmission paths for calls as well as invoke and manage call features that provide a specific variation on the way a call is treated.

In this chapter, our discussion of IS-41-C call processing encompasses the following categories of functions:

1. *Call origination functions*—the intersystem functions that support the mobile-originated calling capabilities, including the handling of call origination features

2. *Call termination functions*—the intersystem functions that support the mobile-terminated calling capabilities, including the handling of call termination features

3. *Feature control functions*—the intersystem functions that support the subscriber's ability to modify certain feature parameters (e.g., feature activation status, call-forwarding forward-to number)

Where Are Call Processing Functions Specified in IS-41-C?

Call processing functions are specified in three parts of IS-41-C:

1. IS-41.5-C provides the required formats—the bit-by-bit encoding—of all the IS-41-C operation components, including those used for call processing. IS-41.5-C defines both the messages (e.g., LocationRequest Invoke) and the message parameters (e.g., MobileIdentificationNumber).

2. IS-41.6-C provides algorithmic descriptions of the procedures that are associated with sending and receiving most IS-41-C messages, including those used for call processing. This part of IS-41-C also includes a significant level of detail on the application process logic associated with call origination, call termination, and voice feature control.

3. IS-41.3-C steps back from the protocol details contained in parts 5 and 6 and attempts to explain, using information flow diagrams and step-by-step descriptions, how the operations are used individually and together to accomplish call processing application process tasks.

General Issues Associated with Call Processing

Many call processing functions are beyond the scope of the IS-41 standard, and many aspects of call processing are closely coupled to the basic IS-41 automatic roaming and intersystem handoff functions described in Chaps. 9 and 10. These aspects include:

- Delivering calls to a roaming subscriber. Location and status information needs to be obtained, involving the location management and MS state management functions described in Chap. 10.
- Controlling call origination from a roaming subscriber. The subscriber's authorized calling capabilities must be provided to the serving system, involving the service qualification functions described in Chap. 10.
- Controlling inter-MSC circuits during intersystem handoff, involving the various intersystem handoff functions described in Chap. 9.

There are also some general issues that impact a number of the call processing functions including these:

1. All features are optional, and many are not supported everywhere. Subscribers that transit to a new system via either roaming or intersystem handoff may lose features not supported by that system.
2. Feature codes used to control features are not standardized. A feature code used to activate call forwarding, for instance, may not be the same in a visited system as in the home system.
3. Feature interaction.

Feature interaction

Feature interaction is a complex issue. It is a situation that arises when a subscriber is authorized for multiple features. Four aspects to feature interactions need to be analyzed:

1. Definition of a feature interaction
2. Definition of feature precedence

3. Definition of feature treatment

4. Which functional entities control an interaction

A *feature interaction* simply means that if multiple features are simultaneously active, these features need to be analyzed in order to apply the appropriate call treatment. At this point, it does not matter which feature is analyzed first; they are all reviewed, and a logical decision is made as to how to treat the call.

Feature precedence usually refers to one feature superseding another. That is, if two features are active and one feature takes precedence over the other, the subordinate feature need not be factored into the algorithm to determine call treatment; i.e., the fact that it is active can be ignored.

Feature treatment means that certain actions are taken on a call based on calling and called party numbers, active features, and other events that can affect the call. For example, if call forwarding—no answer (CFNA) is active and conditions are such that CFNA is invoked (e.g., the call is not answered), then CFNA treatment means to forward the call.

Depending on the individual features involved in the interaction, the MSC, the HLR, or a combination of the MSC and HLR controls the interaction. Control of the interaction includes determining feature precedence as well as the call treatment to apply.

An example of an HLR-controlled feature interaction with precedence is the interaction between call forwarding—unconditional (CFU) and call waiting (CW). If a mobile-terminated call arrives, the HLR controls the interaction between CFU and CW. If CFU is active, CW is never even checked for activation. This means that CFU takes precedence over CW. It does not matter whether the subscriber has the feature, or if it is active. It is simply never analyzed in this case.

An example of an MSC-controlled feature interaction with precedence is the interaction between call transfer (CT) and three-way call (3WC). If a subscriber has established a three-way call and then disconnects, the normal 3WC feature processing calls is to disconnect all the parties involved in the call. However, if CT is also active, the MSC transfers the call to the two other parties in the three-way call when the controlling subscriber disconnects. For this aspect of the interactions between the CT and 3WC features (i.e., disconnect treatment), CT takes precedence over 3WC.

Feature interaction analysis is a very important function to maintain seamlessness. It is necessary to address each call scenario involving every feature for proper call processing. Some of the feature interactions are specified in IS-41-C. Other feature interactions are implementation-dependent; i.e., those that are HLR-controlled can be seamlessly provided without standardized, feature-specific serving system support. The management of feature interactions, however, is required to ensure proper and efficient logic processing at both the HLR and the serving system.

Call Origination Services

The simplicity of the process of making a mobile call—turning the phone on, entering digits, and pressing the SEND key—masks an enormous amount of call processing that takes place in the mobile telecommunications network. Prior to Revision C, IS-41 had relatively little impact on this processing—the Revision B services were mainly call termination-related (i.e., call delivery, call forwarding, call waiting). This situation changed in Revision C, and will likely continue to change in subsequent revisions as "wireless intelligent network" concepts are applied to the standard and call origination becomes a more network-controlled—rather than a serving system–controlled— process.

In general, IS-41-C call origination services are the network functions that enable, restrict, supplement, or otherwise impact an MS's ability to originate a call while roaming outside the home service area. IS-41-C supports the following call origination services:

- Basic call origination
- Subscriber personal identification number (PIN) access (SPINA)
- Subscriber PIN intercept (SPINI)
- Calling number identification restriction (CNIR)
- Message waiting notification (MWN)
- Voice message retrieval (VMR)
- Call transfer (CT)
- Three-way calling (3WC)
- Conference calling (CC)
- Priority access and channel assignment (PACA)
- Preferred language (PL)
- Voice privacy (VP)

The message waiting notification, preferred language, and voice privacy features are not solely call origination features—they also include call termination support capabilities.

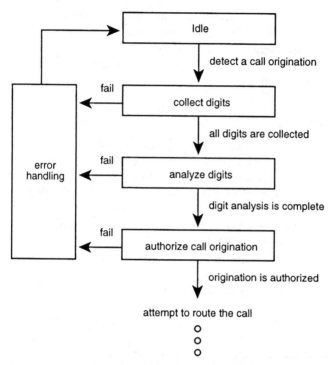

Figure 12.1   Part of a simplified call origination process in a serving system.

Basic call origination

A portion of a simplified call origination process in a serving system is shown in Fig. 12.1. The protocol and procedures in IS-41-C impact this model in two areas: digit analysis and call origination authorization.

IS-41-C impact on the digit analysis process.   IS-41-C defines two basic categories of *triggers* that can be applied to the digit analysis process, call origination triggers and feature request triggers. The call origination triggers are "armed" when the serving system receives the IS-41-C OriginationTriggers parameter. This parameter is part of the MS's service profile information provided by the home system during the service qualification process (see Chap. 10). Some of the potential triggers are:

- All call originations
- Call origination to an international destination
- Call origination to the subscriber's own number, termed a *revertive call*

- Call origination to a number beginning with the * digit
- Call origination to a number beginning with the # digit
- Call origination matching a criterion set in an agreement between the home and serving systems
- Call origination with n digits, where n may be set from 0 to 14 digits
- Call origination with 15 or more digits

When a trigger condition is detected during the digit analysis process, the serving system sends an OriginationRequest Invoke (**ORREQ**) message to the MS's HLR, containing (among other things) the digits received from the MS. This allows the HLR to evaluate the dialed digits and to provide routing instructions to the serving system via the OriginationRequest Return Result (**orreq**) message (see Fig. 12.2). Generally, the evaluation and routing procedures in the HLR are associated with some subscriber-specific features, such as subscriber PIN intercept or voice message retrieval. Thus, the OriginationTriggers parameter and OriginationRequest operation can be considered to be the building blocks for other call processing features.

The basic "feature request trigger" does not require explicit arming; i.e., when the serving system detects a call origination to a number beginning with the * digit—this is called a *feature code string*—it sends a FeatureRequest Invoke (**FEATREQ**) message to the MS's HLR. However, note that the corresponding origination trigger takes

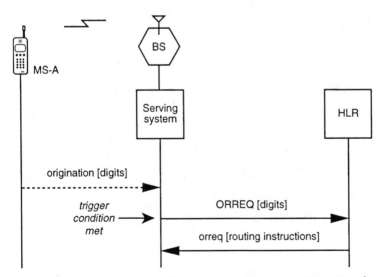

Figure 12.2 An example of a call origination trigger processing using the OriginationRequest operation.

precedence over the feature request trigger; i.e., if the origination trigger for the * digit is armed, then the serving system sends an **ORREQ** message rather than a **FEATREQ** message to the HLR. See "Feature Control Services" for additional discussion of the feature request processes.

IS-41-C impact on the call origination authorization process. The key IS-41-C parameters that impact the basic call origination authorization process are:

- AuthorizationDenied
- AuthorizationPeriod
- OriginationIndicator

This information is obtained through the service qualification processes described in Chap. 10 (e.g., using the Registration-Notification or QualificationDirective operation). The HLR sends either the AuthorizationDenied or the AuthorizationPeriod parameter to the serving system, but not both. If authorization is denied, then the subscriber receives access denial treatment (e.g., a special tone or announcement) from the serving system when call origination is attempted. If authorization is allowed, the AuthorizationPeriod parameter indicates the period for which the authorization applies:

- Per call
- For n calls, hours, days, or weeks, where n ranges from 0 to 255
- For a period set in an agreement between the home and serving systems
- Indefinitely (i.e., until denied or deregistered)

When the authorization period expires, for example, after 24 h, the next call origination or other system access request by the MS triggers the service qualification process in the serving system (see Fig. 12.3).

The OriginationIndicator (ORIGIND) parameter specifies the types of calls the MS is allowed to originate:

- No calls (i.e., call origination not allowed)
- Call types as defined by agreement between the home and serving systems
- Local calls only
- Calls to a specified NPA-NXX only
- Calls to a specified NPA-NXX and local calls only

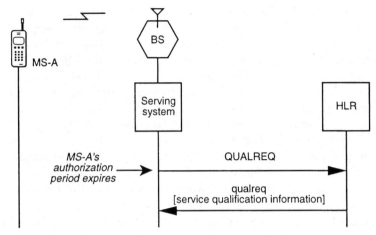

Figure 12.3 An example of call origination authorization using the QualificationRequest operation.

- Calls within World Zone 1 (or, more generally, the serving country)
- All calls; i.e., local, serving country, and international
- Calls to a specified NPA-NXX-XXXX only

This last call type supports a *hotline* service; i.e., regardless of the number received from the MS, the serving system establishes a call to the single number provided by the HLR.* Also note that each serving system may define locally allowed calls (e.g., to numbers such as 911, *911) that are not affected by the value of the OriginationIndicator parameter.

Subscriber PIN access

The subscriber PIN access (SPINA) feature supports a limited form of subscriber control over the IS-41-C OriginationIndicator (ORIGIND) parameter value; i.e., the subscriber—by performing either a single-step or multistep feature control procedure that includes the entry of a personal identification number—can indirectly toggle the value of the ORIGIND parameter between no calls (SPINA active) and the subscriber's normally assigned origination authorization level (SPINA

---

*While this value of the OriginationIndicator has been defined since IS-41 Revision A, it is only in IS-41-C that the serving system processing for this case is explicitly defined; therefore, serving system implementations of prior versions of IS-41 may not operate the same as IS-41-C products.

inactive). The HLR usually conveys the revised ORIGIND parameter to the serving system by using the QualificationDirective operation (see Fig. 12.4). In effect, SPINA provides a network-based MS locking mechanism that is presumably more secure than MS-based locks.

While the SPINA feature is described in IS-53-A and IS-41-C primarily as a subscriber-controlled on/off switch for call originations, it is also possible for the home service provider to use the feature in an *automatic lock* mode as a fraud-deterrent measure. In other words, the SPINA feature is automatically activated by the service provider and must be deactivated on a per-call basis by the subscriber (in effect, the feature becomes a derivative of the subscriber PIN intercept feature described in the next section). However, this application of SPINA diminishes in importance when authentication—a much more effective fraud-deterrent measure—is ubiquitously deployed in the network.

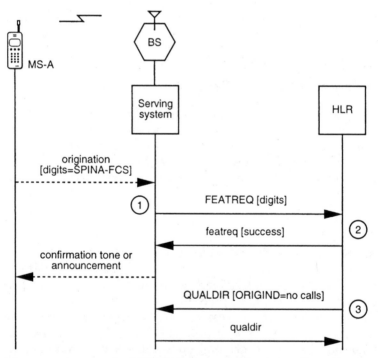

Figure 12.4 An example of SPINA feature activation using the single-step feature control procedure: (*1*) The subscriber enters the SPINA activation feature code string (SPINA-FCS) and presses the SEND key; the SPINA-FCS is in the form *FC + PIN. The serving system detects the FCS and sends the received digits to the HLR in the **FEATREQ** message. (*2*) The HLR verifies that the PIN is valid and signals operation success in the **featreq** message. The serving system notifies the subscriber that the request was successful. (*3*) The HLR modifies the subscriber's service qualification to disable call origination, using the **QUALDIR** message.

Subscriber PIN intercept

The subscriber PIN intercept (SPINI) feature is an application of the OriginationTriggers parameter and OriginationRequest operation building blocks already described. SPINI allows the home service provider to set the conditions, i.e., triggers, under which the subscriber may be requested to enter a valid PIN before the network agrees to complete the subscriber's call origination request.

Figure 12.5 illustrates the SPINI feature operation when (a) the

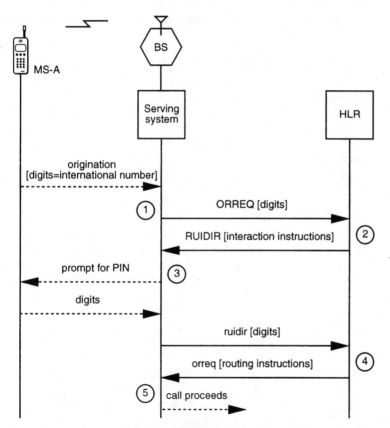

Figure 12.5 An example of SPINI feature invocation using the RemoteUserInteractionDirective operation: (1) The subscriber enters an international phone number and presses the SEND key. The "call origination to an international destination" trigger is set; therefore, the serving system sends an **ORREQ** message to the HLR including the dialed digits. (2) The HLR chooses to query the subscriber for a PIN before allowing the call to proceed; therefore, it sends interaction instructions to the serving system in the **RUIDIR** message. (3) The serving system connects the call to a system capable of user interaction (e.g., an interactive voice response unit), prompts the subscriber for a PIN, collects the digits entered by the subscriber, and sends them to the HLR in the **ruidir** message. (4) The HLR verifies that the PIN is valid and returns call routing instructions in the **orreq** message. (5) The serving system continues with call origination, as directed by the HLR.

origination trigger for international calls is set in the serving system, (*b*) the subscriber attempts to originate an international call, and (*c*) the interaction with the subscriber is directed by the HLR, using the IS-41-C RemoteUserInteractionDirective operation.

It is also possible for the serving system to control the SPINI-related interaction with the subscriber, thereby cutting down on the network signaling traffic between the serving and home systems. For this, the serving system must:

- Inform the HLR that it is capable of local SPINI feature control.

- Obtain the subscriber's SPINI PIN and SPINI-related call origination triggers.

The serving system accomplishes this information transfer in the course of qualifying the MS for service. The serving system's local SPINI capability is conveyed in the IS-41-C TransactionCapability parameter (TRANSCAP), and the subscriber's SPINI PIN and triggers are returned to the serving system in the SPINIPIN and SPINITriggers parameters, respectively. With this information, the serving system can control the SPINI-related call origination processing (see Fig. 12.6).

Calling number identification restriction

The calling number identification restriction (CNIR) feature allows a subscriber to prevent the presentation of the subscriber's calling number identification (CNI) information to the call-terminating party. The subscriber may be the party who originated the call or may be a party who is redirecting the call (e.g., the subscriber is busy and the call forwarding—busy feature is active).

As with many features in IS-41-C, protocol support for the CNIR feature has three components:

1. *Conveying the subscriber's default CNIR feature activity status to the serving system.* The IS-41-C CallingFeaturesIndicator parameter contains this information; the HLR sends this to the serving system during the service qualification process.

2. *Feature control.* The CNIR feature may be activated or deactivated on a per-call origination basis, using the per-call feature control procedure described in "Feature Control Services."

3. *Feature invocation,* i.e., marking the subscriber's CNI information to indicate the status of the CNIR feature for the call.

For the purpose of the third function, each of the following IS-41-C

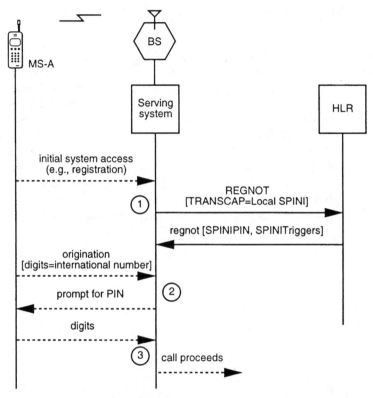

Figure 12.6 An example of local SPINI feature handling by the serving system: (*1*) The subscriber registers in a local SPINI-capable serving system. The HLR provides the subscriber's SPINI PIN and call origination triggers to the serving system. (*2*) The subscriber enters an international phone number and presses the SEND key. The "call origination to an international destination" trigger is set; therefore, the serving system connects the call to a system capable of user interaction (e.g., an interactive voice response unit), prompts the subscriber for a PIN, collects the digits entered by the subscriber, and verifies that they match SPINI PIN digits received from the HLR. (*3*) The serving system allows the call origination to proceed.

number identification parameters includes a CNIR indicator field (i.e., presentation allowed or presentation restricted):

- CallingPartyNumberDigits1
- CallingPartyNumberDigits2
- CallingPartyNumberString1
- CallingPartyNumberString2
- RedirectingNumberDigits

- RedirectingNumberString

When the called party is an MS served by a system that is connected to the subscriber's serving system using IS-41-C, the subscriber's CNI parameters can be conveyed to the serving system by using IS-41-C messages (see Fig. 12.7); otherwise, a call control protocol with number identification support (e.g., the SS7 ISDN user part, ISUP) is required between the subscriber's system and the called party's system to transport the CNI information.

Finally, IS-41-C supports a nonstandard (i.e., not defined from a stage 1 perspective in IS-53-A) capability whereby a legally authorized called subscriber (e.g., an emergency services operator) can override the calling subscriber's CNIR feature and gain access to the CNI information—even when CNIR is activated for the calling subscriber. This is referred to as the *CNIR override* feature.

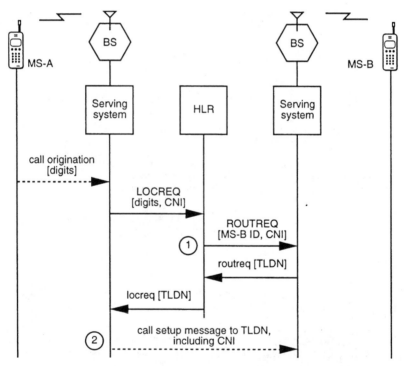

Figure 12.7 Methods of conveying calling number identification (CNI) information: (*1*) The CNI information can be sent from the originating system (i.e., MS-A's serving system) to the terminating system (i.e., MS-B's serving system) by using IS-41-C parameters in the normal call delivery messages **LOCREQ** and **ROUTREQ**. (*2*) The CNI information can also be sent from the originating system to the terminating system by using an appropriate call control protocol (e.g., ISUP).

Message-waiting notification

Message-waiting notification (MWN) is not solely a call origination feature; in fact, the MWN feature supports call origination, call termination, and non-call-associated notification options.

For notification at the time of call origination or termination, the MWN *pip tone* option is used. A pip tone consists of an audible tone (i.e., four 100-ms bursts of a 480-Hz signal) applied by the serving system on the voice path to the MS when the subscriber attempts to originate or terminate a call. The MWN pip tone feature may be deactivated on a per-call origination basis by using the per-call feature control procedure; this capability can be used to ensure that a modem-driven data call is not interrupted by the pip tone.

There are two options for notification when the MS is idle (i.e., not involved in a call):

1. An *alert pip tone* nominally consisting of the same four bursts of signal as the pip tone, but generated by an audible annunciator in the MS under the command of the serving system

2. A *message-waiting indication* on the MS—it may be a simple message-waiting lamp or a digital display of the number of messages waiting for retrieval—that is driven based on signaling messages from the serving system.

The HLR conveys the MWN information to the serving system in the IS-41-C MessageWaitingNotificationCount (MWNCOUNT) and MessageWaitingNotificationType (MWNTYPE) parameters as part of the service qualification process. The serving system uses this information to provide the appropriate MWN indication to the MS.

In the case of call handoff, the revised MWN information may be sent down the handoff chain from the anchor system to the serving system by using the IS-41-C InformationForward Invoke (**INFOFWD**) message (see Fig. 12.8). However, the anchor system will initiate this operation only if the MWN is of the message-waiting indication type described above; the pip tone and alert pip tone methods do not apply in the case of an active call that has been handed off.

Voice message retrieval

The voice message retrieval feature allows service providers to offer subscribers easy access to their voice mailbox for message retrieval purposes. Two forms of subscriber access are supported:

1. The subscriber dials a feature code string (for example, *VM).
2. The subscriber dials the subscriber's own number (i.e., a *revertive* call).

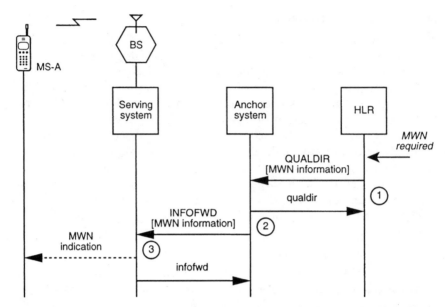

Figure 12.8 An example of message-waiting notification (MWN): (*1*) The HLR deter-mines that MWN is required for MS-A. The HLR sends the MWN information (i.e., MWNTYPE and MWNCOUNT parameters) to the anchor system. (*2*) MS-A is involved in a call that has been handed off; therefore, the anchor system sends the MWN infor-mation to the serving system in the **INFOFWD** message. (*3*) The serving system noti-fies MS-A via the appropriate notification method.

The IS-41-C implementation of the first access method is based on the feature control with call routing procedure described in "Feature Control Services" (see Fig. 12.9). Implementation of the second access method is based on the OriginationTriggers parameter and OriginationRequest operation described above (see Fig. 12.10).

The VMR feature also allows the service provider to route sub-scriber calls to the voice message system by using a form of call deliv-ery and temporary local directory numbers (TLDNs), rather than associating each subscriber's mailbox with an individual directory number (see Figs. 12.9 and 12.10).

Three-way calling and call transfer

The three-way calling and call transfer features allow a subscriber involved in a two-party call (i.e., between the subscriber and party A) to:

1. Place party A on hold—i.e., the subscriber presses the SEND key.

2. Call another party B—i.e., the subscriber enters party B's phone number and presses the SEND key.

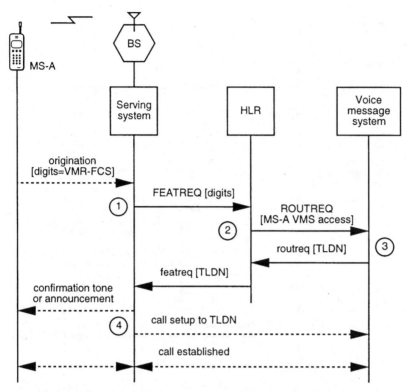

Figure 12.9 An example of VMR feature invocation using the feature code procedure: (*1*) The subscriber enters the VMR invocation feature code string (VMR-FCS), which may include the VMR password, and presses the SEND key. The serving system detects the FCS and sends the received digits to the HLR in the **FEATREQ** message. (*2*) The HLR sends a **ROUTREQ** message to the subscriber's voice message system (VMS), identifying the subscriber (e.g., MS-A's voice mailbox number and VMR password) requesting VMR routing information. (*3*) The VMS returns a TLDN to the HLR, and the HLR relays the TLDN to the serving system. (*4*) The serving system establishes a call to the VMS by using the TLDN, and the subscriber is connected to the voice mailbox.

3. Press the SEND key to

■ Connect the subscriber with both parties A and B (if the subscriber has 3WC active)

■ Release party B and reconnect the subscriber to party A (if the subscriber has CT active but not 3WC)

However, the subscriber presses the END key—rather than the SEND key—to release the subscriber and connect party A to B (if the subscriber has CT active).

IS-41-C does not define procedures that apply solely to the three-

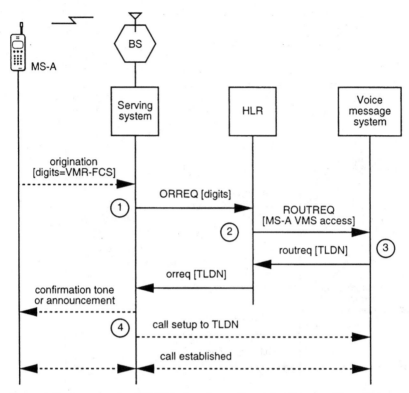

Figure 12.10   An example of VMR feature invocation using the revertive call procedure: (*1*) The subscriber enters his or her own mobile directory number (MDN), possibly along with the VMR password, and presses the SEND key. The serving system detects the revertive call. The "revertive call" trigger is set; therefore, the serving system sends the received digits to the HLR in the **ORREQ** message. (*2*) The HLR sends a **ROUTREQ** message to the subscriber's voice message system, identifying the subscriber (e.g., MS-A's voice mailbox number and VMR password) requesting VMR routing information. (*3*) The VMS returns a TLDN to the HLR, and the HLR relays the TLDN to the serving system. (*4*) The serving system establishes a call to the VMS by using the TLDN, and the subscriber is connected to the voice mailbox.

way calling and call transfer features, stating that these features "are controlled by the call processing in the Anchor MSC in a manner consistent with IS-53."

However, the 3WC and CT features require IS-41-C support in a couple of generic areas:

1. To convey the subscriber's 3WC and CT feature activity status to the serving system. The IS-41-C CallingFeaturesIndicator parameter contains this information; the HLR sends this to the serving system during the service qualification process.

2. Feature invocation under call handoff conditions. Since these features are controlled by the anchor system (this is the IS-41 philosophy), the serving system must notify the anchor system when the subscriber either presses the SEND key (possibly with other dialed digits) or presses the END key.

The serving system uses the IS-41-C FlashRequest Invoke (**FLASHREQ**) message to forward the notification to the anchor system when the SEND key was pressed by the subscriber (see Fig. 12.11); this message also includes any other digits dialed by the subscriber. When the subscriber presses the END key, the serving system initiates the call release process; i.e., it sends a FacilitiesRelease Invoke (**FACREL**) message to the anchor system.

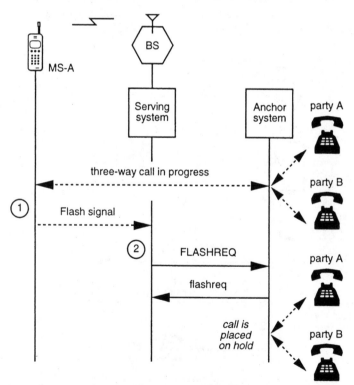

Figure 12.11   An example of the use of the FlashRequest operation for three-way calling: (*1*) The subscriber is involved in a three-way call that has been handed off from the anchor system to a nonanchor serving system. The subscriber presses the SEND key. (*2*) The serving system sends a **FLASHREQ** message to the anchor system. The anchor system places party A and party B on hold.

Conference calling

Conference calling (CC) is another multiparty feature, like 3WC and CT; however, it requires specific IS-41-C protocol elements to support two key capabilities of the feature:

1. The ability of the service provider to define the maximum number of conferees allowed for any particular conference call, rather than fix the number for all calls and all subscribers. For this purpose, the conference call is initiated with a feature code string provided by the subscriber—either at the start of a call (see Fig. 12.12) or during an active two-party call—and conveyed from the serving system to the HLR in the **FEATREQ** message, as described in "Feature Control Services." The **featreq** message contains the IS-41-C ConferenceCallingIndicator (CCI) parameter that specifies the maximum number of conferees for the call. The presence of the CCI parameter in the **featreq** message informs the serving system that the subscriber has invoked the CC service and to process the call accordingly.

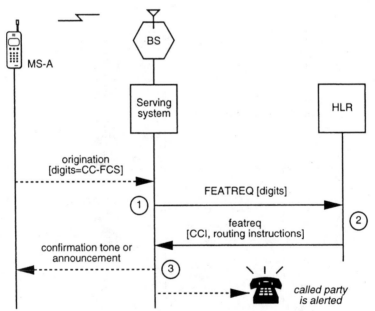

Figure 12.12 An example of conference call initiation: (*1*) The subscriber enters the CC invocation code string (CC-FCS), which includes the phone number of the first party to add to the call, and presses the SEND key. The serving system detects the FCS and sends the received digits to the HLR in the **FEATREQ** message. (*2*) The HLR authorizes the conference call and returns the ConferenceCallingIndicator (CCI) parameter and call-routing instructions in the **featreq** message. (*3*) The serving system notifies the subscriber that the request was successful, begins CC processing, and routes the call as directed by the HLR.

2. The ability of the subscriber to drop the last party added to the conference call. This is useful if a busy signal, wrong number, or voice message system is reached during the attempt to add a party to a conference call. This function is also initiated with a feature code string from the subscriber—the "drop last party" feature code string—which the serving system sends to the HLR in a **FEATREQ** message. The HLR returns a **featreq** message to the serving system, containing the IS-41-C ActionCode parameter set to the value *conference calling drop last party*. The serving system executes the drop-last-party operation, and the conference call continues.

Additionally, the feature is defined so that once the conference call has been authorized by the HLR, new parties can be added to the call under either HLR control or serving system control. The HLR-controlled case is shown in Fig. 12.12; i.e., each new party's number is preceded by a feature code that directs the serving system to query the HLR for routing instructions. With serving system control:

1. The subscriber presses SEND to place the current conferees on hold.
2. The subscriber enters the new party's number—without a feature code—and presses SEND.
3. The serving system immediately establishes a call to the new party; no query (i.e., **FEATREQ** message) is sent to the HLR.
4. The subscriber presses SEND to add the held conferees back onto the call.

This approach may be supplemented with the call origination triggers to give the HLR selected control over adding new parties. For example, if the international destination trigger is armed and the subscriber dials an international number as the next party to add to the conference call, the serving system will launch an **ORREQ** message to the HLR, allowing the HLR the opportunity to explicitly authorize the addition.

Finally, CC works under call handoff conditions in the same manner as the 3WC and CT features; i.e., the serving system uses the **FLASHREQ** message or **FACREL** message to forward the SEND key or END key notifications, respectively, to the anchor system.

Priority access and channel assignment

The priority access and channel assignment (PACA) feature enables a mobile subscriber to have priority access to voice or traffic channels on call origination. When channels are not available, the subscriber is queued at the radio system on a first-come first-served and a priority basis. There can be up to 16 priority levels, with 1 being the highest.

The subscriber is considered to be busy while a call is queued. Note that this feature is not intended to be a pay feature, such that some mobile subscribers can essentially request higher-priority service over others. The feature is meant to provide service priority for emergency response personnel (e.g., police, fire, medical) or extraordinarily important individuals, such as government emergency officials. The PACA feature requires special MS support, not currently defined in any air interface standard.

The PACA feature has two modes of activation which affect the IS-41-C protocol support procedures:

1. The feature is automatically requested on each call origination.

2. The feature is manually requested by the subscriber.

The HLR conveys this subscriber profile attribute to the serving system in the IS-41-C PACAIndicator parameter; the HLR sends this to the serving system during the service qualification process. With manual activation, the subscriber must precede the destination address with the PACA activation feature code; the serving system sends the entire string to the HLR for PACA authorization in the **FEATREQ** message. If approved, the HLR sends a single-call PACA invocation indicator to the serving system in the IS-41-C OneTimeFeatureIndicator (OTFI) parameter, as described in "Feature Control Services." With automatic activation, the serving system treats each call by the subscriber as a potential PACA call; i.e., if PACA conditions exists, e.g., an emergency situation, the call is given the treatment prescribed for the subscriber's priority level (e.g., placed into a queue for the given priority level or assigned an available channel from a set of channels reserved for the given priority level).

Preferred language

The preferred language (PL) feature provides a standardized means of notifying the serving system of a visiting subscriber's preferred language for recorded announcements, directory assistance, and other serving system interactions.

The HLR conveys this subscriber attribute to the serving system in the IS-41-C PreferredLanguageIndicator parameter; the HLR sends this to the serving system during the service qualification process. Normally, this attribute is fixed in the HLR's database; however, IS-41-C supports the ability to change the value for an authorized MS— e.g., an MS used in a phone rental operation—using the single-step feature control mechanism described in "Feature Control Services."

Voice privacy and signaling message encryption

The voice privacy (VP) and signaling message encryption (SME)* features provide an enhanced degree of privacy by encrypting the signals transmitted over the radio channel between the MS and the serving base station. The VP feature provides encryption of the subscriber's conversations, while the SME feature encrypts selected parameters in the signaling messages sent on an analog voice channel or digital traffic channel, e.g., a signaling message that conveys a subscriber's personal identification number from the MS to the BS. The VP feature is only available in the digital modes of operation (i.e., TDMA and CDMA); the SME feature can be applied to both digital and analog modes, provided authentication is supported in the MS and the network. Although we discuss the VP and SME features together, the subscriber has no control over SME—it is a service provider option—and the two features may be independently provided to the subscriber.

To support these features, both the authentication controller[†] and the MS generate a voice privacy mask and a signaling message encryption key—using the cellular authentication and voice encryption (CAVE) algorithm—on a per-call origination or termination basis (see Fig. 12.13). The MS and the serving BS use these parameters to encode and decode the signals transmitted on the radio channel.

IS-41-C support for these features falls into three categories:

1. *Conveying the subscriber's VP feature activity status to the serving system.* The IS-41-C CallingFeaturesIndicator parameter contains this information; the HLR sends this to the serving system during the service qualification process. If the MS is subscribed to VP and requests the feature, then it is invoked—provided the serving system can support it. Since the SME feature is not subscriber-controlled, it is not part of the subscriber's service profile information.

2. *Conveying the appropriate VP and SME masks to the anchor system.* The procedures in IS-41.6-C state that the authentication controller generates the VP and SME masks when the MS attempts a call origination or termination and successfully passes the global authentication challenge. The masks are conveyed from the authentication controller to the anchor MSC in the AuthenticationRequest Return Result (**authreq**) message in the following IS-41-C parameters:

---

*The signaling message encryption acronym SME should not be confused with the same acronym used for short message entity in Chap. 13.

[†]As described in Chap. 11, the authentication controller is the serving system (i.e., the VLR) if SSD is shared or the authentication center (AC) if SSD is not shared.

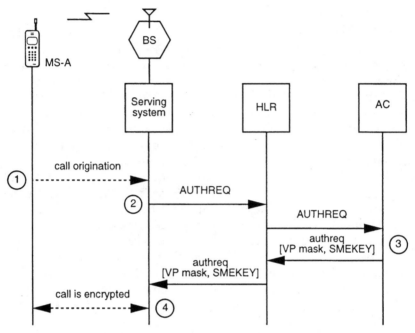

Figure 12.13 Generating the voice privacy mask and SMEKEY when SSD is not shared: (*1*) The subscriber originates a call. The MS generates a voice privacy mask and SMEKEY that it will use for the call. (*2*) The serving system initiates MS authentication, as described in Chap. 11. Since SSD is not shared, the serving system sends an AuthenticationRequest Invoke (AUTHREQ) message to the HLR, which relays the message to the AC. (*3*) The AC performs the authentication functions described in Chap. 11. Since the system access is a call origination, it also generates a voice privacy mask and SMEKEY for the call and sends these to the HLR. If the MS is subscribed to the VP feature, the HLR relays the voice privacy mask to the serving system; otherwise, the HLR discards the voice privacy mask. (*4*) The serving system uses the voice privacy mask and SMEKEY received from the HLR to encrypt and decrypt the information exchanged with the MS over the radio channel.

- For TDMA operation, the voice privacy mask is in the VoicePrivacyMask (VPMASK) parameter.

- For CDMA operation, the voice privacy mask is in the CDMAPrivateLongCodeMask (CDMAPLCM) parameter.

- For both modes of operation, the signaling message encryption key is in the SignalingMessageEncryptionKey (SMEKEY) parameter.

3. *Conveying the appropriate VP mask and SME key to the serving system under call handoff conditions.* To maintain support for the VP and SME features when a call handoff occurs, the parameters are passed to the new serving system in each of the following IS-41-C messages:

- FacilitiesDirective Invoke

- FacilitiesDirective2 Invoke
- HandoffBack Invoke
- HandoffBack2 Invoke
- HandoffToThird Invoke
- HandoffToThird2 Invoke
- InterSystemSetup Invoke

The IS-41-C ConfidentialityModes (CMODES) parameter is also passed in these messages to inform the serving system that VP, SME, both, or neither is desired for the call; the response message from the serving system also includes the CMODES parameter, set to indicate whether the features can be supported by the serving system for the call.

Call Termination Services

In general, IS-41-C call termination services are the network functions that enable, restrict, supplement, or otherwise impact an MS's ability to receive a call while roaming outside the home service area. IS-41-C supports the following call termination services:

- Call delivery (i.e., basic call termination)
- Call forwarding—unconditional (CFU)
- Call forwarding—default (CFD)
- Call forwarding—no answer (CFNA)
- Call forwarding—busy (CFB)
- Call waiting (CW)
- Do not disturb (DND)
- Calling number identification presentation (CNIP)
- Selective call acceptance (SCA)
- Password call acceptance (PCA)
- Mobile access hunting (MAH)
- Flexible alerting (FA)

Call delivery

The call delivery (CD) feature can be considered the most fundamental form of call redirection supported by IS-41-C.* With CD, the call is redirected from its original called number—the called MS's mobile

---

*Call delivery has been supported by IS-41 since Revision A. Prior to the availability of automatic call delivery, calls to roaming subscribers required the calling party to dial a special "roamer port" number to receive a second dial tone; then the calling party would dial the MS's directory number.

directory number* (MDN)—to another telephone number temporarily assigned to the MS by the serving system, specifically for call delivery purposes. The system that performs the call delivery redirection is called the *originating system*. The originating system is often the MS's home system. However, another system may be specifically provisioned to recognize calls to mobile subscribers and to initiate the call delivery processing on behalf of the home system; such a system is generally termed a *gateway system*.

The originating system relies on three key IS-41-C functions for normal call delivery:

1. The location management functions (see Chap. 10) that keep track of the MS's location, i.e., the serving system identification

2. The MS state management functions (see Chap. 10) that keep track of the MS's ability to receive incoming calls (i.e., the MS is either active or inactive)

3. The call routing functions that provide the originating system with the routing information—at a minimum, a telephone number called the temporary local directory number—needed to get the call to the current serving system. These functions are provided by the IS-41-C LocationRequest and RoutingRequest operations (see Fig. 12.14).

IS-41-C also allows an authorized subscriber to activate and deactivate the CD feature by using the single-step feature control mechanism.

Unsuccessful call delivery. There are a number of opportunities for failure in the call delivery process (failures that can be mitigated with the other call termination features, such as call forwarding, described in this chapter). In some cases, when the HLR receives the LocationRequest Invoke (**LOCREQ**) message, it can make an immediate decision to deny the call delivery attempt, i.e., when:

- The directory number is not recognized.
- The MS's current location is not known.
- The MS is not authorized for call delivery.
- The MS is inactive.
- Termination to the MS is otherwise denied (e.g., due to a delinquent account).

---

*The MS's mobile directory number is the number to call to reach the MS. In current practice, this number normally corresponds to the MS's mobile identification number (MIN) that is programmed into the mobile station equipment; however, IS-41-C allows these two numbers to be different.

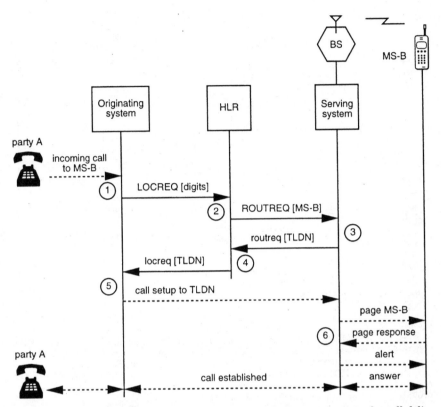

Figure 12.14 Use of the LocationRequest and RoutingRequest operations for call delivery: (*1*) Party A dials a phone number (i.e., MS-B's mobile directory number) that is routed via the PSTN to the originating system. The originating system launches a **LOCREQ** message to the HLR to determine how to route the call. (*2*) The HLR determines that the call is for MS-B and sends a **ROUTREQ** message to MS-B's current serving system. (*3*) The serving system allocates a TLDN for the call and returns it to the HLR in the **routreq** message. (*4*) The HLR relays the TLDN to the originating system in the **locreq** message. (*5*) The originating system routes the call to the TLDN. (*6*) The serving system associates the incoming call to the TLDN with the previous **ROUTREQ** message identifying MS-B; therefore, it pages MS-B. MS-B responds to the page; therefore, the serving system directs MS-B to alert the subscriber. When the subscriber answers, the call is established between the calling party A and MS-B.

In this case, the HLR immediately sends a LocationRequest Return Result (**locreq**) message to the originating system, including the IS-41-C AccessDeniedReason (ACCDEN) parameter with an indication of the reason for denying access to the called MS (see Fig. 12.15). The message may also include an AnnouncementList parameter. This parameter specifies one or more tones or announcements that may be provided to the calling party by the originating system (e.g., a tone followed by "The customer you have called is not accessible. Please try your call again later").

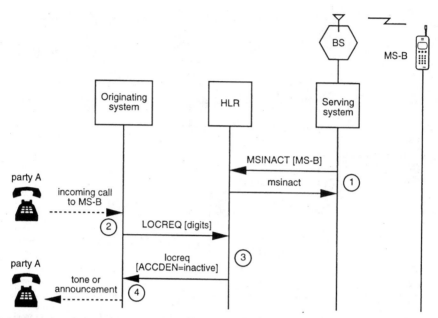

Figure 12.15   An example of unsuccessful call delivery (MS is inactive): (*1*) The serving system declares MS-B inactive and notifies the HLR by using the MSInactive Invoke (**MSINACT**) message. (*2*) Party A dials a phone number (i.e., MS-B's mobile directory number) that is routed via the PSTN to the originating system. The originating system launches a **LOCREQ** message to the HLR to determine how to route the call. (*3*) Since MS-B is inactive, the HLR denies the call delivery attempt. (*4*) The originating system provides a tone or announcement to the calling party.

Otherwise, the HLR sends a RoutingRequest Invoke (**ROUTREQ**) message to the serving system. At this point, the serving system can make the decision to deny the call delivery attempt if:

- The MS is inactive, based on the serving system's determination.
- The MS is busy.
- The MS does not respond to paging (assuming the system performs paging at this point).
- The MS is otherwise unavailable* (e.g., turned off).

In this case, the serving system sends a RoutingRequest Return Result (**routreq**) message to the HLR, including the IS-41-C AccessDeniedReason parameter with the reason for denying access to

---

*Classifying the MS as *unavailable* versus *inactive* is a serving system decision. However, if the MS is labeled inactive, the HLR may suppress call delivery attempts to the MS; therefore, the serving system must explicitly notify the HLR when the MS becomes active again, using the RegistrationNotification operation (see Chap. 10). On the other hand, if the MS is labeled unavailable, the HLR will continue to attempt call delivery as it is notified of each new call to the MS.

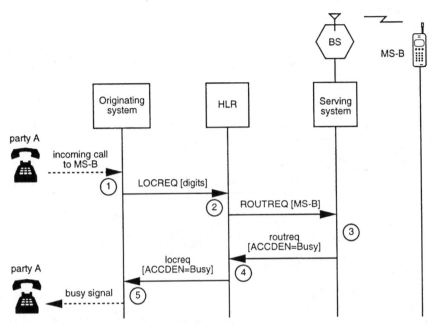

Figure 12.16   An example of an unsuccessful call delivery (MS is busy): (*1*) Party A dials a phone number (i.e., MS-B's mobile directory number) that is routed via the PSTN to the originating system. The originating system launches a **LOCREQ** message to the HLR to determine how to route the call. (*2*) The HLR determines that the call is for MS-B and sends a **ROUTREQ** message to MS-B's current serving system. (*3*) Since MS-B is busy, the serving system denies the call delivery attempt, returning the AccessDeniedReason (ACCDEN) parameter set to the value *busy*. (*4*) The HLR relays the ACCDEN parameter to the originating system in the **locreq** message. (*5*) The originating system provides a busy signal to the calling party.

the called MS. The HLR forwards the AccessDeniedReason parameter to the originating system in the **locreq** message (see Fig. 12.16).

Finally, even when the MS is considered available, a TLDN is provided to the originating system, and a call is established to the serving system, call delivery can still fail under a number of circumstances, including:

- When the MS becomes inactive, busy, or unavailable after the serving system provides the TLDN to the HLR but before the call to the TLDN is received by the serving system
- When the MS does not respond to paging or does not answer the incoming call alert
- When the MS fails authentication

In these cases, the calling party will be given the appropriate call treatment by the serving system; e.g., a busy signal, tone, or announcement (see Fig. 12.17).

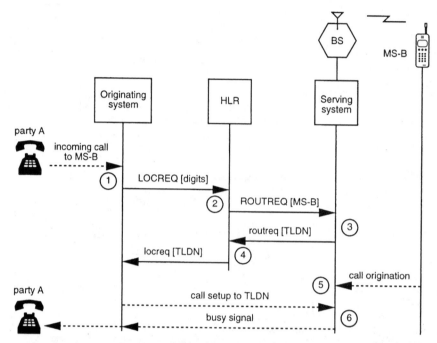

Figure 12.17 An example of unsuccessful call delivery (MS is busy when call arrives at the serving system): (*1–4*) Same as in Fig. 12.14. (*5*) MS-B originates a call prior to the serving system's receiving the call to the TLDN. (*6*) When the call to the TLDN arrives, the serving system provides a busy signal to the calling party.

Call delivery in border cell situations.   The border cell situations discussed in Chap. 10 may also impact call delivery if the MS sends a page response to a border system, even though the page was issued from the serving system. This situation can occur if:

1. The MS hears the page in the serving system, then rescans and chooses the border system as having the strongest control channel signal.

2. The MS has just registered in the border system, and the location update process has not yet been completed in the HLR.

3. The MS's location is otherwise incorrect in the HLR.

IS-41-C includes a number of capabilities designed to alleviate these problems. The solutions fall under the following general categories: *predelivery paging solutions and postdelivery paging solutions.* Predelivery paging solutions apply to serving systems that page the MS when the **ROUTREQ** message is received from the HLR; these methods use the IS-41-C InterSystemPage or UnsolicitedResponse operations. The postdelivery solution can be used by serving systems that page the MS when the actual call to the TLDN is received from the originating system. We believe postdelivery paging is the easier approach and describe this border cell solution in Fig. 12.18.

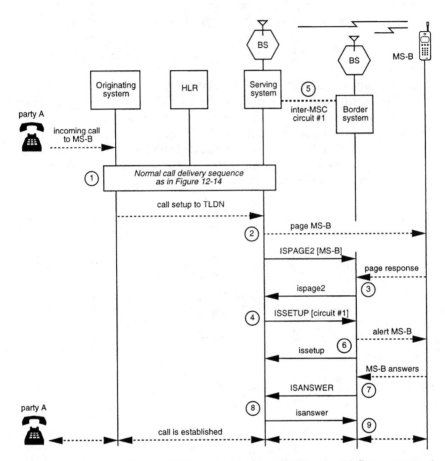

Figure 12.18 An example of call delivery in a border cell situation: (*1*) Same as steps 1 through 4 in Fig. 12.14. (*2*) When the call to the TLDN arrives, the serving system pages MS-B, but also sends an InterSystemPage2 Invoke (**ISPAGE2**) message to one or more border systems. The **ISPAGE2** message instructs the border system either to page MS-B and listen or to just listen (as shown) for a page response from MS-B. (*3*) The border system receives the page response from MS-B; the border system notifies the serving system by using the InterSystemPage2 Return Result (**ispage2**) message. (*4*) The serving system initiates call handoff to the border system by using the InterSystemSetup Invoke (**ISSETUP**) message, specifying an inter-MSC circuit between the two systems for handoff purposes. (*5*) The inter-MSC circuit is established. (*6*) The border system alerts MS-B and responds to the serving system via the InterSystemSetup Return Result (**issetup**) message. (*7*) The subscriber answers the call; the border system notifies the serving system by using the InterSystemAnswer Invoke (**ISANSWER**) message. (*8*) The serving (now anchor) system connects the call path from the originating system to the inter-MSC circuit and sends an InterSystemAnswer Return Result (**isanswer**) acknowledgment to the border system. (*9*) The call is now established between the calling party A and MS-B.

Call forwarding—unconditional

The call forwarding—unconditional feature enables the subscriber to have all incoming calls redirected either to a selected forward-to number or to the subscriber's voice mailbox. The basic CFU feature is supported by the IS-41-C LocationRequest operation (see Fig. 12.19). The subscriber may also be given an alert indication when CFU is invoked to forward a call, e.g., to remind the subscriber that the feature is active. This reminder feature is enabled by the IS-41-C InformationDirective operation.

IS-41-C allows an authorized subscriber to activate and deactivate the CFU feature and to register a new CFU forward-to number by using the single-step feature control mechanism. A *courtesy call* may be provided to the newly registered CFU forward-to number, permitting the subscriber to verify the validity of the forward-to number, by using the feature control with call-routing procedures described in "Feature Control Services."

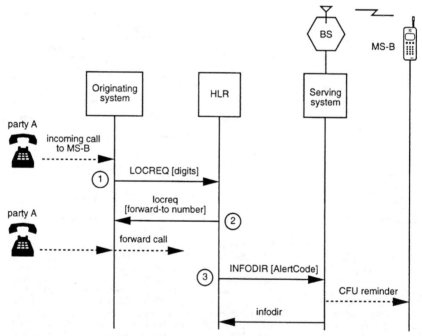

Figure 12.19 An example of CFU: (*1*) Party A dials a phone number (i.e., MS-B's mobile directory number) that is routed via the PSTN to the originating system. The originating system launches a **LOCREQ** message to the HLR to determine how to route the call. (*2*) Since MS-B has CFU active, the HLR directs the originating system to forward the call to the CFU forward-to number. (*3*) The HLR may also send an InformationDirectiveInvoke (**INFODIR**) message to the serving system. The AlertCode parameter in the message directs the serving system to provide a CFU reminder alert to the subscriber.

Call forwarding—busy and call forwarding—no answer

The call forwarding—busy and call forwarding—no answer features enable the subscriber to have all incoming calls that encounter a busy or a no-answer condition, respectively, to be redirected either to a selected forward-to number or to the subscriber's voice mailbox. While the busy condition is easily defined—the subscriber is already engaged in a call or service—the no-answer condition encompasses a number of situations, including these:

- The MS does not respond to paging.
- The subscriber does not respond to MS "alerting" (i.e., the phone ringing).
- The MS's current location is not known.
- The MS is inactive.
- The MS is otherwise inaccessible (e.g., the call delivery feature is inactive).

Checking the MS's busy status requires querying the serving system; however, in some cases the HLR can make an immediate CFNA invocation decision (e.g., if the MS is listed as inactive in the HLR's database). In these cases, the CFNA feature requires only the IS-41-C LocationRequest operation, and the **locreq** message contains the CFNA forward-to number (see Fig. 12.20).

Otherwise, the HLR sends a **ROUTREQ** message to the serving system, attempting normal call delivery. The serving system may respond with a **routreq** message, including the IS-41-C Access-DeniedReason parameter with the reason for denying access to the called MS. The HLR directs the originating system to forward the call (1) to the CFB forward-to number if the AccessDeniedReason parameter value is busy (see Fig. 12.21) or (2) to the CFNA forward-to number if the value of the parameter is any one of these:

1. Unavailable
2. No page response
3. Inactive
4. Unassigned directory number
5. Termination denied

Finally, even when the MS is considered available, a TLDN is provided to the originating system, and a call established to the serving system, CFB, or CFNA can still be invoked under a number of circumstances, including these:

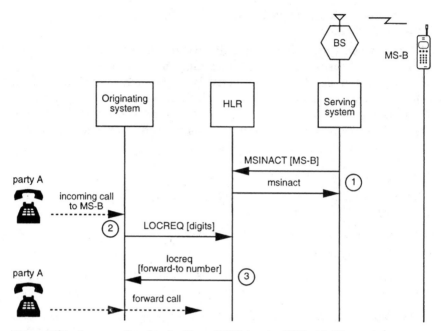

Figure 12.20   An example of immediate CFNA by the HLR: (1) The serving system determines that MS-B is inactive; the HLR is notified by using the MSInactive operation. (2) Party A dials a phone number (i.e., MS-B's mobile directory number) that is routed via the PSTN to the originating system. The originating system launches a **LOCREQ** message to the HLR to determine how to route the call. (3) Since MS-B is inactive and has CFNA active, the HLR immediately directs the originating system to forward the call to the CFNA forward-to number.

- When the MS becomes busy after the serving system provides the TLDN to the HLR but before the call to the TLDN is received by the serving system (this case is referred to as a *call collision* in IS-41-C)

- When the MS becomes inactive or unavailable after the serving system provides the TLDN to the HLR

- When the MS does not respond to paging or does not answer the incoming call alert

- When the MS fails authentication (IS-41-C does not explicitly allow or disallow this case)

In these cases, the serving system initiates a set of call redirection processes that normally involve the steps shown in Fig. 12.22.*

---

*If the originating system does not accept the redirection request, IS-41 specifies that the serving system may query the HLR for the appropriate forward-to number (via the TransferToNumberRequest operation) and forward the call itself. However, this approach is less efficient in terms of the number of trunks involved since it includes the path between the originating and serving systems.

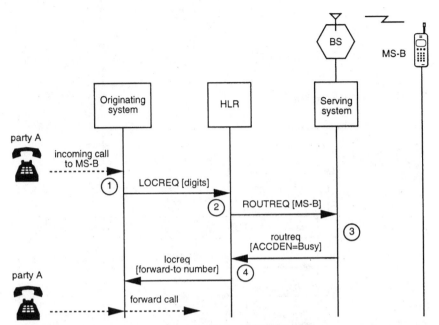

Figure 12.21   An example of CFB by the HLR: (*1*) Party A dials a phone number (i.e., MS-B's mobile directory number) that is routed via the PSTN to the originating system. The originating system launches a **LOCREQ** message to the HLR to determine how to route the call. (*2*) The HLR determines that the call is for MS-B and sends a **ROUTREQ** message to the MS-B's current serving system. (*3*) Since MS-B is busy, the serving system denies the call delivery attempt, returning the AccessDeniedReason (ACCDEN) parameter set to the value *busy*. (*4*) Since MS-B is busy and has CFB active, the HLR directs the originating system to forward the call to the CFB forward-to number.

Call forwarding—default

The call forwarding—default feature enables a called subscriber to have the network send all incoming calls addressed to the subscriber's directory number to another directory number (i.e., the forward-to number) when a busy or no-answer condition (described in the previous section) is encountered.

From the IS-41-C protocol perspective, a subscriber activation of the CFD feature is essentially equivalent to activating both the CFNA and CFB features simultaneously; registration of a CFD forward-to number registers a single forward-to number for both CFNA and CFB features. The only reference to the CFD feature in an IS-41-C message or parameter is in the DMH_Redirection Indicator parameter, which is primarily used for call-recording purposes. Therefore, refer to the previous descriptions of the CFB and CFNA features for an understanding of IS-41-C's support of CFD.

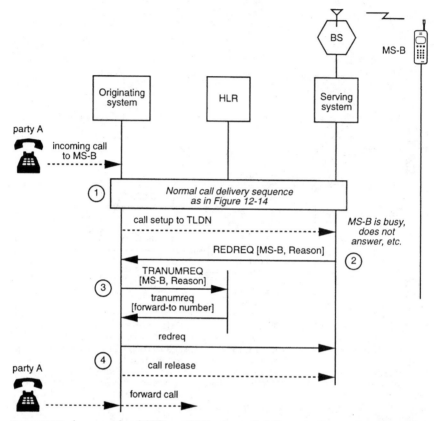

Figure 12.22   An example of CFB or CFNA after call delivery to the serving system: (*1*) Same as steps 1 through 4 in Fig. 12.14. (*2*) When the call to the TLDN arrives, the serving system determines that CFB or CFNA should be invoked (see text for reasons). The serving system sends a RedirectionRequest Invoke (**REDREQ**) message to the originating system, requesting that the call be redirected. (*3*) The originating system uses the TransferToNumberRequest operation to obtain the forward-to number for the given redirection reason (i.e., busy or no answer) from the HLR. (*4*) The originating system notifies the serving system that it accepts the redirection request, releases the call leg to the serving system, and then forwards the call to the forward-to number provided by the HLR.

The CFD feature is generally considered to be the basic form of call forwarding that would be provided to each subscriber. It enables call completion, typically to a voice mail system, when the subscriber cannot be reached, for just about any reason. With this basic—or default—forwarding feature in effect, the subscriber can add the CFB and CFNA features to provide special handling of these conditions. For example, a subscriber has CFD to voice mail but wants calls temporarily redirected to a coworker if the subscriber is engaged in a call, e.g., to handle particularly important calls. For this purpose, the subscriber activates the CFB feature and registers the coworker's telephone number as the CFB forward-to number.

Call waiting

The call-waiting feature allows an MS that is already engaged in a call to be informed of an additional incoming call and to "toggle" between the two calls—placing one call on hold and getting connected to the other—by pressing the SEND key.

IS-41-C protocol support for the CW feature has three components:

1. *Conveying the subscriber's CW feature activity status to the serving system.* The IS-41-C CallingFeaturesIndicator parameter contains this information; the HLR sends this to the serving system during the service qualification process.

2. *Feature control.* The CW feature may be activated and deactivated by an authorized MS by using the single-step feature control procedure and may be deactivated on a per-call basis—either at call origination time or during an active call—by using the per-call feature control procedures described in "Feature Control Services."

3. *Feature invocation.* This is described below.

Feature invocation begins in the anchor system when a **ROUTREQ** message is received from the HLR for an MS that is already engaged in a call (we refer to this as *call-A*). Rather than return the AccessDeniedReason parameter set to indicate that the MS is busy, the anchor system (the call may have been handed off to a new serving system) returns a normal TLDN. When the anchor system receives the incoming call (we refer to this as *call-B*) to the TLDN, it attempts to notify the MS. Depending on the anchor system's and serving system's capabilities and on those of the called MS, a number of CW notification methods are possible:

1. If the MS is capable of only an *in-band notification* (i.e., a tone injected momentarily into the voice path to the called MS), then this is provided by the anchor system, even if call-A has been handed off. The tone is applied once and then again in 15 s if call-B is not answered (i.e., by the subscriber's pressing the SEND key).

2. If call-A has not been handed off and the MS and anchor system are capable of an *out-of-band notification* (i.e., a signal sent outside of the voice path to the called MS that directs the MS to generate a CW alert), then the anchor system orders the MS to apply the CW tone.

3. If call-A has been handed off and the MS, anchor, and serving systems are capable of an out-of-band notification, then the anchor system sends an InformationForward Invoke (**INFOFWD**) message to the serving system. The **INFOFWD** message includes the AnnouncementList parameter that directs the serving system to

order the MS to apply the CW tone. The subscriber may then press
the SEND key to place call-A on hold and answer call-B. This triggers
the serving system to send a **FLASHREQ** message to the anchor system. The anchor system places call-A on hold and connects the subscriber to call-B (see Fig. 12.23).

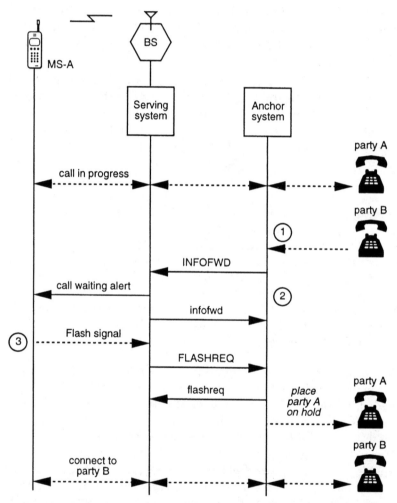

Figure 12.23 An example of call-waiting alert after handoff: (*1*) The subscriber
is involved in a call that has been handed off from the anchor system to a
nonanchor serving system. A new call arrives for the subscriber from party B.
(*2*) Since the subscriber has CW active, the anchor system sends an **INFOFWD**
message to the serving system, directing the serving system to order the MS to
apply the CW tone. (*3*) The subscriber presses the SEND key. The serving system sends a **FLASHREQ** message to the anchor system. The anchor system
places party A on hold and connects the subscriber to party B.

A little-known feature supported by the IS-41-C protocol and procedures is the priority call waiting (PCW) feature. This is actually a call
origination feature although it impacts the CW feature: It allows calls
originated by a PCW subscriber to be assigned a higher-than-normal
priority level (this is indicated in the "Call waiting for incoming call"
field of the OneTimeFeatureIndicator parameter). Once a call with
this priority level is completed to a CW subscriber (MS-A), CW will be
invoked only if another PCW subscriber attempts to reach MS-A; calls
from subscribers without PCW are given busy treatment.

Do not disturb

When a subscriber activates the do not disturb (DND) feature, the
MS is essentially placed in an inactive state for call delivery purposes. Incoming calls are provided the treatment described in Fig. 12.15;
i.e., the calling party receives a call refusal tone, announcement, or
both. The feature also blocks all other audible alerting of the called
party, such as the message-waiting notification indications and the
CFU alert.

The DND feature may be activated and deactivated by an authorized MS by using the single-step feature control procedure.

Calling number identification presentation

The calling number identification presentation (CNIP) feature allows
the subscriber to obtain—and usually view in some manner—the calling number identification information from the serving system when
the incoming call is received and the subscriber is alerted. The CNI
information can include up to two calling party numbers, a redirecting number, as well as calling and redirecting subaddress information. The ability to present two calling party numbers is optional. A
user-provided calling party number is intended to represent the actual address of the calling party, useful when the calling party is served
by a private branch exchange (PBX). The network-provided calling
party number is intended to provide another number representing
network address information that may be helpful to the called party.
For example, the (user-provided) calling party's *direct inward dialing*
(DID) number can be provided as well as the (network-provided) *listed directory number* for a given corporate office.

With normal call delivery, the CNI information gets to the serving
system in the following manner (refer to Fig. 12.14 which illustrates
call delivery):

1. The originating system must receive the CNI information from the
   PSTN or other source of the call.

2. The originating system includes the CNI information in the **LOCREQ** message it sends to the HLR. The CNI information is carried in the following IS-41-C parameters in the **LOCREQ** message:

- CallingPartyNumberDigits1 (i.e., the network-provided number)
- CallingPartyNumberDigits2 (i.e., the user-provided number)
- CallingPartySubaddress
- RedirectingNumberDigits
- RedirectingSubaddress

3. The HLR provides the CNI information to the anchor system in the **ROUTREQ** message. The CNI information is carried in the following IS-41-C parameters in the **ROUTREQ** message, mapped from the corresponding parameters in the **LOCREQ** message:

- CallingPartyNumberString1
- CallingPartyNumberString2
- CallingPartySubaddress
- RedirectingNumberString
- RedirectingSubaddress

The numbers—but not the subaddresses—are converted from one format in the **LOCREQ** message (e.g., CallingPartyNumberDigits1) to another format in the **ROUTREQ** message (e.g., Calling-PartyNumberString1) for the following reasons:

1. The numbers are generally received from the PSTN in binary-coded decimal (BCD) format, so that is the format that is conveyed to the HLR in the **LOCREQ** message.

2. The numbers are generally delivered to the MS over the radio channel in the International Alphabet No. 5 (IA5) format, so that is the format that the HLR conveys to the anchor system in the **ROUTREQ** message.

3. Using the IA5 format in the **ROUTREQ** message to the anchor system gives the HLR the option to substitute letters (e.g., a name) for the numeric information.

At the serving system, the number information is analyzed. If the presentation indicator within a number parameter indicates that presentation to the called party is restricted (i.e., transmission of the number over the radio channel is not allowed or the number is not available), then the serving system shall not provide the number to the MS. Otherwise, the CNI information is transmitted to the MS.

Selective call acceptance

The selective call acceptance (SCA) feature also makes use of the CNI information received from the PSTN or other source of the call to the subscriber. The SCA feature uses the CNI information to perform call screening at the HLR, i.e., checking the calling party number against a list of allowed numbers. Calls from numbers on the list are allowed to proceed as normal call delivery attempts (see Fig. 12.14). Incoming calls from parties not on the screening list—including calls that do not provide CNI information—are given one of two possible call refusal treatments:

1. If the subscriber has registered an SCA call refusal treatment, by using the single-step feature control procedure described in "Feature Control Services," then refused calls are redirected either to the selected SCA forward-to number (see Fig. 12.24) or to the subscriber's voice mailbox.

2. Otherwise, the call is given the default system call refusal treatment; this could include a tone and announcement to the calling party (e.g., "The number you have dialed does not accept incoming calls").

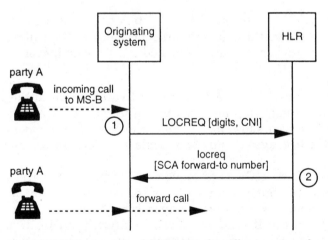

Figure 12.24 An example of SCA when the caller is not on the screening list: (1) Party A dials a phone number (i.e., MS-B's mobile directory number) that is routed via the PSTN to the originating system. The originating system launches a **LOCREQ** message to the HLR to determine how to route the call. The **LOCREQ** message includes party A's CNI information. (2) Since MS-B has SCA active and party A's calling number is not on the SCA screening list, the HLR directs the originating system to forward the call to the SCA forward-to number.

The SCA feature may be activated and deactivated by an authorized MS using the single-step feature control procedure. It is also possible for an authorized subscriber to add to and remove numbers from the screening list, again using the single-step feature control procedure.

Password call acceptance

The password call acceptance (PCA) feature can be used by itself or to augment the SCA feature just described. For example, if the calling party's regular number is on the SCA screening list but the party is calling from a hotel room, the SCA feature will refuse the call. However, if PCA is active, the calling party will be prompted for a password and—if the entered password is verified—call delivery to the subscriber will proceed as usual.

Incoming calls from parties not able to provide a valid password are given one of two possible call refusal treatments:

1. If the subscriber has registered a PCA call refusal treatment, using the single-step feature control procedure described in "Feature Control Services," then refused calls are redirected either to the selected PCA forward-to number or to the subscriber's voice mailbox.

2. Otherwise, the call is given the default system call refusal treatment; this could include a tone and announcement to the calling party (e.g., "The number you have dialed does not accept incoming calls").

The PCA feature may be activated and deactivated by an authorized MS using the single-step feature control procedure. It is also possible for an authorized subscriber to add to and remove numbers from the password list, again using the single-step feature control procedure.

IS-41-C describes two methods of implementing the PCA feature invocation procedures which we refer to as the *remote interaction* approach and the *home-based interaction* approach. The remote interaction approach uses the IS-41-C RemoteUserInteractionDirective operation to allow the HLR to remotely control the user interaction (i.e., the prompt for the password and digit collection) that is provided by the originating system (see Fig. 12.25). The home-based interaction approach has the originating system route the call to the home system where, presumably, a more customized user interaction "experience" is provided, e.g., voice entry of the password, rather than digit entry via the telephone keypad (see Fig. 12.26).

In fact, the two approaches can coexist. If the originating system is

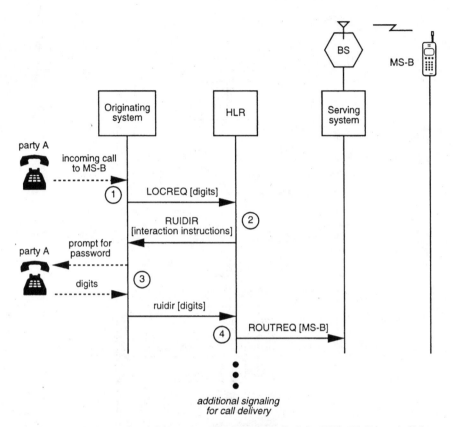

Figure 12.25 An example of the remote interaction method for PCA: (*1*) Party A dials a phone number (i.e., MS-B's mobile directory number) that is routed via the PSTN to the originating system. The originating system launches a **LOCREQ** message to the HLR to determine how to route the call. (*2*) Since MS-B has PCA active, the HLR must query the caller for a valid password before allowing the call to proceed; therefore, it sends interaction instructions to the originating system in the **RUIDIR** message. (*3*) The originating system connects the call to a system capable of user interaction (e.g., an interactive voice response unit), prompts party A for a PIN, collects the digits entered, and sends them to the HLR in the **ruidir** message. (*4*) The HLR verifies that the PIN is valid and then proceeds with normal call delivery.

not capable of interacting with the user via the remote interaction approach—this is indicated in the IS-41-C TransactionCapability parameter supplied by the originating system in the **LOCREQ** message—the HLR can fall back on the home-based interaction approach.

In current practice, the originating system is normally the home system and is often colocated with the HLR; therefore, the two approaches appear identical (i.e., messages between the originating system and the HLR are internal). However, it is quite likely that in

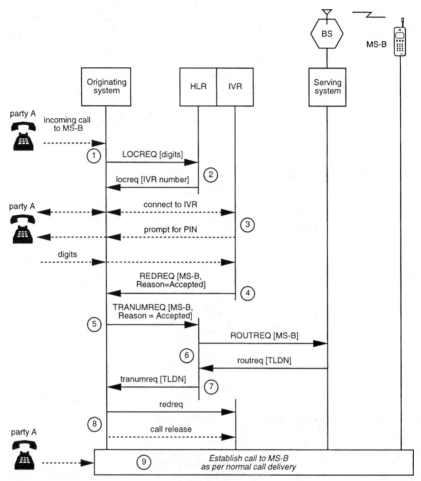

Figure 12.26 An example of the home-based interaction method for PCA: (*1*) Party A dials a phone number (i.e., MS-B's mobile directory number) that is routed via the PSTN to the originating system. The originating system launches a **LOCREQ** message to the HLR to determine how to route the call. (*2*) Since MS-B has PCA active, the HLR must query the subscriber for a valid password before allowing the call to proceed; therefore, it sends instructions to the originating system to route the call to a home-based interactive voice response (IVR) unit. The HLR communicates with the IVR to prepare for the incoming call. (*3*) The originating system routes the call to an IVR unit. The IVR unit prompts party A for a PIN, collects the digits entered, and verifies that the PIN is valid. (*4*) The IVR sends a **REDREQ** message to the originating system, requesting that the call to MS-B be redirected; the RedirectionReason (Reason) parameter is set to the value *Call accepted*. (*5*) The originating system uses the **TRANUMREQ** message to obtain the forward-to number for the given redirection reason (i.e., *Call accepted*) from the HLR. (*6*) The HLR handling for the *Call accepted* forwarding case is to process the request as a normal call delivery attempt. Therefore, the HLR obtains a TLDN from the serving system by using the RoutingRequest operation. (*7*) The HLR returns the TLDN to the originating system in the **tranumreq** message. (*8*) The originating system notifies the IVR that it accepts the redirection request, releases the call leg to the IVR, and then forwards the call to the TLDN provided by the HLR. (*9*) The call is then established to the subscriber as a normal call delivery.

the future the HLR and originating system functions will reside in physically separate equipment and locations, particularly if wire line switches develop IS-41-C originating system functionality, i.e., when the wire line calling party's central office switch can detect a call to a mobile subscriber and launch the **LOCREQ** message to the HLR. In this case, the remote interaction approach appears more efficient than routing the call home only to find that the call is refused.

Mobile access hunting

The call flows and procedures in IS-41-C that define the mobile access hunting (MAH) feature may look complex, but the feature has a simple objective: to complete a call to one of a group of telephone numbers by proceeding sequentially through the group, one member at a time, until one of the group members answers. The MAH process is initiated when a party calls the MAH pilot directory number.

Two types of MAH groups are defined which impact the MAH feature processing:

- *Single-user type.* This type of MAH group is considered busy when any one of the group members is detected as busy. It is useful for the individual who wants to set up a group including an MS, an office phone, and a business phone.

- *Multiple-user type.* This type of MAH group is considered busy when all the group members are detected as busy. It is useful for the company that wants to set up a group including multiple employees.

Authorized MS members of the MAH group can enter and leave the group and change their position in the group (e.g., go to either first or last) by using the single-step feature control procedure.

Figure 12.27 illustrates the basic MAH call completion process for a single user group composed of mobile stations MS-A and MS-B and a land line telephone:

1. The call to the MAH pilot directory number starts out as a normal call delivery attempt; i.e., the originating system sends a **LOCREQ** message to the HLR.

2. The HLR initiates MAH processing and selects the first member of the group, MS-A, which is roaming in another network. The HLR sends a **ROUTREQ** message to the serving system and receives a TLDN in return.

3. The HLR sends the TLDN to the originating system and instructs the system that if it cannot complete the call (e.g.,

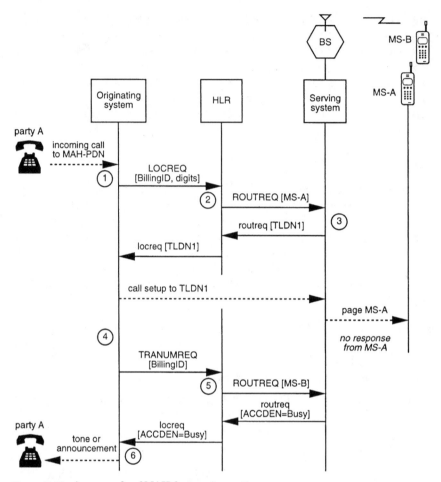

Figure 12.27   An example of MAH feature invocation.

because the MS is busy or does not answer), it must notify the HLR by using the **TRANUMREQ** message. This capability is provided by the IS-41-C TerminationTriggers (TERMTRIG) parameter and is critical in that it allows the HLR to maintain control of the call in case of unsuccessful termination to an MAH member. The HLR may also provide the AnnouncementList parameter, directing the originating system to inform the calling party that the call is being forwarded.

4. The originating system establishes a call to the serving system, but MS-A does not answer. The originating system waits a period of time indicated by the HLR in the NoAnswerTime parameter (for

example, 10 s) and then sends a **TRANUMREQ** message to the HLR.
The message includes either the BillingID parameter used in the
original **LOCREQ** message for the call or the PilotNumber parameter, containing the MAH pilot directory number; either of these parameters allows the HLR to relate the **TRANUMREQ** message to the
MAH call.

  5. The HLR selects the next member of the group, MS-B, which is
also roaming in another network. The HLR sends a **ROUTREQ** message to the serving system and receives an AccessDeniedReason parameter, indicating that MS-B is busy. Since the MAH group is a single-
user group and a member is busy, the entire group is considered busy.
Therefore, the HLR returns the AccessDeniedReason parameter to
the originating system in the **tranumreq** message and terminates
the MAH processing, without attempting to complete the call to the
last member of the group, the land line telephone.

  6. The originating system notifies the calling party that the call
attempt was unsuccessful and completes its processing of the call.

Flexible alerting

Like MAH, flexible alerting (FA) is another complex-looking feature
with a simple objective: to complete a call to one of a group of telephone numbers by attempting to establish a call to each member of
the group—in parallel—until one of the group members answers.
The FA process is initiated when a party calls the FA pilot directory
number.

  Two types of FA group are defined which impact the FA feature processing:

- *Single-user type.* This type of FA group is considered busy when any
  one of the groups members is detected as busy. It is useful for the
  individual who wants to set up a group including an MS, an office
  phone, and a business phone.

- *Multiple-user type.* This type of FA group is considered busy when
  all the group members are detected as busy. It is useful for the company that wants to set up a group including multiple employees.

  Authorized MS members of the FA group can enter and leave the
group using the single-step feature control procedure.

  Figure 12.28 illustrates the basic FA call completion process for a
single-user group composed of two mobile stations MS-A and MS-B
with numbers MDN-A and MDN-B, respectively, and a land line telephone with directory number POTS1:

  1. The call to the FA pilot directory number starts out as a normal

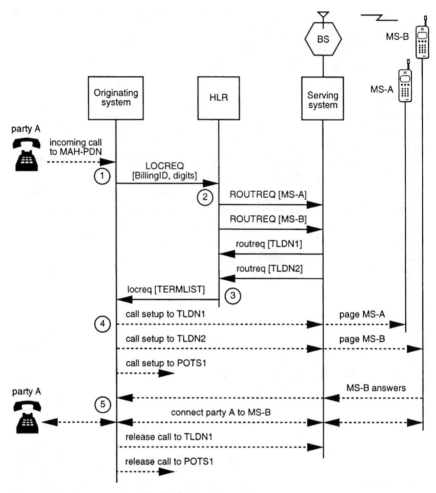

Figure 12.28   An example of the FA feature invocation.

call delivery attempt; i.e., the originating system sends a **LOCREQ** message to the HLR. The originating system includes the TransactionCapability parameter, indicating whether it can support the multiple terminations required for the FA feature; this parameter is required in the IS-41-C **LOCREQ** message.

2. The HLR initiates FA processing. Both mobile members of the group, MS-A and MS-B, are roaming in another network. The HLR sends two **ROUTREQ** messages to the serving system, one for each of the two MSs, and receives two temporary local directory numbers (TLDN1 and TLDN2) in return.

3. The HLR sends the two TLDNs and the land line number

POTS1 to the originating system in the IS-41-C TerminationList parameter. The HLR may also provide the AnnouncementList parameter, directing the originating system to inform the calling party that the call is being forwarded. The HLR may then terminate its FA call processing.

4. The originating system establishes calls—in parallel—to each of TLDN1, TLDN2, and POTS1.

5. MS-B answers. Therefore, the originating system connects the calling party to MS-B and releases the calls to the other two members of the FA group. This completes the originating system's processing of the FA call.

Feature Control Services

In general, IS-41-C feature control services are the network functions that allow the subscriber to activate, deactivate, and (in some cases) invoke features as well as register and "deregister" (i.e., delete) the information that enables the subsequent operation of the feature (e.g., forward-to numbers). These services effectively implement the feature control capabilities that are defined in the IS-53-A standard. In particular, IS-41-C seeks to standardize these services for subscribers who are roaming outside their home service areas.

IS-41-C supports the following feature control services:

- Basic single-step feature control
- Feature control with call routing
- Per-call feature control
- Multistep feature control
- Remote feature control

We discuss these after introducing the feature code concept.

Feature codes

Feature codes are most commonly used to modify, invoke, or cancel a cellular feature that has been subscribed to (refer to *TIA/EIA IS-52-A* and *TIA/EIA IS-53-A* for the standard specifications of feature codes).

Different feature code digit sequences are specified by the service provider for use with particular features. The sequence is known as a *feature code string* that consists of a preceding asterisk * or double asterisk ** followed by a series of numeric digits (0 through 9); a pound sign # may be used as a delimiter between separate sequences of digits. For example, the feature code string:

$$*, 72, 4085550303$$

could mean that a call forwarding forward-to number is being registered. In this example, *72 indicates that the call-forwarding feature
is being accessed. The digit sequence 4085550303 indicates the forward-to number. A # is not necessary since only one sequence of digits
is used. The subscriber presses SEND (or some other key with similar
functionality) on the mobile station to transmit the feature code
string to the network.

One problem with feature codes is that they are not standardized. A
feature code string that controls the use of a feature in one serving system may not be identical to that in another. The differences in feature
codes are due to original disparate implementations in early cellular
systems. Network operators are reluctant to change feature codes that
have been in use for years since they have become familiar mechanisms for subscribers. However, IS-41 supports seamless operation by
providing the protocol to convey the subscriber-entered feature code
string to the home system for interpretation; in this manner, home system feature codes may take precedence over local serving system codes.
We describe these protocol mechanisms in the following sections.

Basic single-step feature control

The basic *single-step feature control* service comprises the following
steps (see Fig. 12.29):

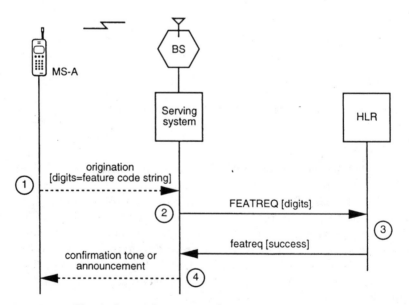

Figure 12.29 The single-step feature control process.

1. The subscriber enters a feature code string into the MS and presses the SEND key.

2. The serving system recognizes the received digits as a feature code string and sends them to the HLR in a FeatureRequest Invoke (**FEATREQ**) message.

3. The HLR processes the feature control request and updates its database, if necessary. The HLR returns a success or failure indication to the serving system in the FeatureRequest Return Result (**featreq**) message.

4. The serving system provides the appropriate indication to the MS.

Following completion of a successful feature control procedure, the HLR may invoke the service qualification process to convey the subscriber's new profile to the serving system.

When the call has been handed off and the serving system receives a feature code string from an MS, it uses the **FLASHREQ** message to forward the digits received from the MS to the anchor system (see Fig. 12.30).

This process works flawlessly—that is, as long as the serving system cooperates and sends the received feature code string to the HLR

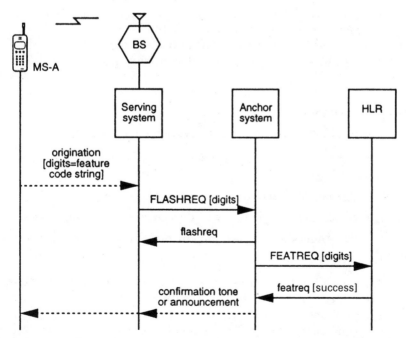

Figure 12.30 The single-step feature control process under call handoff conditions. Note that the anchor system provides the confirmation tone or announcement.

for processing. Unfortunately, because the serving system may not be able to distinguish a visiting MS from a home MS—or because the feature code string is interpreted as a local feature request—the feature code string may never leave the serving system. IS-41-C procedures essentially direct the serving system to give the home system "first shot" at interpreting the feature code string. Then if the HLR does not recognize the feature code string, the digits are returned to the serving system in the **featreq** message along with an indication that the HLR was unsuccessful in the FeatureResult parameter. At this point, the serving system may—if it is able—attempt to interpret the feature code string and provide the requested feature. The alternative method of accessing a local feature in a visited system, as defined in the IS-52-A dialing plan standard, is for the subscriber to precede the feature code string with an additional * digit; i.e., to access the local feature that is associated with the string *123, the visiting subscriber dials **123. This form of string should trigger immediate local processing in the serving system, though currently this is also not guaranteed!

Feature control with call routing

The *feature control with call routing* service is an extension of the single-step feature control service. In addition to the feature result indication, the HLR includes routing information, e.g., the Digits parameter, in the **featreq** message it sends to the serving system. The serving system gives the feature result indication to the MS and then attempts to establish a call to the number provided by the HLR (see Fig. 12.31).

Per-call feature control

Some features allow activation or deactivation on a per-call basis, i.e.:

- Calling number identification restriction may be activated or deactivated for a single call.

- Call waiting may be deactivated for a single call.

- Message-waiting notification may be deactivated for a single call.

- Priority access and channel assignment may be activated for a single call.

IS-41-C supports this capability by supplementing the basic feature control procedures with the OneTimeFeatureIndicator (OTFI) parameter; this provides a *per-call feature control* service. An example of the process—for deactivating CW for a call—is described below (see Fig. 12.32):

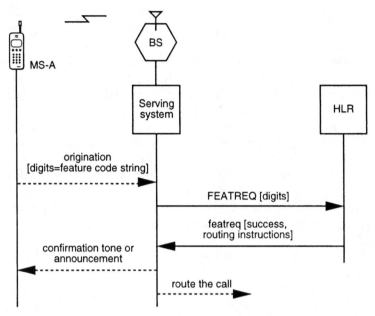

Figure 12.31 The feature control with call routing process.

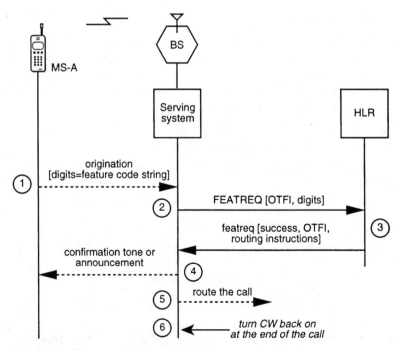

Figure 12.32 The per-call feature control process.

278 IS-41 Revision C Explained

1. The subscriber enters a feature code string, including the cancel-call-waiting feature code and a called directory number, into the MS and presses the SEND key.

2. The serving system recognizes the received digits as a feature code string and sends the string to the HLR in a **FEATREQ** message. IS-41-C specifies that the serving system also initializes the OTFI parameter and includes it in the **FEATREQ** message.

3. The HLR processes the feature control request. The subscriber is authorized for demand CW cancellation; therefore, the HLR deactivates CW for a single call by setting the Call Waiting for Future Incoming Call (CWFI) field in the OTFI to the value *No CW*. The HLR returns a success indication to the serving system in the **featreq** message, along with the revised OTFI parameter and routing information corresponding to the called directory number received in the **FEATREQ** message.

4. The serving system provides the feature confirmation indication to the MS.

5. The serving system turns off CW for the call and establishes a call to the specified directory number.

6. At the end of the call, the serving system turns CW back on for the MS.

Multistep feature control

The *multistep feature control* service is used when additional information—beyond the initial feature code string—is required to effect a feature control operation. The most common additional information is a password or a personal identification number. Currently, the only standardized features that require this capability are:

- Subscriber PIN access, where it is required for PIN entry during registration, activation, and deactivation

- Subscriber PIN intercept, where it is required for PIN entry during registration

- Voice message retrieval, where it is required for VMR password entry during feature invocation

However, the capability is generic and may be used for future standard features as well as nonstandard feature control purposes.

The process works as follows (see Fig. 12.33):

1. The subscriber enters a feature code string into the MS and presses the SEND key.

2. The serving system recognizes the received digits as a feature code string and sends them to the HLR in a **FEATREQ** message.

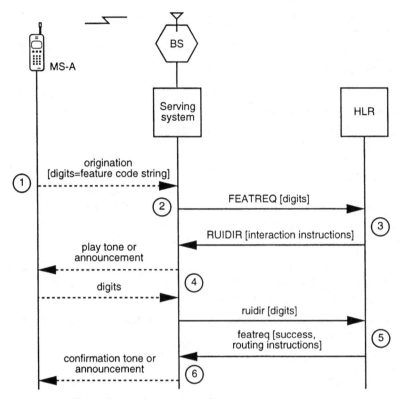

Figure 12.33 The multistep feature control process.

3. The HLR processes the feature control request. The HLR sends a RemoteUserInteractionDirective Invoke (**RUIDIR**) message to the serving system.

4. The **RUIDIR** message directs the serving system to

   ■ Connect the caller to a unit capable of voice prompts and digit collection.

   ■ Play the tone or announcement indicated in the Announcement List parameter.

   ■ Collect digits entered by the subscriber; the digits may be sent as in-band DTMF signals or in an out-of-band signaling message.

   ■ Send the digits to the HLR in the RemoteUserInteraction Directive Return Result (**ruidir**) message.

5. The HLR processes the digits and updates its database, if necessary. Additional RemoteUserInteractionDirective exchanges may

be executed. The HLR returns a success or failure indication to the serving system in the **featreq** message.

6. The serving system provides the appropriate indication to the MS and releases the call.

Following completion of a successful feature control procedure, the HLR may invoke the service qualification process to convey the subscriber's new profile to the serving system.

Remote feature control

Remote feature control (RFC) service is similar in concept to the multistep feature control service. However, the RFC service may be accessed from any land line or mobile station, not just the subscriber's own mobile station.* The service works as follows (see Fig. 12.34):

1. The subscriber calls a special RFC directory number (say, 1-800-SYNACOM). The PSTN delivers the call to the RFC access system which acts as the originating system for the call.

2. The RFC access system launches a **LOCREQ** message to the HLR.

3. The HLR sends a **RUIDIR** message to the RFC access system.

4. The **RUIDIR** message directs the RFC access system to

   - Connect the subscriber to a unit capable of voice prompts and digit collection.
   - Prompt the subscriber for its mobile directory number.
   - Collect digits entered by the subscriber.
   - Send the digits to the HLR in the **ruidir** message.

5. The HLR processes the digits and sends another **RUIDIR** message to the RFC access system.

6. This **RUIDIR** message directs the RFC access system to

   - Prompt the subscriber for its personal identification number.
   - Collect digits entered by the subscriber.
   - Send the digits to the HLR in the **ruidir** message.

7. The HLR processes the digits and sends another **RUIDIR** message to the RFC access system.

---

*Of course, the capability to access the service from other than the subscriber's (authenticated) mobile station opens the remote feature control service to potential fraud. It would appear advisable to deploy this feature with caution.

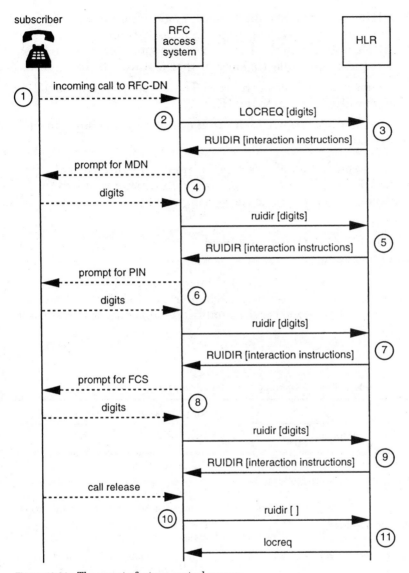

subscriber

RFC access system

HLR

① incoming call to RFC-DN

② LOCREQ [digits]

③ RUIDIR [interaction instructions]

prompt for MDN

④ digits

ruidir [digits]

⑤ RUIDIR [interaction instructions]

prompt for PIN

⑥ digits

ruidir [digits]

⑦ RUIDIR [interaction instructions]

prompt for FCS

⑧ digits

ruidir [digits]

⑨ RUIDIR [interaction instructions]

call release

⑩ ruidir [ ]

⑪ locreq

Figure 12.34 The remote feature control process.

8. This **RUIDIR** message directs the RFC access system to

- Prompt the subscriber for a feature request (i.e., a feature code string).
- Collect digits entered by the subscriber.
- Send the digits to the HLR in the **ruidir** message.

9. The HLR processes the digits as it would a normal feature code string, possibly updating its database. It then sends another **RUIDIR** message to the RFC access system, confirming the feature request and requesting any further feature control requests.

10. The subscriber releases the call. The serving system sends an empty **ruidir** message to the HLR.

11. The HLR responds with the **locreq** message, completing the transaction.

Following completion of a successful remote feature control procedure, the HLR may invoke the service qualification process to convey the subscriber's new profile to the serving system.

Summary of IS-41-C Operations Used for Call Processing

Table 12.1 summarizes the IS-41-C operations used by the call processing functions described in this chapter. Note that most functions are also supported by the service qualification operations described in Chap. 10.

TABLE 12.1   Use of IS-41-C Operations for Call Processing

Function	IS-41-C operations used for the function
Basic call origination	OriginationRequest, FeatureRequest
Call delivery (CD)	LocationRequest, RoutingRequest, FeatureRequest
Call delivery in border cell situations	CD operations plus InterSystemPage, InterSystemPage2, InterSystemSetup, InterSystemAnswer, UnsolicitedResponse
Call forwarding—busy	CD operations plus RedirectionRequest, TransferToNumberRequest
Call forwarding—default	CD operations plus RedirectionRequest, TransferToNumberRequest
Call forwarding—no answer	CD operations plus RedirectionRequest, TransferToNumberRequest
Call forwarding—unconditional	CD operations plus InformationDirective
Call transfer	FeatureRequest, FlashRequest
Call waiting	FeatureRequest, FlashRequest, InformationForward
Calling number identification presentation	CD operations plus InformationForward

TABLE 12.1   Use of IS-41-C Operations for Call Processing (Continued)

Function	IS-41-C operations used for the function
Calling number identification restriction	CD operations
Conference calling	FeatureRequest, FlashRequest
Do not disturb	LocationRequest, FeatureRequest
Flexible alerting	CD operations
Message-waiting notification	FeatureRequest, InformationForward
Mobile access hunting	CD operations plus RedirectionRequest, TransferToNumberRequest
Password call acceptance	CD operations plus RemoteUserInteractionDirective or (RedirectionRequest and TransferToNumberRequest)
Preferred language	FeatureRequest
Priority access and channel assignment	FeatureRequest
Selective call acceptance	CD operations
Subscriber PIN access	FeatureRequest, RemoteUserInteractionDirective
Subscriber PIN intercept	FeatureRequest, RemoteUserInteractionDirective, OriginationRequest
Three-way calling	FlashRequest
Voice message retrieval	CD operations plus OriginationRequest
Voice privacy and signaling message encryption	AuthenticationRequest, FacilitiesDirective, FacilitiesDirective2, HandoffBack, HandoffBack2, HandoffToThird, HandoffToThird2, InterSystemSetup
Single-step feature control	FeatureRequest, FlashRequest
Feature control with call routing	FeatureRequest
Per-call feature control	FeatureRequest, FlashRequest
Multistep feature control	FeatureRequest, FlashRequest, RemoteUserInteractionDirective
Remote feature control	LocationRequest, RemoteUserInteractionDirective

Chapter

13

Short Message Service Functions

In this chapter, we discuss the IS-41-C mobile telecommunications network functions related to short message service (SMS). These functions are divided into the following categories:*

- Short message entity (SME†) service qualification

- SME location management

- SME state management

- Short message processing

We also examine the IS-41-C application processes that support these network functions. In the course of describing the processes, we identify the IS-41-C mobile application part (MAP) operations (e.g., SMSRequest) that are used to accomplish SMS process tasks. We summarize this information at the end of the chapter. Note that we employ the IS-41-C convention for operation component acronyms; i.e., the Invoke component acronym is in all-capital letters (e.g., **SMSREQ**), while the Return Result component acronym is in all-lowercase letters (e.g., **smsreq**). Refer to the IS-41-C standard for the descriptions of the individual IS-41-C operations and parameters. Also, consult the Glossary for a description of general terms (e.g., *serving system*) that are not explicitly defined in this chapter.

---

*The list of SMS functions reflects the authors' subjective categorization of certain IS-41-C functions. Likewise, the process descriptions are the authors' subjective interpretation of the IS-41-C procedures.

†The short message entity acronym, SME, should not be confused with the same acronym used for signaling message encryption in Chap. 12.

Throughout this chapter we consider the serving system as a single entity encompassing the mobile switching center (MSC) and visitor location register (VLR) functional entities. This simplifies the descriptions of the IS-41-C SMS functions and also is representative of a large percentage of the IS-41 implementations currently in service. However, keep in mind that the potential for separation of the MSC and VLR exists and is fully defined in IS-41-C.

What Is Short Message Service?

IS-41-C specifies a set of data services collectively known as the short message service (SMS). SMS includes services that are specially designed for the mobile environment. Most traditional data services are not appropriate for this environment since they require bulky terminals compared to the size of a handheld mobile station (MS). That is, data services are generally not designed for an integrated implementation on mobile telephones since they usually require a keyboard and a reasonably large display screen. However, SMS supports the transmission and reception of simple messages that are convenient for display on mobile terminals.

The IS-41-C SMS is designed as a generic short-message transport mechanism. The following design imperatives were applied to the SMS:

- Support a variety of teleservice applications.
- Make use of commonly implemented transport protocols.
- Incorporate a flexible addressing scheme.
- Easily interwork with other packet-switched data networks.
- Be compatible with electronic mail services, paging services, and other commonly used messaging services.

The SMS is categorized into the following two types of services: SMS bearer service and SMS teleservices. These services require two functional entities in addition to those used for basic mobile telecommunications: the message center (MC) and the short message entity (SME). These services and functional entities are described in detail in the following sections.

SMS bearer service

The SMS bearer service is the basic transport mechanism to convey a short message as a packet of data between two points on the network. The short message may have length up to 200 octets.

The SMS bearer service does not contain many features itself. This allows for the flexibility of custom applications (i.e., teleservices) to use a simple transport mechanism. The bearer service supports the standardized teleservices described in the next section.

The SMS bearer service is designed to use any one of the following transport protocols:

- Signaling System No. 7 (SS7)

- X.25

- Internet protocol (IP)

In addition, IS-41-C does not preclude the use of a proprietary data protocol for transport of the SMS bearer service messages (see Fig. 13.1).

The SMS bearer service is described in *TIA/EIA IS-53-A* as the "Short Message Delivery—Point-to-Point Bearer Service." The points are short message entities, which are essentially applications that can both send and receive short messages. Thus, the bearer service is bidirectional and symmetrical, and there is no technical service differentiation (aside from signaling procedures) between mobile-originated SMS and mobile-terminated SMS.

The bearer service always attempts to deliver a short message to an MS-based SME whenever the MS is registered, even if the mobile sta-

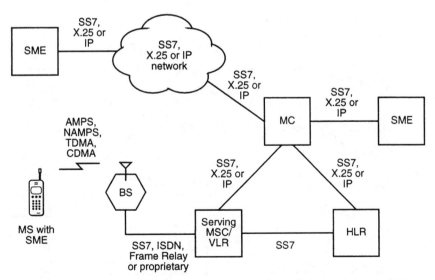

Figure 13.1 The basic network architecture supporting IS-41 SMS. SS7, X.25, and IP can be used for transport of short messages. If a packet-switched protocol other than SS7 is used, the serving MSC requires an interworking function to support that protocol.

tion is engaged in a call. The network is informed whether a message has been received by the MS. This allows the messages to be held by the sender in cases of unsuccessful delivery and then retransmitted to the destination when possible.

SMS teleservices

An *SMS teleservice* is defined as an SMS service that provides the complete application capability, including terminal equipment functions, for communication between SMS users (i.e., SMEs) according to application protocols established between the users. The SMS teleservices are designed as special applications that use the SMS bearer service as a transport mechanism (see Fig. 13.2). There are currently two TIA-standardized teleservices:

1. *Cellular paging teleservice* (CPT). The CPT specifies a short message–based teleservice to provide paging-like services to an MS. CPT is based upon a minimal character set, including uppercase letters A through Z and digits 0 through 9, with a maximum message length of 63 characters.

2. *Cellular messaging teleservice* (CMT). The CMT specifies a short message–based teleservice to provide a generic short messaging application to an MS. CMT is based upon an extensive character set and a maximum message length of 200 characters.

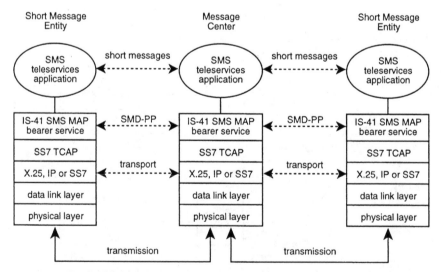

Figure 13.2 Layered protocol model showing the SMS teleservices as applications supported by the SMS bearer service (i.e., SMS point-to-point delivery service). Note that X.25, IP, or SS7 can be used for network transport of short messages.

Although stage 1 of CPT and CMT is standardized in TIA IS-53-A, stages 2 and 3 are designed to be network-independent and are not specified in IS-41-C.

Message center

The message center (MC) provides the *store-and-forward* function for most (see below) mobile-originated short messages and for all mobile-terminated short messages. The MC is typically implemented as a physically separate network entity, but it may be combined with other functional entities. Many MCs may be connected on a single network, or a single MC may be connected to many networks.

SMS subscribers are associated with an MC, known as the *home* MC, in the MS's home system. The MC maintains the mobile identification number (MIN) address information of the MSs that it serves for mobile-terminated messages. The MC is addressable by the directory number addresses of the MSs that it serves for mobile-terminating messages. In general, an MC provides the following capabilities:

- Forward messages to the addressed MS-based SME.

- Store short messages for unavailable MS-based SMEs.

- Apply originating and terminating SMS supplementary services to short messages.

- Optionally, provide interworking between different transport protocols.

IS-41-C supports bypassing the MC for mobile-originated short messages, so that messages can be sent directly to the destination SME from the serving MSC. However, no originating supplementary services can be applied to these messages.

Short message entity

The short message entity is a functional entity capable of composing (originating) and decomposing (receiving) a short message. An SME may be located in a fixed network (outside the IS-41 network), in a mobile station, or within the IS-41 network. An SME is generally considered to be an application entity that represents the originator and recipient of short messages provided via the point-to-point short message delivery service. In general, an SME has the following capabilities:

- Compose short messages.

- Dispose of, act upon, or display short messages.

- Request supplementary services.

- Store received short messages.
- Manage stored messages.

The methods for performing these tasks are implementation-dependent and are not addressed by IS-41-C.

Where Are SMS Functions Specified in IS-41-C?

SMS functions are specified in three parts of IS-41-C:

1. IS-41.5-C provides the required formats of all the IS-41-C operation components, including those used for SMS. IS-41.5-C defines both the messages (e.g., SMSRequest Invoke) and the message parameters (e.g., SMS_Address).

2. IS-41.6-C provides algorithmic descriptions of the procedures that are associated with sending and receiving most IS-41-C messages, including those used for SMS. This part of IS-41-C also includes an example of an air interface definition for SMS in the informative (i.e., not technically a requirement of the standard) Annex C. This information is intended to illustrate the assumptions regarding the air interface SMS support that were made in the design of the IS-41-C SMS network protocols and procedures.

3. IS-41.3-C steps back from the protocol details contained in parts 5 and 6 and attempts to explain, using information flow diagrams and step-by-step descriptions, how the operations are used individually and together to accomplish SMS application process tasks.

Issues Associated with SMS

The following issues need to be addressed in the implementation of SMS:

- *Delivering short messages to roaming subscribers.* Location, status, and address information needs to be obtained from the serving system to deliver a mobile-terminated short message.

- *Delivering short messages during intersystem handoff.* Short messages being delivered during intersystem handoff must be delivered to the subscriber intact.

- *Methods for originating short messages from an MS.* Typical mobile stations do not have an adequate keypad for entering message information to be transmitted. In a basic implementation, a message may be selected from a predetermined set of messages stored in the MS; alternatively, operator assistance or portable computers can be used to generate messages.

- *Subscribers may roam into areas where SMS is not supported.* The subscriber may not have access to SMS in all cellular coverage areas. Subscribers may not be aware that they have lost service.

- *Different air interface protocols support different message lengths.* NAMPS supports transmission of only 14 characters at a time, whereas the digital protocols support short messages of at least 140 characters. Different systems may also support varying maximum-message lengths, possibly limited by the service provider to conserve bandwidth.

- *SMS interconnection to other data networks.* A primary consideration in the design of SMS is support of standardized transport protocols for easy access to packet data networks such as the Internet or X.25, in addition to SS7.

SME Service Qualification

The IS-41-C SME service qualification function encompasses the processes which establish an SME's financial accountability and service capabilities in a serving system. There are two types of SMEs defined for SMS: MS-based SME and fixed SME. IS-41-C defines the service qualification procedures for MS-based SMEs only; qualification procedures for fixed SMEs are determined by individual service providers.

The IS-41-C MS-based SME service qualification processes are tightly coupled to the service qualification of the host MS itself; i.e., if the MS is qualified, generally the SME is also qualified. These procedures are described in Chap. 10 and use the following IS-41-C operations:

- RegistrationNotification

- QualificationRequest

- QualificationDirective

The only significant SME-related addition to MS service qualification is that the HLR sends the SME service profile information—in the form of the IS-41-C SMS_OriginationRestrictions and SMS_Termination-Restrictions parameters—to the SMS-capable serving system, along with the other MS-related profile parameters (see Fig. 13.3).

SME Location Management

Like the associated MS function described in Chap. 10, the SME location management function comprises two components:

1. SME location update processes have the effect of creating or modifying the SME-related location information in an MS's temporary

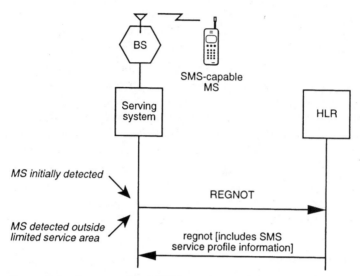

Figure 13.3   An example of SMS service qualification using the RegistrationNotification operation.

record in a visited system and updating the SME location information in an MS's record in the HLR.

2. SME location cancellation processes have the effect of deleting SME-related location information in an MS's temporary record in a visited system and updating the SME location information in an MS's record in the HLR.

The IS-41-C MS-based SME location management processes are tightly coupled to the associated processes of the host MS itself; i.e., the location of the MS is the location of the SME as well. These procedures are described in Chap. 10 and use the following IS-41-C operations:

1. To update location:

  ▪ RegistrationNotification

2. To cancel location:

  ▪ MSInactive, including the DeregistrationType parameter
  ▪ BulkDeregistration
  ▪ RegistrationCancellation
  ▪ UnreliableRoamerDataDirective

Note that the serving system also includes a temporary routing address—the SMS_Address parameter—in the Registration Notification Invoke message. This is used for routing short messages to the serving system on their way to the visiting SME. However, we consider this address to be functionally associated with short message termination (described later in this chapter) rather than the SME location update process—somewhat analogous to the serving system's providing a temporary number (i.e., the TLDN) for call termination purposes (see Chap. 12). Note also that, based on this approach, the MC does not maintain "location information" for the SME. In fact, the MC does not need to know the mobile network location for the SME— merely an address that the MC can use to get a message to the SME's location.

SME State Management

The IS-41-C SME state management function encompasses the processes by which the short message termination availability state of an MS-based SME is maintained in the HLR.

Note that there is no formal definition of an *SME state* in IS-41-C. There are, however, procedures that define how the HLR responds to the MC's requests for short-message termination routing information. We use the concept of an SME state, having value either *available* or *unavailable,* to aid our explanation of these procedures. Generally, if the SME's state is *unavailable,* the HLR will indicate this to the MC; otherwise, the HLR either will provide the MC with the routing information currently stored in its database or will attempt to obtain more up-to-date routing information from the serving system. The HLR may consider an MS-based SME *unavailable* because:

1. The MS is not "registered." When the HLR does not have a valid location for the MS, the associated MS-based SME is considered *unavailable.*

2. The MS is registered on an SMS-incapable system.

3. The SME is not authorized for SMS service on the current serving system, although the SME is generally authorized for SMS service on other systems.

4. The host MS is out of radio contact. The MS may have missed autonomous registration events due to a loss of radio contact and been designated *inactive* by the serving system. The SME state in this case is system-dependent.

5. The host MS is intentionally inaccessible. The MS may have gone into a mode (e.g., sleep mode) whereby it is intentionally inaccessible. The SME state in this case is also system-dependent.

Note that the SME state is not necessarily the same as the MS state described in Chap. 10; an *inactive* MS may be *available* for short message service deliveries, based on system-dependent algorithms.

The HLR SME state management process makes use of information provided by the serving system and HLR location management processes. Generally, the SME state is set to *available* by the HLR after a successful location update for the MS. Likewise, after a successful location cancellation, the HLR sets the SME state to *unavailable*; the state may be reset to *available* if the location cancellation is nested within a location update process. In this sense, the HLR SME state management process makes use of the same operations as the location update and location cancellation processes, i.e., RegistrationNotification, RegistrationCancellation, MSInactive, BulkDeregistration, and UnreliableRoamerDataDirective. Additionally, the SME state may be explicitly set to *unavailable* in the HLR in the following ways:

1. When the HLR receives a valid IS-41-C **REGNOT** message including a valid AvailabilityType parameter from the serving system, it may set the SME's state to *unavailable* (see Fig. 13.4).

2. When the HLR receives a valid IS-41-C **MSINACT** message from the serving system, it may set the SME's state to *unavailable* (see Fig. 13.5). Note that this message may also result in location cancellation for the MS if the DeregistrationType parameter is included.

3. When the HLR receives a valid IS-41-C RoutingRequest RETURN RESULT (**routreq**) message from the serving system, including

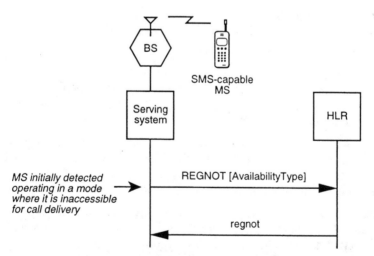

Figure 13.4   Setting the SME's state to unavailable by using the RegistrationNotification operation.

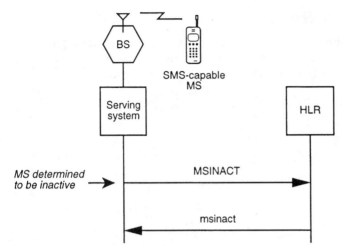

Figure 13.5 Setting the SME's state to unavailable by using the MSInactive operation.

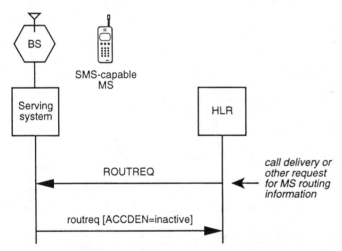

Figure 13.6 Setting the SME's state to unavailable by using the RoutingRequest operation.

the AccessDeniedReason parameter set to *inactive,* the HLR may set the SME's state to *unavailable* (see Fig. 13.6).

Once set, the SME's state remains *unavailable* until a serving system—for which the SME is SMS-authorized—sends a valid **REGNOT** message to the HLR that includes the SMS_Address parameter but does not include the AvailabilityType parameter.

Short Message Processing

The IS-41-C short message processing functions encompass the processes that enable, restrict, supplement, or otherwise impact an SME's ability to originate and terminate a short message.

Figure 13.7 illustrates a basic message origination and termination sequence for message transfer between two MS-based SMEs; SME-A is the originator, and SME-B is the destination. The key SMS elements that apply to the scenario shown in Fig. 13.7 are:

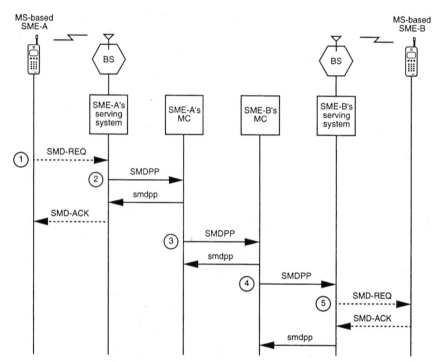

Figure 13.7 A basic message origination and termination sequence for short message transfer between two MS-based SMEs: (1) MS-based SME-A sends an air interface message, SMD-REQUEST (SMD-REQ), to the serving system. (2) The serving system routes the short message to SME-A's MC, using the IS-41-C SMSDeliveryPointToPoint Invoke (**SMDPP**) message. Each of the **SMDPP** messages shown in this scenario may be routed by using the same SS7 signaling network as is used for routing other IS-41-C messages; alternatively, a separate network, based on TCP/IP or some other network protocol, may be employed. When the acknowledgment (i.e., the **smdpp** message) is received from the MC, the serving system converts it to an air interface acknowledgment, the SMD-ACK message. (3) SME-A's MC may apply an originating supplementary service to the short message (this is not currently defined in IS-41-C); the **SMDPP** message is then routed to the destination SME's MC. (4) SME-B's MC may apply a terminating supplementary service to the short message (this is not currently defined in IS-41-C); the **SMDPP** message is then routed to the destination SME's serving system. (5) The serving system forwards the short message to the destination SME by using the air interface SMD-REQ message. SME-B responds with an automatic acknowledgment (SMD-ACK) to signal acceptance of the SMD-REQ message.

- Short message addressing and routing
- Short message barring, i.e., enforcing messaging restrictions
- Applying SMS supplementary services

Short message addressing and routing

Because of the store-and-forward nature of the short message transfer process, messages may take a circuitous path from the originator to the final destination. The addressing mechanisms defined in IS-41-C provide for this.

Refer to Fig. 13.7. IS-41-C allows the originating SME (SME-A) to provide up to four pieces of address information in the air interface (e.g., TDMA or CDMA) equivalent of the SMD-REQ message:

1. OriginalOriginatingAddress

2. OriginalDestinationAddress

3. OriginalOriginatingSubaddress

4. OriginalDestinationSubaddress

In general, the OriginalOriginatingAddress information is required for message termination but is not necessarily included for message origination; likewise, the OriginalDestinationAddress information is required for message origination but is not necessarily included for message termination.

IS-41-C defines six SMS address parameters for the SMSDelivery-PointToPoint Invoke (**SMDPP**) message:

1. SMS_OriginalOriginatingAddress

2. SMS_OriginalOriginatingSubaddress

3. SMS_OriginatingAddress

4. SMS_OriginalDestinationAddress

5. SMS_OriginalDestinationSubaddress

6. SMS_DestinationAddress

The air interface OriginalOriginatingSubaddress and Original-DestinationSubaddress information is optional and, if provided, is passed transparently from end to end by the SMS point-to-point bearer service in the SMS_OriginalOriginatingSubaddress and SMS_OriginalDestinationSubaddress parameters, respectively.

Various numbering formats are supported for the other SMS address parameters, including:

- ITU-T E.164 format

- ITU-T X.121 format
- Private numbering plan formats
- Internet protocol (IP) address format

A probable scenario is for each MS-based SME in Fig. 13.7 to be addressed by its host MS's MIN, MIN-A (for SME-A) and MIN-B (for SME-B); the MIN-based addresses are encoded by using the E.164 format. Table 13.1 summarizes the relationship between the air interface address values in the SMD-REQ messages and the IS-41-C SMS address values in the **SMDPP** messages for each step identified in Fig. 13.7. For the purposes of illustration, we assume that all possible address parameters—with the exception of the subaddress parameters—are included in each message.

Most of the mappings between the air interface and IS-41-C address elements are straightforward:

- The air interface OriginalOriginatingAddress parameter maps to the IS-41-C SMS_OriginalOriginatingAddress parameter.
- The air interface OriginalDestinationAddress parameter maps to the IS-41-C SMS_OriginalDestinationAddress parameter.

However, the values of the SMS_OriginatingAddress and SMS_DestinationAddress parameters vary depending on the particular information flow in Fig. 13.7 and are set to ensure correct routing of the message:

- In step 2, the SMS_DestinationAddress parameter is set to MIN-A, rather than to MIN-B, to route the **SMDPP** message to SME-A's MC.
- In step 3, the SMS_OriginatingAddress parameter is set to MIN-A, to identify the source of the message as SME-A's MC; the

TABLE 13.1   Relationship between Air Interface and IS-41-C SMS Address Information

Step	Original Originating Address	Original Destination Address	SMS_Original Originating Address	SMS_ Originating Address	SMS_Original Destination Address	SMS_ Destination Address
1	MIN-A	MIN-B				
2			MIN-A	MIN-A	MIN-B	MIN-A
3			MIN-A	MIN-A	MIN-B	MIN-B
4			MIN-A	MIN-B	MIN-B	MIN-B
5	MIN-A	MIN-B				

SMS_DestinationAddress parameter is set to MIN-B, to route the **SMDPP** message to SME-B's MC.

- In step 4, the SMS_OriginatingAddress parameter is set to MIN-B, rather than to MIN-A, to identify the source of the message as SME-B's MC.

To route the message to SME-A's MC in step 2, the serving system can maintain a table of MIN-to-MC addresses (e.g., MIN to SS7 destination point code), as is often done today in IS-41 networks for routing IS-41 messages to an MS's HLR. If messages are transported by using an SS7 signaling network, the serving system can use the SS7 global title translation (GTT) capability. In this case, the serving system creates a global title address containing the SMS_DestinationAddress parameter value and requests a MIN-to-MC global title translation.

These same routing possibilities exist for SME-A's MC in step 3; i.e., to route the message to SME-B's MC, either a fixed MIN-to-MC address table or GTT on MIN-B may be employed, although the latter approach is much more likely (i.e., it is far easier to maintain the MIN-to-MC routing information in the SS7 network than in each and every accessible MC).

Terminating short messages to MS-based SMEs. Steps 2 and 3 in Fig. 13.7 involve message routing between fixed points. In step 4, the destination SME (SME-B) is mobile; therefore, if it does not already have one, SME-B's MC must get a valid routing address for the system currently serving SME-B. IS-41-C provides the SMSRequest operation specifically for this purpose; this routing requirement also impacts the RegistrationNotification operation, as explained below (see Fig. 13.8):

1. When a new SMS-capable MS is detected, the serving system sends a RegistrationNotification Invoke (**REGNOT**) message to the HLR. If the serving system is SMS-capable, the message includes the SMS_Address parameter that is used to route short messages to the serving system for delivery to the MS-based SME. For example, if the short-message transport network is SS7-based, the SMS_Address parameter may contain an SS7 point code and subsystem number; when the serving system receives an **SMDPP** message addressed to this point code and subsystem number, it assumes the message is intended for a visiting MS-based SME. The specific destination MS is identified by the address parameters within the **SMDPP** message.

2. When the MC requires a current routing address for an MS-based SME, it sends an SMSRequest Invoke (**SMSREQ**) message to the MS's HLR.

3. The HLR then makes a decision: Is the SMS_Address received in step 1 sufficiently current for SMS delivery purposes, or should the

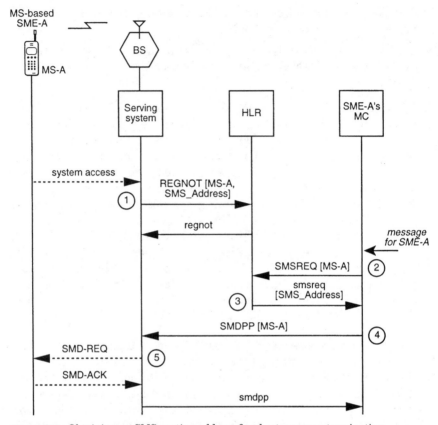

Figure 13.8   Obtaining an SMS routing address for short message termination.

HLR attempt to obtain a new address? If the SMS_Address is considered valid, the HLR returns it to the MC in the SMSRequest Return Result (**smsreq**) message; otherwise (not shown in Fig. 13.8), the HLR may send an **SMSREQ** message to the serving system, requesting an update on the MS's accessibility and routing information.

4. The MC uses the SMS_Address to route the **SMDPP** message to the serving system.

5. The serving system sends the message to the MS-based SME identified in the **SMDPP** message.

SMS delivery pending flag (SMSDPF) management.   The MC may become aware of an MS-based SME's unavailability under three circumstances:

1. The HLR returns the SMS_AccessDeniedReason (SMSACCDEN) parameter in the **smsreq** message in response to the MC's request for a current routing address for the MS-based SME.

2. The serving system returns the SMSACCDEN parameter in the **smdpp** message in response to the MC's request for delivery of a short message to the MS-based SME.

3. The HLR or serving system sends the SMSACCDEN parameter in an SMSNotification Invoke message to the MC.

The MC can handle the MS-based SME unavailability condition by taking one of two approaches:

1. *The polled approach.* The MC periodically resends the **SMSREQ** message to the HLR or the **SMDPP** message to the serving system until it is successful. This approach is entirely MC-controlled and requires no special IS-41-C protocol support.

2. *The interrupt-driven approach.* The MC requests notification from the HLR or serving system when the MS-based SME becomes *available.* For this approach, the MC includes the SMS_ NotificationIndicator (SMSNOTIND) parameter in the **SMSREQ** or **SMDPP** message, set to the value *notify when available.* If (1) the notification request in the **SMSREQ** message is accepted by the HLR (or serving system if the **SMSREQ** is relayed to the serving system) or (2) the notification request in the **SMDPP** message is accepted by the serving system, then the SMSACCDEN parameter is returned to the MC with value *postponed;* otherwise, the SMSACCDEN parameter is returned to the MC with value *denied.*

Once the notification request is accepted by a system (either the serving system or the HLR), the system sets the SMS delivery pending flag (SMSDPF). If the state of the MS-based SME changes, the system sends an SMSNotification Invoke (**SMSNOT**) message to the MC. The **SMSNOT** message is the "interrupt" signal for the MC at which point it evaluates whether the pending message is still valid or has already been discarded.

Figure 13.9 illustrates an example of the interrupt-driven approach when the HLR provides the notification to the MC, while Fig. 13.10 illustrates the case when the serving system provides the notification. Once the SMSDPF is set in the serving system, special procedures are necessary to ensure that the short-message pending status is preserved when, e.g., the serving system changes. The SMS_- MessageWaitingIndicator (SMSMWI) parameter is designed for this purpose (see Figs. 13.11 and 13.12).

Short message routing with call handoff.   One thing not shown in Fig. 13.7 is IS-41-C's support for short message origination and termination under call handoff conditions. Two operations are defined:

1. SMSDeliveryBackward conveys the mobile-originated short mes-

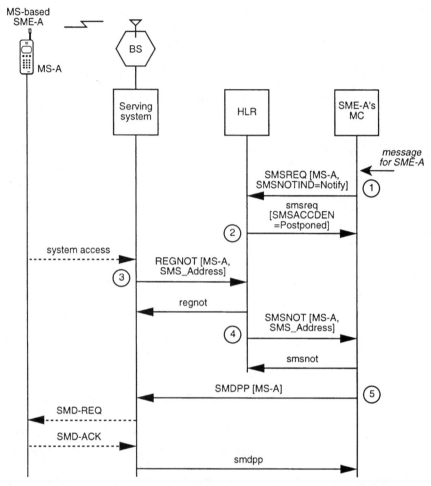

Figure 13.9 An example of the interrupt-driven approach to SME availability notification by the HLR: (*1*) When MS-based SME-A's MC requires a current routing address for SME-A, it sends an **SMSREQ** message to the MS's HLR. It includes the SMSNOTIND parameter set to the value *Notify when available*. (*2*) The HLR considers SME-A to be *unavailable*. The HLR accepts the notification request by including the SMSACCDEN parameter set to the value *Postponed*. The HLR sets the SMSDPF. (*3*) MS-A accesses the serving system. The HLR marks SME-A as *available* and stores the current SMS routing address (SMS_Address). (*4*) Since the SMSDPF is set, the HLR launches an **SMSNOT** message to SME-A's MC, informing the MC of the routing address for SME-A. (*5*) The MC uses this address to route the short message to SME-A.

sage from the serving system, possibly via one or more tandem systems, to the anchor system.

2. SMSDeliveryForward conveys the mobile-terminated short message from the anchor system, possibly via one or more tandem systems, to the serving system.

These operations are illustrated in Fig. 13.13 and Fig. 13.14.

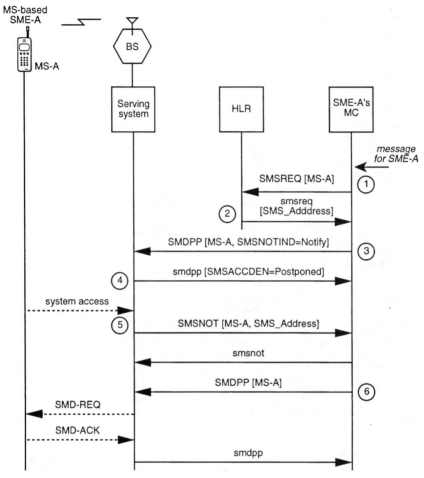

Figure 13.10 An example of the interrupt-driven approach to SME availability notification by the serving system: (*1*) When MS-based SME-A's MC requires a current routing address for SME-A, it sends an **SMSREQ** message to the MS's HLR. (*2*) The HLR considers SME-A to be *available* and provides the current routing address for the SME in the SMS_Address parameter. (*3*) The MC uses this address to route the short message to SME-A. It includes the SMSNOTIND parameter set to the value *Notify when available*. (*4*) SME-A is temporarily unavailable for short message delivery. The serving system accepts the notification request by including the SMSACCDEN parameter set to the value *Postponed*. The serving system sets the SMSDPF. (*5*) MS-A accesses the serving system. Since the SMSDPF is set, the serving system launches an **SMSNOT** message to SME-A's MC, informing the MC of the routing address for SME-A. (*6*) The MC uses this address to route the short message to SME-A.

A final word on short message addressing and routing. The addressing and routing mechanisms in IS-41-C provide an extremely flexible set of tools for short message exchange. Part of the design requirement was to support multiple transport technologies—SS7, TCP/IP, and X.25—and to allow interworking between these technologies. The

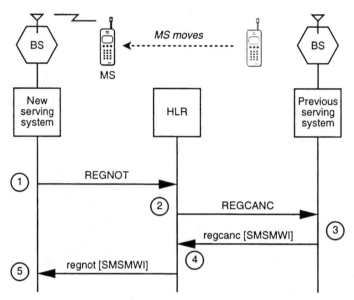

Figure 13.11 Transferring the state of the SMSDPF by using the SMSMWI parameter during the location update and location cancellation processes: (*1*) A new MS moves into its service area; therefore, the serving MSC requests location update in the HLR. (*2*) The HLR has stored location information for the MS; therefore, it requests location cancellation from the previous serving system. (*3*) The previous serving MSC deletes the temporary MS record and sends an acknowledgment to the HLR. It includes the SMSMWI parameter if the SMS-DPF is set. (*4*) The HLR acknowledges the location update and relays the SMSMWI parameter to the new serving system. (*5*) When the new serving system receives the SMSMWI parameter, it sets the SMSDPF.

messages, parameters, and procedures in IS-41-C support this requirement.

For example, Table 13.2 provides another example of the relationship between the air interface address values in the SMD-REQ messages and the IS-41-C SMS address values in the **SMDPP** messages for each step identified in Fig. 13.15. This time, SME-B is addressed by an IP address IP-B. For the purposes of illustration, we assume that no subaddress information is provided for SME-A or SME-B.

Referring to Table 13.2, routing proceeds as per our first example in Table 13.1 until step 3. At this point, SME-A's MC must send the **SMDPP** message to SME-B's MC. Since SME-B is identified with an IP address, SME-A's MC sends the **SMDPP** message via a TCP/IP network, where it terminates on SME-B's MC. Noting that the mes-

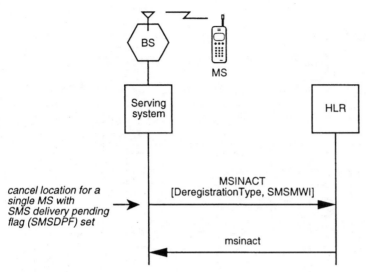

Figure 13.12 Transferring the state of the SMSDPF by using the SMSMWI parameter during location cancellation via the MSInactive operation. The HLR sets its SMSDPF when it receives the SMSMWI parameter.

TABLE 13.2   Another Example of IS-41-C SMS Address Information

Step	Original Originating Address	Original Destination Address	SMS_Original Originating Address	SMS_ Originating Address	SMS_Original Destination Address	SMS_ Destination Address
1	MIN-A	IP-B				
2			MIN-A	MIN-A	IP-B	MIN-A
3			MIN-A	MIN-A	IP-B	IP-B
4			MIN-A	IP-B	IP-B	MIN-B
5	MIN-A	IP-B				

sage is intended for IP-B, and that IP-B is currently associated with MS-B, SME-B's MC sends the **SMDPP** message to MS-B's current serving system (e.g., via an SS7 transport network using the SMS_Address provided by the serving system). SME-B's MC informs the serving system that the original destination address is IP-B, while the "intermediate" destination address is MIN-B (the MIN associated with MS-B). The serving system should, at this point, send the message to MS-B, identifying the destination address as IP-B.

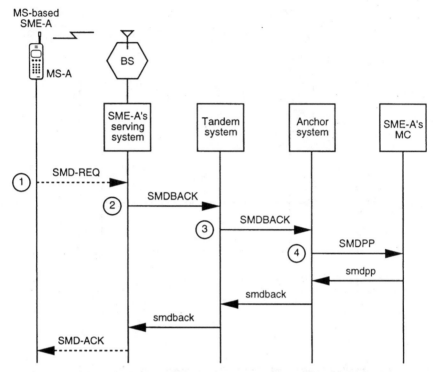

Figure 13.13 MS-based SME message origination under call handoff conditions: (1) MS-based SME-A sends an air interface message, SMD-REQ, to the serving system. (2) The call has previously been handed off from the anchor system to a tandem system and then to the current serving system; therefore, the serving system routes the short message to the anchor system by using the SMSDeliveryBackward Invoke (**SMDBACK**) message. (3) The tandem system forwards the **SMDBACK** message to the anchor system. (4) The anchor system then sends the message to SME-A's MC, using the **SMDPP** message. Acknowledgment messages are relayed forward through the handoff chain to SME-A.

SMS barring

IS-41-C provides a number of SMS origination and termination barring capabilities, conveyed to the serving system during the service qualification process for the host MS in the following IS-41-C parameters: SMS_OriginationRestrictions and SMS_TerminationRestrictions. The SMS_OriginationRestrictions parameter defines the type of short messages the MS is allowed to originate:

- No messages; i.e., block all messages.
- All messages; i.e., allow all messages.
- Block message if direct routing is requested (see below).

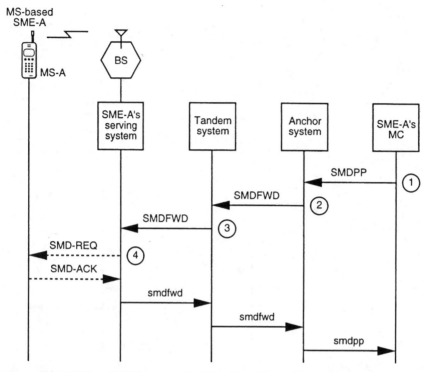

Figure 13.14 MS-based SME message termination under call handoff conditions: (*1*) A message destined for MS-based SME-A is sent by SME-A's MC to the anchor system. (*2*) The call has previously been handed off from the anchor system to a tandem system and then to the current serving system; therefore, the anchor system routes the short message to the serving system by using the SMSDeliveryForward Invoke (**SMD-FWD**) message. (*3*) The tandem system forwards the **SMDFWD** message to the serving system. (*4*) The serving system sends the air interface message SMD-REQ to SME-A. Acknowledgment messages are relayed back through the handoff chain to the MC.

- Force indirect routing for all messages (see below).

In the examples of the previous section, we showed messages routed via the originating SME's MC; this is called *indirect routing*. IS-41-C allows MS-based SMEs to request that originated messages bypass the originator's MC; this is called *direct routing*. Figure 13.16 and Table 13.3 illustrate how our first example from the previous section (i.e., in Fig. 13.7 and Table 13.1) changes when the originating SME requests direct routing of the message. Note the change in the SMS-DestinationAddress parameter in step 2.

The SMS_TerminationRestrictions parameter defines the type of short messages the MS is allowed to receive:

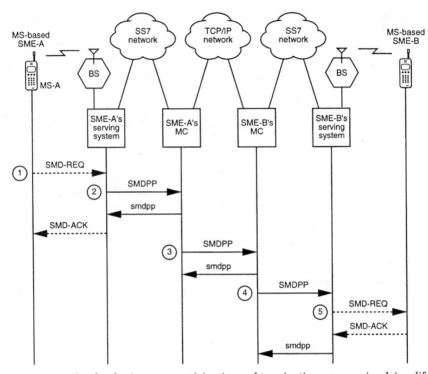

Figure 13.15 Another basic message origination and termination sequence involving different network technologies: (1) MS-based SME-A sends the air interface message SMD-REQ to the serving system. (2) The serving system routes the short message to SME-A's MC, using the IS-41-C **SMDPP** message. The **SMDPP** message is routed by using the same SS7 signaling network as is used for routing other IS-41-C messages. When the acknowledgment (i.e., the **smdpp** message) is received from the MC, the serving system converts it into an air interface acknowledgment, the SMD-ACK message. (3) SME-A's MC may apply an originating supplementary service to the short message (this is not currently defined in IS-41-C); the **SMDPP** message is then routed to the destination SME's MC by using a TCP/IP network. (4) SME-B's MC may apply a terminating supplementary service to the short message (this is not currently defined in IS-41-C); the **SMDPP** message is then routed to the destination SME's serving system by using the SS7 signaling network. (5) The serving system forwards the short message to the destination SME by using the air interface SMD-REQ message. SME-B responds with an automatic acknowledgment (SMD-ACK) to signal acceptance of the SMD-REQ message.

- No messages; i.e., block all messages.
- All messages; i.e., allow all messages.
- Block message if it is charged to destination (see below).

The SMS_ChargeIndicator parameter, when it is included in the **SMDPP** message, indicates whether the originator of the message has requested that the destination be charged for the message delivery.

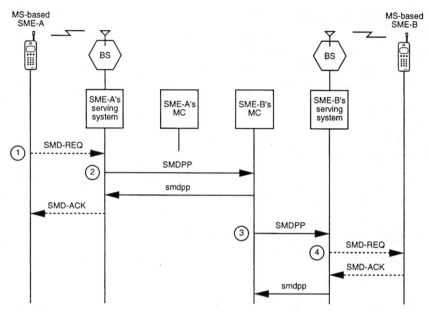

Figure 13.16 Another basic message origination and termination sequence involving direct routing: (*1*) MS-based SME-A sends an air interface message, SMD-REQUEST (SMD-REQ), to the serving system, requesting direct routing. (*2*) The serving system routes the short message directly to SME-B's MC, using the IS-41-C SMSDeliveryPointToPoint Invoke (**SMDPP**) message. When the acknowledgment (i.e., the **smdpp** message) is received from the MC, the serving system converts it to an air interface acknowledgment, the SMD-ACK message. (*3*) SME-B's MC may apply a terminating supplementary service to the short message (this is not currently defined in IS-41-C); the **SMDPP** message is then routed to the destination SME's serving system. (*4*) The serving system forwards the short message to the destination SME by using the air interface SMD-REQ message. SME-B responds with an automatic acknowledgment (SMD-ACK) to signal acceptance of the SMD-REQ message.

TABLE 13.3   Addressing When Direct Routing Is Provided

Step	Original Originating Address	Original Destination Address	SMS_Original Originating Address	SMS_ Originating Address	SMS_Original Destination Address	SMS_ Destination Address
1	MIN-A	MIN-B				
2			MIN-A	MIN-A	MIN-B	MIN-B
3			MIN-A	MIN-B	MIN-B	MIN-B
4	MIN-A	MIN-B				

SMS supplementary services

Routing messages through the originator's and destination's MCs allows supplementary services to be applied to the basic SMS bearer service at one or both of the originating and terminating MCs. Possible services include:

- Delayed delivery
- Repeated delivery
- Delivery to a distribution list

These potential supplementary services, and many others, are left as system implementation issues in the IS-41-C specification.

Summary of IS-41-C Operations Used for SMS

Table 13.4 summarizes the IS-41-C operations used by the SMS functions described in this chapter.

TABLE 13.4   Use of IS-41-C Operations for SMS

Function	IS-41-C operations used for the function
SME service qualification	RegistrationNotification, QualificationRequest, QualificationDirective
SME location management	RegistrationNotification, MSInactive, BulkDeregistration, RegistrationCancellation, UnreliableRoamerDataDirective
SME state management	SME location management operations plus RoutingRequest
Short message processing	RegistrationNotification, MSInactive, RegistrationCancellation, SMSDeliveryBackward, SMSDeliveryForward, SMSDeliveryPointToPoint, SMSNotification, SMSRequest

Chapter

14

Operations, Administration, and Maintenance Functions

In this chapter, we discuss the IS-41-C mobile telecommunications network functions related to operations, administration and maintenance (OA&M). The subset of OA&M functions supported by IS-41-C is limited to inter-MSC circuit management. Circuit management can be considered a subset of network management functionality. There are many functions used to manage and control inter-MSC circuits; however, the IS-41 intersystem operations only support the most primary functions:

1. Inter-MSC circuit blocking (and unblocking)

2. Inter-MSC circuit reset

3. Inter-MSC circuit testing

These functions are necessary to remove trunks from service, place them back into service, test them for transmission and reset them to an idle state when necessary. Other functions such as traffic load control and collection of traffic statistics are beyond the scope of IS-41.

We describe each of the functions and provide examples of how the IS-41-C mobile application part (MAP) operations (e.g., TrunkTest) are used to accomplish inter-MSC circuit management tasks. We summarize this information at the end of the chapter. Note that we employ the IS-41-C convention for operation component acronyms; i.e., the Invoke component acronym is in all capital letters (e.g., **TRUNKTEST**), while the Return Result component acronym is in all lowercase letters (e.g., **trunktest**). Refer to the IS-41-C standard for the descriptions of the individual IS-41-C operations and parameters.

Also, the reader should consult the Glossary for the description of general terms (e.g., like *serving system*) that are not explicitly defined in this chapter.

Throughout this chapter we consider the serving system as a single entity encompassing the mobile switching center (MSC) and visitor location register (VLR) functional entities. This simplifies the descriptions of the IS-41-C OA&M functions and also is representative of a large percentage of the IS-41 implementations currently in service. However, the reader should keep in mind that the potential for separation of the MSC and VLR exists and is fully defined in IS-41-C.

Where Are OA&M Functions Specified in IS-41-C?

OA&M functions are specified in two parts of IS-41-C:

1. IS-41.5-C provides the required formats—the bit-by-bit encoding—of all the IS-41-C operation components, including those used for OA&M. IS-41.5-C defines both the messages (e.g., TrunkTest Invoke) and the message parameters (e.g., InterMSCCircuitID).

2. IS-41.4-C provides algorithmic descriptions of the procedures that are associated with sending and receiving the IS-41-C OA&M-related messages; these procedures were considered sufficiently unique to warrant separating them from the main procedures in IS-41.6-C. IS-41.4-C also includes stage 2 style information for OA&M.

Inter-MSC Circuit Blocking

The IS-41-C circuit blocking function—which also encompasses circuit unblocking—allows the MSCs at each end of an inter-MSC circuit to coordinate:

1. Removal of the inter-MSC circuit from service (*blocking*)

2. Reinstatement of the inter-MSC circuit into service (*unblocking*)

Circuit blocking may be used for fault management or testing purposes. Since the inter-MSC circuits are used solely for handoff, circuit blocking has important consequences for the intersystem handoff functions described in Chap. 9. IS-41-C defines the concept of a *blocking state* to describe the impact on handoff of circuit blocking. We describe the IS-41-C blocking state in terms of the inter-MSC circuit model shown in Fig. 14.1.

Each MSC maintains a record of the blocking state for each inter-MSC circuit. For our modeling purposes, *Sxy* is the blocking state of circuit x at MSC-y. In general, SxA is not necessarily equal to SxB.

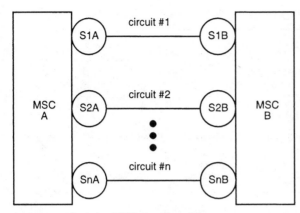

Figure 14.1 An inter-MSC circuit model.

TABLE 14.1 Possible Values for S1A, the Blocking State of Circuit 1 at MSC-A

Blocking state, S1A	Symbol	Meaning
Active	ACT	Handoff can be initiated on circuit 1 by either MSC-A or MSC-B
Locally blocked	LB	Handoff can be initiated on circuit 1 only by MSC-A
Remotely blocked	RB	Handoff can be initiated on circuit 1 only by MSC-B
Locally and remotely blocked	LRB	Circuit 1 cannot be used for handoff

The blocking state for each circuit can take on one of four values at any given point in time. Table 14.1 describes the possible blocking state values for circuit 1 at MSC-A.

The blocking state for a circuit makes a transition from one state value to another based on *events*. Events include the sending or receiving of certain IS-41-C messages. IS-41.4-C specifies dozens of events, including:

- Sending and receiving blocking and unblocking messages
- Sending and receiving intersystem handoff messages
- Sending and receiving other OA&M messages (e.g., ResetCircuit Invoke)

Figure 14.2 illustrates a small portion of the overall *state transition diagram* for S1A, the blocking state of circuit 1 at MSC-A, when we consider only the events associated with receiving and transmitting

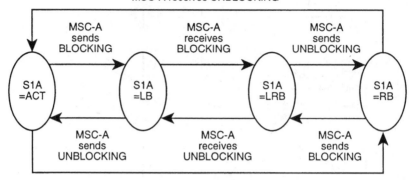

Figure 14.2   A portion of the overall state transition diagram for S1A, the blocking state of circuit 1 at MSC-A (see Fig. 14.1).

the Blocking Invoke (**BLOCKING**) and Unblocking Invoke (**UNBLOCKING**) messages.

Inter-MSC Circuit Reset

The IS-41-C inter-MSC circuit reset function may be used for two related purposes:

1. To initialize an inter-MSC circuit to a known state when it is placed into service.

2. To re-initialize an inter-MSC circuit to a known state after a failure that may have resulted in a loss of circuit state information.

The latter function addresses the sticky issue that arises when something in the "real world" is modeled as an object with a limited set of states and events that cause transitions between states—what do you do when the object's current state information is lost? The IS-41-C ResetCircuit operation solves this problem, as illustrated in Fig. 14.3 and described below:

Step 1. MSC-A and MSC-B have exchanged blocking messages and the state of inter-MSC circuit 1 is set to *locally and remotely blocked* at both ends.

Step 2. MSC-A suffers a data failure.

Step 3. When MSC-A recovers from the fault, it does not know the current state of inter-MSC circuit 1. It sends a ResetCircuit Invoke (**RESET**) message to MSC-B; effectively, this message tells MSC-B:

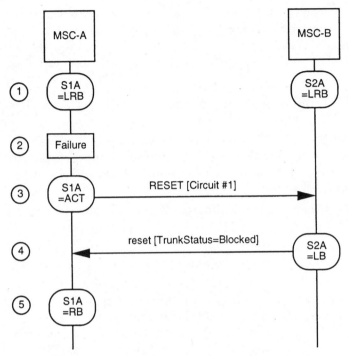

Figure 14.3   An example of the use of the IS-41-C Reset operation.

"I consider circuit 1 to be in the *active* state. What is the state from your perspective?"

Step 4. MSC-B responds with the ResetCircuit Return Result (**reset**) message, indicating that it considers the circuit to be blocked. This is conveyed in the IS-41-C TrunkStatus parameter. MSC-B then changes the circuit blocking state to *locally blocked,* since the circuit is blocked from MSC-B's end and active from MSC-A's end.

Step 5. When MSC-A receives the **reset** message, it sets the blocking state to *remotely blocked.* The circuit state is now re-initialized.

Inter-MSC Circuit Testing

The IS-41-C inter-MSC circuit testing function supports *loop-back* testing of inter-MSC circuits. This capability is important to verify end-to-end path continuity and quality, especially when analog trunks are used to provide inter-MSC circuits. Otherwise, the only time a signal appears on the circuit is when it is used for handoff purposes—not the time to find out that the trunk is defective. With digital trunks—e.g., based on a T-1 carrier system—there are alternate methods of monitoring the integrity of the physical facility but loop-back testing is still effective.

The circuit testing process is illustrated in Fig. 14.4 and described below:

Step 1. MSC-A initiates circuit testing by sending a TrunkTest Invoke (**TTEST**) message to MSC-B.

Step 2. MSC-B decides to accept the test request. It places the selected circuit in loop-back mode; this involves connecting the circuit's receive-audio-path to the transmit-audio-path, with appropriate signal level adjustment.

Step 3. MSC-B then sends a TrunkTest Return Result (**ttest**) message to MSC-A.

Step 4. MSC-A applies either a test tone or a termination to the circuit to test transmission level or noise level on the circuit, respectively.

Step 5. When MSC-A wishes to end testing, it sends a TrunkTestDisconnect Invoke (**TTESTDISC**) message to MSC-B.

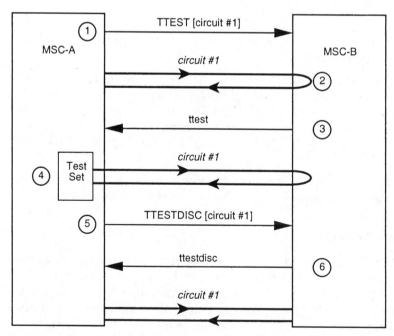

Figure 14.4 An example of the IS-41-C inter-MSC circuit-testing process.

Step 6. MSC-B ends testing, removes the loop-back, and sends a TrunkTestDisconnect Return Result (**ttestdisc**) message to MSC-A.

Summary of IS-41-C Operations Used for OA&M

Table 14.2 summarizes the IS-41-C operations used by the OA&M functions described in this chapter.

TABLE 14.2   Use of IS-41-C Operations for OA&M

Function	IS-41-C operations used for the function
Inter-MSC circuit management	Blocking, Unblocking
Inter-MSC circuit reset	ResetCircuit
Inter-MSC circuit testing	TrunkTest, TrunkTestDisconnect

15

Looking beyond IS-41-C

IS-41-C represents a significant advance in the intersystem functionality available to roaming mobile telecommunications network subscribers—but there is certainly more to come. The protocol provides ready tools for extending its capabilities, in both standard and nonstandard manners. Even as we write, the TIA standards committees are hard at work developing new capabilities that will drive the standard into the future.

Protocol Extension Mechanisms

IS-41-C provides a number of mechanisms for extending the protocol. These mechanisms offer the potential for new operations, new parameters, and new parameter values. Furthermore, section 7 of IS-41.5-C describes guidelines to be applied when the protocol is extended so as to maintain a level of backward and forward compatibility.

New operations

Each of the 54 operations defined in IS-41-C (e.g., LocationRequest) is identified by a number known as the *operation specifier* (see Chap. 8). Operation specifier values are used as listed in Table 15.1.

As Table 15.1 shows, IS-41-C provides 32 operation specifier values for use in defining nonstandard protocol extensions. Individual vendors may use these values to identify new operations with the assurance that future revisions of the standard will not use these operation specifiers, forcing the vendor to modify its implementation to remain

TABLE 15.1   Use of IS-41-C Operation Specifier Values

Operation specifier values	Use of operation specifiers
0	Not used
1 through 40	Used by operations in IS-41-C
41	Reserved
42 through 55	Used by operations in IS-41-C
56 through 223	Reserved for future standard operations
224 through 255	Reserved for nonstandard protocol extensions

standard-compliant. These proprietary operations can be defined and implemented to provide differentiating capabilities for a vendor. Of course, protocol extension in this manner runs the risk of conflicts between two manufacturers' use of the same specifier.

New parameters

Each of the 167 parameters defined in IS-41-C (e.g., Mobile-IdentificationNumber) is identified by a number known as the *parameter identifier* (see Chap. 8). Parameter identifier values are used as listed in Table 15.2.

As Table 15.2 shows, IS-41-C provides 128 parameter identifier values for use in defining nonstandard protocol extensions. Individual vendors may use these values to identify new parameters with the assurance that future revisions of the standard will not use these parameter identifiers, forcing the vendor to modify its implementation to remain standard-compliant.

TABLE 15.2   Use of IS-41-C Parameter Identifier Values

Parameter identifier values	Use of parameter identifiers
0	Not used
1 through 124	Used by parameters in IS-41-C
125 through 127	Reserved
128 through 169	Used by parameters in IS-41-C
170 through 16255	Reserved for future standard operations
16256 through 16383	Reserved for nonstandard protocol extensions

TABLE 15.3   Values of the IS-41-C ActionCode Parameter

ActionCode parameter values	Meaning
0	Not used
1 through 7	Meaning is defined in IS-41-C
8 through 95	Reserved for future standard parameter values
96 through 127	Reserved for nonstandard parameter values
128 through 223	Reserved for future standard parameter values
224 through 255	Reserved for nonstandard parameter values

New parameter values

Many IS-41-C parameters have ranges of values (e.g., 224 to 255) explicitly reserved for private protocol extension purposes. For example, the values of the IS-41-C ActionCode parameter are listed in Table 15.3.

As Table 15.3 shows, IS-41-C provides the ranges from 96 to 127 and from 224 through 255 for use in defining nonstandard parameter values. Individual vendors may use these values to identify new actions associated with the ActionCode parameter with the assurance that future revisions of the standard will not use these parameter values, forcing the vendor to modify its implementation to remain standard-compliant.

Capabilities Planned for Future Versions of IS-41

As of the end of 1996, IS-41-C had become a full-fledged ANSI standard, ANSI/TIA/EIA-41; therefore, there will be no "IS-41 Revision D." However, work continues in the TR45.2 standards committee on enhancements to the standard that will likely show up in future revisions of the ANSI standard (e.g., ANSI/TIA/EIA-41 Revision A) or as Telecommunications Systems Bulletins (TSBs, see Chap. 2). These capabilities include:

1. Enhanced emergency services
2. Circuit-switched asynchronous data and group III fax
3. Network-directed system selection
4. Over-the-air activation
5. Advanced digital features

6. International mobile station identifier (IMSI)

7. Subscriber confidentiality

8. Lawfully authorized electronic surveillance

9. PCS operation and interworking

10. Wireless intelligent network (WIN)

Enhanced emergency services

Emergency 911 and related services constitute a high priority for future versions of the IS-41 standard. Capabilities under consideration at the time of writing the book include these:

1. *Emergency services (911)* permits a subscriber to press 911 plus the SEND key and be connected to a public safety answering point (PSAP) to request an emergency response from the appropriate agency (e.g., fire, police, ambulance, poison control center, or suicide prevention center).

2. *Emergency services callback* permits a PSAP to call back a subscriber that had previously dialed 911.

3. *Emergency services reconnect* attempts to reconnect a 911 call that was in progress when radio contact was lost with an MS.

4. *Emergency area congestion control* allows the cellular system to automatically limit the 911 calls originating from the same area to control call traffic to the PSAP.

5. *Emergency call attempt reporting* to the appropriate PSAP.

6. *Emergency location information delivery* informs the PSAP of the location (e.g., latitude, longitude) of an MS invoking 911.

7. *Emergency subscriber information delivery* informs the PSAP of information concerning the calling subscriber, such as name, address, home phone number, other subscriber information (e.g., medical condition), serving wireless system ID, MIN, ESN, and callback number.

8. *Emergency system congestion control* allows the cellular system to screen the 911 calls, based on requests from the PSAP.

Circuit-switched asynchronous data and group III fax

While analog air interfaces already support these capabilities, special digital radio interface support is needed for the asynchronous data and group III fax services due to the distorting effect of digital voice coders (vocoders) on standard modem tones. Network capabilities include

- Separate directory numbers for voice and data with a single MIN
- Intersystem transport of data after handoff
- Interworking between air interface data formats and standard modem formats
- Data privacy using encryption of the air interface signal

Network-directed system selection

The network-directed system selection feature is a network capability that allows a service provider, based on various customer and service provider–specified criteria, to automatically direct a subscriber to a desired serving system. The selected serving system could be any system available to the MS, regardless of frequency band (cellular A/B or PCS bands A/B/C/D/E/F) or technology (analog or digital).

Over-the-air activation

The over-the-air activation feature allows the service provider to activate cellular service for a potential subscriber without the need for a third party (e.g., customer service personnel in a retail outlet). These procedures would consist of a call to the service provider customer service center, over-the-air programming of number assignment modules (NAMs), and potentially manufacturer-specific parameters (e.g., lock code, call timer). Cryptographic techniques would be employed to secure the information exchange.

Advanced digital features

The latest members of the digital family of AMPS-based standards, IS-136 for TDMA and IS-95-A for CDMA, include a number of capabilities that are not fully supported in IS-41-C:

- Sleep mode, which provides enhanced battery conservation capabilities
- Network access from private and residential phone systems (see Chap. 17)
- Special community of interest, or user group, services
- Selection of alternate voice coders

International mobile station identifier

The international mobile station identifier (IMSI), defined in ITU-T Recommendation E.212, will be the successor to the mobile identification number (MIN) as the standard subscription identifier for AMPS-based systems. The IMSI is supported in both IS-136 and IS-95-A; as of mid-1996, work was progressing on modifying IS-41-C to also support the IMSI. During the transition to IMSI-only identification, the network signaling protocols will be required to support multiple identification options, say, MSs that use MIN only, IMSI only, or both MIN and IMSI.

Subscriber confidentiality

Subscriber confidentiality affords some protection against cloning fraud by allowing the wireless system to assign the subscriber a temporary mobile station identity (TMSI). Once assigned, the TMSI is used in the subsequent communications over the air interface with the MS, rather than the MIN and ESN.

Lawfully authorized electronic surveillance

Lawfully authorized electronic surveillance (LAES) provides law enforcement agencies access to a cellular subscriber's communications and call-identifying information when lawfully authorized.

PCS operation and interworking

Personal communication service (PCS) operation and interworking entail ensuring that features and other network capabilities operate in the PCS band (1800 MHz) just as they do in the cellular band (800 MHz), and that roaming and handoff between the two bands are also possible. The required changes to IS-41-C are specified in TSB76.

Wireless intelligent network

The wireless intelligent network effort addresses the desire on the part of service providers to define and implement custom features in a multivendor environment without the need for individual feature standardization. The WIN model for feature definition relies on a few key building blocks:

- An *originating basic call state machine* that models the call control activities associated with call origination in the MSC
- A *terminating basic call state machine* that models the call control activities associated with call termination in the MSC

- *Mobility management state machines* that model the mobility management activities associated with a roaming subscriber in the MSC, VLR, and HLR

- *Triggers* (i.e., detection points) that are defined at various points in the originating, terminating, and mobility management state machine processing

- *Standardized messaging* to "arm" the triggers, communicate the occurrence of trigger firing events between functional entities involved, and remotely control the operation of the originating, terminating, and mobility management state machines

As can be readily imagined, this is a sizable undertaking. To focus the WIN effort, work is initially targeted on the following features: voice-controlled services, incoming call screening, and calling name identification presentation.

Voice-controlled services. Voice-controlled services are based on voice recognition technology and allow the cellular user to control features and services by using spoken commands. These services include:

1. Voice-controlled dialing
2. Voice-controlled feature control
3. Voice-based user identification
4. Text-to-speech conversion

Incoming call screening. Incoming call screening provides for alternate routing, blocking, or allowing of specified incoming calls. Incoming calls to a subscriber may be given one of the following terminating treatments:

1. Call is terminated normally to the subscriber with normal alerting.
2. Call is terminated to the subscriber with distinctive alerting or call-waiting tone.
3. Call is forwarded to voice mail.
4. Call is forwarded to another number.
5. Call is routed to a subscriber-specified announcement.
6. Call is blocked (e.g., call routed to a standard announcement or tone).

Calling name identification presentation. Calling name identification presentation gives the name of the calling party and the date and time of the call. The information is presented to the called user during the first ringing cycle.

Big Bang versus Piecemeal Evolution

Revision C can be considered a "big bang" in the evolution of the IS-41 standard. Overnight (actually, over several nights), the standard mushroomed in size and capabilities. Based on the current mood in the TR45.2 subcommittee, it is unlikely that such a dramatic change will occur in the standard again, at least not in the near future. Instead, a more "piecemeal" approach is being attempted, with individual features or groups of related features planned for separate standardization. In fact, according to David Crowe's *Cellular Networking Perspectives* newsletter, "The scope and purpose of IS-41 Rev. D are currently under review. Proposals have been made to limit IS-41 to a 'core' protocol, and publish features in standalone documents."*

Whatever the outcome of the debate, we can be certain of one thing: IS-41 will continue to reflect the network interoperability needs of carriers and manufacturers well into the next millennium.

Cellular Networking Perspectives, vol. 5, no. 8, August 1996.

IS-41 Network Implementations

16

IS-41 Interoperation
with Other Networks

IS-41 cellular networks interoperate with each other to provide nationwide roaming services to mobile subscribers. They also interoperate with wire line networks to support calls between cellular and wire line parties. These calls can be traditional voice calls or data calls that can convey circuit-switched or packet-switched data. IS-41 network interoperation involves the interconnection of trunks and signaling links between the networks as well as business agreements between the service providers. This chapter describes the interconnection of IS-41 networks with other cellular and wire line networks.

Roaming Agreements

A roaming agreement is simply a legal and business contract entered into by two cellular network service providers. This contract defines the methods by which one cellular network provides service to subscribers from the other cellular network. It also defines the tariffs and the methods for billing of roaming subscribers. Roaming agreements are also the basis for proper routing of signaling information between the two networks. Based on these agreements, provisionable tables in the network equipment are populated with information that enables subscribers to register and receive services from a visited system (see Fig. 16.1).

Roaming agreements permit the IS-41 signaling communications between cellular networks that provides the mobility management and call processing functions. Without these agreements, a subscriber may be prohibited from registering on a visited system and receiving

cellular network X

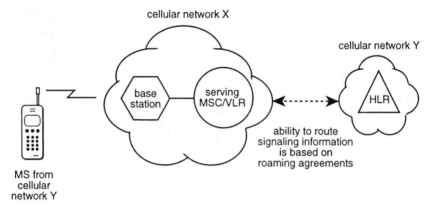

Figure 16.1 The ability to route signaling information and provide mobility management functions (e.g., registration, location management, service qualification) is based on a roaming agreement. An MS from cellular network Y can only receive seamless roaming service from cellular network X if a roaming agreement exists between the two networks.

any cellular service. Some networks support *indirect* roaming. This type of roaming occurs when there is no explicit roaming agreement between the visited system and the subscriber's home system. The subscriber can still be afforded service, but it may not be seamless. When a subscriber attempts to register on a visited system, that system may not be able to route the registration signaling information to the subscriber's home network (i.e., the HLR), since there is no roaming agreement. In this case, the visited system can perform one of three actions:

1. Deny service to the visiting subscriber.

2. Provide service to the visiting subscriber, by obtaining credit information directly from the subscriber.

3. Route the signaling information about the visiting subscriber to a clearinghouse network, which may have a roaming agreement with the subscriber's home system.

The most undesirable action by the visited system is to deny service to the subscriber. The subscriber suffers, and there is no revenue generated for either the home or the visited network.

When a roaming subscriber attempts call origination and the visited system does not recognize the subscriber's MIN, it can direct the subscriber to a customer service center. This customer service center typically requests credit information from the subscriber so that calls can be originated and billed directly to the subscriber (see Fig. 16.2).

cellular network X

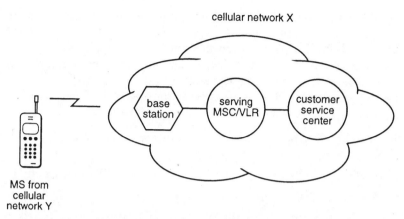

Figure 16.2 If no roaming agreement exists, the serving system can request credit information from the subscriber and allow the subscriber to originate calls; however, mobile-terminated calls will not be delivered to the subscriber.

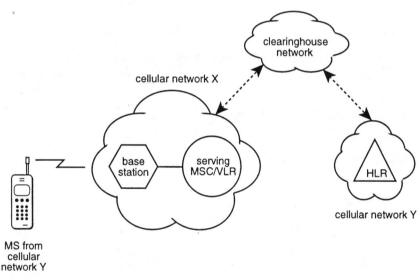

Figure 16.3 A clearinghouse network can provide roaming capabilities to an MS visiting a system that has no roaming agreement with the MS's home service provider system.

In this case, the location management and service qualification processes are not performed, and mobile-terminated calls cannot be delivered to the subscriber.

Another option is based on an agreement the visited system has with a clearinghouse network (see Fig. 16.3). Signaling information for all unrecognized mobile subscribers is routed to the clearinghouse network. If the clearinghouse network has an agreement with the subscriber's home system, it can support the mobility management functions on behalf of the visited system. If the clearinghouse network does not have an agreement with the home system, it can obtain credit information directly from the subscriber.

Roaming agreements define the interactions permitted between cellular systems. Some examples of these interactions are:

- Methods for routing signaling information about roaming subscribers

- Methods for the networks to settle accounting and billing information

- Methods for recording call details to support accounting and billing

- Methods for handling toll calls (i.e., use of interexchange carriers for calls)

- Methods for handling fraud (i.e., responsibility for and methods to avoid lost revenue due to roamer fraud)

- Methods for updating routing tables

- Methods for sharing statistical data about roaming subscribers

Roaming agreements also require that technical information be shared about the cellular networks involved. Examples of this information are:

- MSC NPA-NXX codes supported

- MSC identification codes supported

- SS7 point codes

- Switch makes and models

- Software generic versions

- Feature code values supported

Roaming agreements provide the basis for service providers to offer nationwide cellular service to subscribers. Without these agreements, service providers would need to operate their own cellular networks in every area to provide nationwide coverage for subscribers.

Network Interconnections

Whenever two telecommunications service providers connect to each other to pass signaling information, deliver calls, or do both, they enter into an *interconnection agreement*. The interconnection agreement defines methods used to provide communications between the two networks.

Cellular networks can interconnect with a variety of other networks including the following:

- Public switched telephone networks (PSTNs)
- Other cellular networks
- Personal communications service (PCS) networks
- Public switched packet data networks (PSPDNs)
- Signaling System No. 7 (SS7) *backbone* networks

Cellular networks directly interconnect with the PSTN to support land-to-mobile and mobile-to-land calls. The PSTN may also be used to support mobile-to-mobile calls. The PSTN comprises local exchange carriers (LECs), interexchange carriers (IXCs), and other wire line carrier networks. Some cellular networks directly interconnect with other cellular networks to support mobile-to-mobile calls. These interconnections are sometimes used in highly populated areas where there is a potential for many mobile-to-mobile calls between different cellular service providers serving the same area. Cellular networks will also interconnect with PCS networks. From a network perspective, there is no difference between PCS and cellular network operation if they are both based on IS-41 signaling. Cellular network interconnection with a PSPDN is also possible if the cellular network supports packet data services.

Most cellular networks use SS7 to provide call control and IS-41 signaling. These cellular networks directly interconnect with an SS7 backbone network. This backbone network can be owned and operated by the cellular service provider itself, or it can be provided by a separate independent SS7 network provider.

The interconnection between an individual cellular network and other networks is primarily defined by three criteria:

1. Physical communications facilities
2. Communications protocol supported across the physical facilities
3. The type of information transferred by the communications protocol

The physical communications facilities consist of the lines, trunks, and data links directly provided between the network nodes of the different service providers. The communications protocols supported range from the physical protocol layer through the application protocol layer. For example, a trunk interconnection may support T-1 transmission, SS7, and IS-41 MAP signaling. The type of information transferred across an interconnection is limited by the protocol supported. The protocol, however, may support optional information that is only provided under agreement between the two networks.

A_i-D_i interfaces

The interconnections between a cellular service provider and another network are specified in the standard *TIA/EIA IS-93 Cellular Radio Telecommunications A_i-D_i Interfaces Standard*. This standard provides signaling protocol requirements for interfaces that interconnect a switching system in a cellular network with a switching system in another network. The A_i and D_i interfaces represent the interfaces from the MSC to the PSTN and ISDN, respectively, in the IS-41 network reference model. For all intents and purposes, these two interfaces are the same. The A and the D in the interface names stand for *analog* and *digital*. The distinction between these two interfaces has not been specified and is left over from an anticipated need to differentiate them. Since the PSTN supports both analog and digital protocols, there is no need to distinguish between interfaces to an analog and digital network.

The interconnection types specified in *TIA/EIA IS-93* define generic interfaces that support a number of individual signaling protocols with a variety of signaling methods to provide telecommunications services. The standard specifies two subtle characteristics of the cellular-to-PSTN interface that distinguish it from other PSTN interface specifications:

1. The interfaces are bidirectional and symmetric.

2. The interfaces support the transport of automatic number identification (ANI) information into the cellular network.

Bidirectional and symmetric interfaces mean that a given interface can support the same address and signaling information in both directions (i.e., either entering or exiting the cellular network). The transport of ANI information into the cellular network is an important business issue concerning network interconnections. ANI represents charge number information that can be received with a call from the originating or a mediating network. This information, when available to a terminating network, allows for control over billing and the ability to provide additional services that can be charged for by the terminating network. Note that the ANI information may or may not be the same as the calling party number. Typically, the calling party number represents the number to charge a call to; however, this may not be the case (e.g., calls from businesses with multiple calling party numbers, but one billing number). It should never be assumed that the ANI is always the same as the calling party number or vice versa.

Prior to the publication of *TIA/EIA IS-93*, interconnections to the PSTN were specified by Bellcore. In these specifications, ANI (or charge number) information was never included in the signaling information provided to the cellular networks for mobile-terminating calls. The reason for this is historical and to the advantage of the PSTN and, especially, the local exchange carriers that delivered calls to the cellular network. Cellular networks were traditionally treated as extensions to the PSTN, similar to the way PBXs are treated.

The cellular service providers originally depended on the PSTN to carry mobile-to-land calls, the most common call type. This dependency came from the fact that cellular service providers needed to only provide MSC capabilities in their service areas and then use the existing nationwide switching infrastructure of the PSTN to carry the calls. Because of this dependency, cellular networks were considered subordinate to the PSTN and hence were never given charge number information that would allow them control over services and billing arrangements. The primary purpose of *TIA/EIA IS-93* was to provide cellular service providers with a nationally standardized interface specification that supported the transport of charge number information into their networks. This important change enables the cellular networks to gain a certain degree of "economic equality" with the PSTN and to negotiate interconnection agreements that are favorable to both networks.

TIA/EIA IS-93 specifies the logical interface reference model used for cellular-to-PSTN interconnections (see Fig. 16.4). This model represents a generic interconnection between a cellular network element and any other network element. This enables the application of the interface types between a variety of network elements. The trunk point of interface (POI-T) is designed to convey trunk user traffic and in-band multifrequency (MF) signaling. The SS7 point of interface (POI-S) is used in conjunction with POI-T to represent out-of-band SS7 signaling used on the interface.

The interface types specified in *TIA/EIA IS-93* are distinguished by the telecommunications protocol supported and the information conveyed with that protocol. The information defines the application of the interface type. For example, some interfaces support general trunk access for interexchange carrier calls, and others support emergency services calls. These interconnections require different information to be transferred to provide the appropriate service.

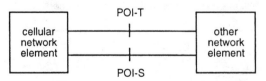

Figure 16.4 Generic interface reference model for cellular interconnection to the PSTN. The trunk point of interface (POI-T) specifies trunk interconnections to convey user traffic, while the SS7 point of interface (POI-S) specifies SS7 interconnections.

Of the interface types specified in *TIA/EIA IS-93,* there are primarily only three in practice:

1. Trunk with line treatment
2. General trunk access
3. Direct trunk access

These three interface types are the same as the original interface types specified by Bellcore (*Compatibility Information for Interconnection of a Wireless Service Provider and Local Exchange Carrier Network TA-NPL-000145*), with one difference: they are bidirectional and symmetric. The other interface types specified in *TIA/EIA IS-93* were added to provide various forms of flexibility to the cellular service providers while allowing them to negotiate and charge a tariff for each type of information that is terminated in their networks. The three primary interface types are really supersets of all the information that is required over a cellular-to-PSTN interconnection. Table 16.1 shows the mapping of these three interface types to the Bellcore-specified interface types.

The trunk with line treatment interface type uses MF signaling to establish a connection to an end office to access valid directory number addresses which are directly connected to that end office. The primary signaling information included in the address signaling sequence to obtain access to network addresses is the called-party number. Charging information is not exchanged over this interface type (see Fig. 16.5).

The general trunk access interface type uses MF or SS7 ISUP signaling to establish a connection to any common carrier switch to access any valid network endpoint accessible by that common carrier switch. The primary signaling information included in the address signaling sequence to obtain access to network addresses is the called-party number, charge number, originating line information, and a carrier identification code (see Fig. 16.6).

TABLE 16.1 Mapping of *TIA/EIA* IS-93 Interface Types to the Bellcore-Defined Interface Types

Interface name	IS-93 type(s)	Bellcore type
Trunk with line treatment	POI-T1	Type 1
General trunk access	POI-T4, POI-T5 & POI-S5	Type 2A
Direct trunk access	POI-T6, POI-T7 & POI-S7	Type 2B

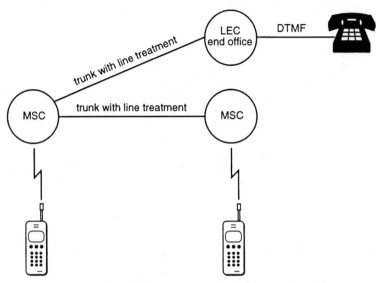

Figure 16.5 The trunk with line treatment interface type. The interconnection allows access to directory numbers directly accessible by the interconnected switches only.

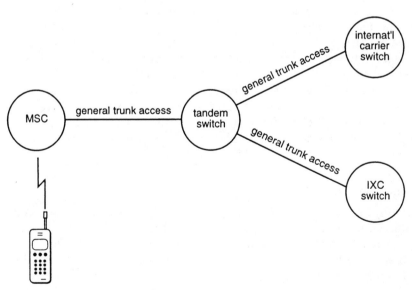

Figure 16.6 The general trunk access interface type. The interconnection allows access to any endpoint accessible by the connected switches.

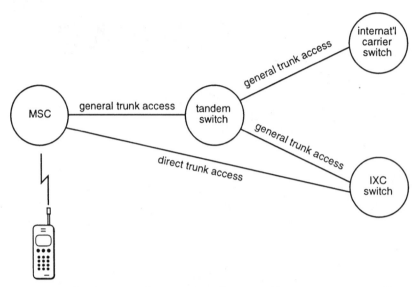

Figure 16.7 The direct trunk access interface type. The interconnection allows direct access to directory numbers served by the connected switch.

The direct trunk access interface type uses MF or SS7 ISUP signaling to establish a direct connection to any common carrier switch to access any valid directory numbers accessible by that common carrier switch. The directory number address space is restricted to local, national, and international directory numbers. This interface type is designed for high-volume traffic routes in conjunction with the general trunk access interface type only. For example, MSCs that carry a large volume of traffic to a single IXC may add this type of trunk specifically for the purpose of accessing directory numbers accessible by that IXC only. The primary signaling information included in the address signaling sequence to obtain access to network addresses is the called-party number (see Fig. 16.7).

SS7 network interfaces

Although some older North American IS-41 networks use the X.25 protocol for signaling message transport, the majority use SS7. And SS7 is not just a signaling and transport protocol; SS7 connectivity defines an architectural strategy to provide a robust, highly reliable, and highly available signaling network.

Basic SS7 network configuration. The basic North American SS7 network configuration is shown in Fig. 16.8. The end signaling points (SPs) represent the signaling capabilities of switches that can originate

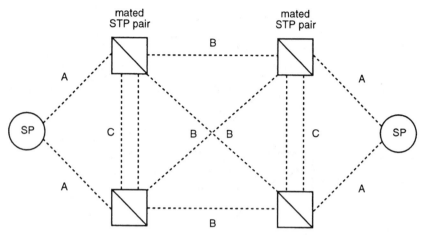

Figure 16.8 Basic SS7 configuration for North America. The architecture uses quasi-associated signaling to create a signaling relationship between end signaling points.

or terminate signaling traffic. The signaling transfer points (STPs) route signaling traffic through the network by transferring that traffic from one signaling link to another to direct it toward a particular destination. This configuration is known as *quasi-associated signaling,* since the signaling relationship between the end SPs is not a direct association (i.e., there are no direct links between end SPs).

The STPs are deployed as a load-sharing mated pair, with each STP carrying half the traffic load. In case of a failure, the remaining STP is capable of carrying the full traffic load. An STP can be implemented as a stand-alone system or can be integrated with a switch.

End signaling points are connected to the STP mated pair by a pair of *A links.* Each of these A links is a member of a *link set.* The pair of link sets containing the A links is known as a *combined link set.* The signaling traffic load is shared equally among the links of the combined link set. Mated pairs of STPs are connected by C links. Connections between STP pairs of different SS7 networks (i.e., gateway STPs) are made with B links. There is functionally no difference between the link types. They simply define different types of traffic that can traverse them since they connect different types of SS7 nodes. Some SS7 networks employ D links, which can define multiple levels of STPs, for example, local and regional STP mated pairs (see Fig. 16.9). E links are used to avoid massive failures or outages of STP mated pairs. They connect an end signaling point to a remote STP pair (in addition to the local pair) that can be within the same SS7 network or in a different network (see Fig. 16.10).

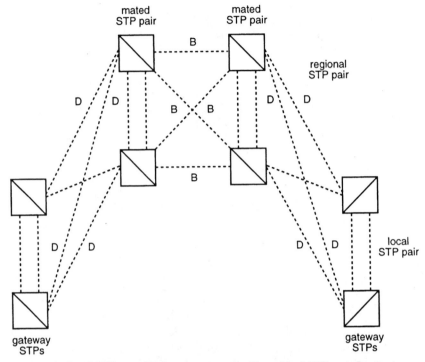

Figure 16.9 Local STP mated pairs can connect with regional STP mated pairs to create a hierarchy for the SS7 network topology.

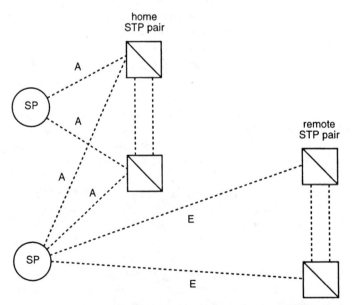

Figure 16.10 End SPs can connect with more than one STP mated pair. The remote pair is used to avoid failures of the local STP pair.

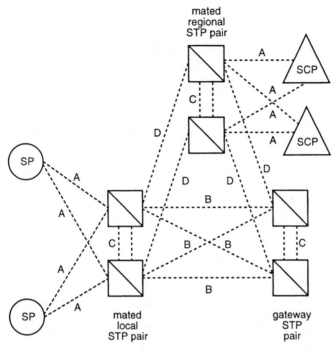

Figure 16.11 SCPs are usually connected to regional STP pairs for concentration of signaling traffic. The SCPs are connected by A links since they are considered to be end signaling points.

Service control points (SCPs) are special types of end signaling points that use the TCAP protocol to perform transaction processing of remote operations. In many cases, the SCP supports a database to perform the required operations. SCPs are typically connected to an STP mated pair by a pair of A links (see Fig. 16.11). SCPs can also be deployed as a redundant pair for higher reliability.

MSCs connect with SS7 networks as end signaling points. An MSC may be part of the cellular service provider's own SS7 network, or it may connect to another service provider's SS7 network. Typically MSCs connect to a local gateway STP mated pair using A links. Analog or digital trunks connect the MSC to common carrier switches for call traffic (see Fig. 16.12).

Independent SS7 networks

Many cellular service providers connect to independent SS7 networks to carry IS-41 signaling traffic, rather than operate their own backbone signaling networks (see Fig. 16.13). The largest cellular service providers in the United States do own and operate their own SS7 net-

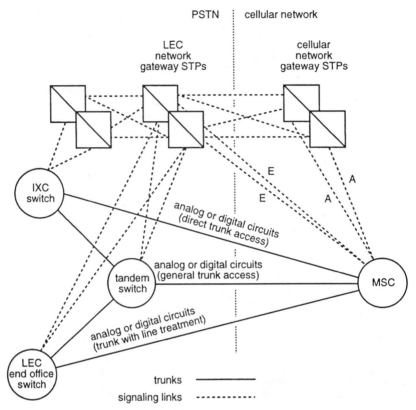

Figure 16.12 MSCs connect to the SS7 network as end SPs. The cellular SS7 backbone network connects to the PSTN SS7 backbone network through gateway STPs. Note the use of the *TIA/EIA IS-93* interface types for trunk connections to the PSTN.

works, but this may not be advantageous for smaller cellular service providers. The independent SS7 network service providers coordinate and manage the SS7 network as well, allowing the cellular service provider to concentrate on managing the cellular infrastructure.

Cellular service providers can interconnect their MSCs directly to gateway STPs of the independent SS7 network (see Fig. 16.13). The cellular service providers may also operate part of an SS7 network themselves and interconnect their local STP mated pair to the gateway STPs of the independent network (see Fig. 16.14). This enables cellular service providers to access signaling destination points that may be inaccessible through their own networks, such as international signaling points.

In July 1993, the Cellular Telecommunications Industry Association (CTIA) named Independent Telecommunications

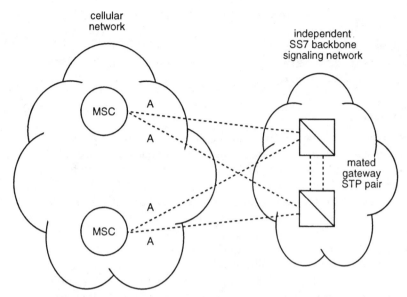

Figure 16.13 Cellular network MSCs can connect to an independent SS7 network directly via A links to the gateway STP.

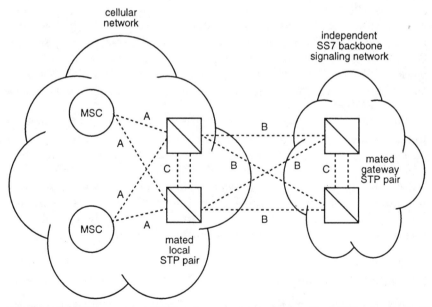

Figure 16.14 Cellular network MSCs can connect to an independent SS7 network via B links between two sets of gateway STPs, one set operated by the cellular network and the other set operated by the independent SS7 network.

Network, Inc. (ITN, now known as Illuminet) the official CTIA SS7/IS-41 backbone signaling network provider for cellular service providers. Of course, there is no requirement to use Illuminet's SS7 network, but they are among the largest SS7 network service providers in North America.

Cellular resellers

Cellular resellers are cellular service providers that do not operate their own networks. Resellers make arrangements to prepurchase a percentage of the available capacity of a service provider's facilities in bulk and at a discount. These facilities generally include radio channels and switching systems, but they can include any location registers and processing centers in the service provider's network. The reseller provides service under a different brand name, using the identical infrastructure as the service provider.

The reseller obtains a block of mobile identification numbers (MINs) from the cellular service provider. These are the numbers that the reseller provides to subscribers of the reseller service. Since the MINs fall within the range already supported by the original service provider, no routing changes are required to provide service to subscribers. The reseller's subscribers may also be supported by the service provider's HLR. The HLR can be partitioned to support the block of the reseller's subscribers. Alternative architectures may include a separate HLR operated by the reseller, but this would affect routing in the network and would require additional roaming agreements beyond those supported by the original service provider. From a network perspective, cellular resellers are typically distinguishable from the original service provider only by different customer service centers and different billing systems.

Cellular reselling is typically very advantageous to the service provider due to guaranteed revenue when the facilities may not have otherwise been in use. It is also advantageous to the reseller, which is afforded the ability to provide cellular service and reap profits without owning and operating a network. The reseller may even compete for subscribers with the service provider providing the facilities, although the agreement between the companies usually prohibits the direct luring of subscribers from each other's service.

International Roaming

With the explosive growth of the global wireless market, it is very desirable to provide seamless automatic roaming internationally. International roaming, from a North American perspective, includes the ability of either a North American subscriber to seamlessly roam to other countries or subscribers from other countries to seamlessly roam in North America. From a cellular network perspective, North America consists generally of the United States and Canada, since these countries both use common administration of numbering and addressing. International roaming in this context refers to roaming between IS-41 networks in different countries.

Numbering problems

Mobile stations in North America are always identified by a 10-digit mobile identification number. The MIN happens to also be used as the 10-digit subscriber directory number which follows the North American Numbering Plan (NANP). The NANP is a subset of the international ISDN numbering plan standardized in *ITU-T Recommendation E.164*. The international E.164 plan includes a country code (CC), a national destination code (NDC), and a subscriber number (SN). Together the NDC and the SN comprise the national significant number (NSN). The 10-digit MIN complies with the international NSN format of E.164 (see Fig. 16.15).

The primary problem arises because the 10-digit MIN follows only a national numbering format and contains no country code. It is impossible to distinguish some 10-digit MINs used in North American IS-41 networks from a 10-digit MIN that includes a country code that is used in other countries. Therefore, it becomes difficult to register an international subscriber in North American networks; the country code of the MIN could be construed as part of the numbering plan area (NPA) or area code, and the wrong HLR could be queried for MS service qualification. Similarly, a North American subscriber roaming internationally would have the first few digits of the MIN construed

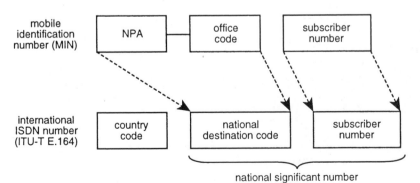

Figure 16.15 The 10-digit MIN complies with the format of the national significant number field of the E.164 international ISDN numbering plan.

as a country code, and thus registration would also be attempted at the wrong HLR.

An excellent example of this problem is with Mexico. The country code in Mexico is 52. However, a valid area code in Arizona is 520. A Mexican roamer in Arizona would attempt a call and be denied, since service qualification would be attempted in Arizona instead of the subscriber's HLR in Mexico. In the short term, Mexico has agreed not to assign any MIN values beginning with 520; however, this problem exists for the country code values of dozens of countries.

This problem has been slightly alleviated by *TIA/EIA TSB29-A,* which defines a numbering plan for mobile stations that are outside of world zone 1, of which the United States and Canada are members. World zones are specified by *ITU-T E.164.* They categorize each of the countries of the world into one of nine zones. The zones cover large contiguous geographic areas.

TIA/EIA TSB29-A specifies that non-world zone 1 mobile stations be programmed with MINs that are formatted like a 10-digit international mobile station identification (IMSI) number as specified in ITU-T Recommendation E.212 (see Chap. 15). *TIA/EIA TSB29-A* specifies that the first three digits of the MIN represent a mobile country code, the fourth digit represents a mobile network code (specified as either 0 or 1), and the last six digits represent the subscriber number. Since the fourth digit is 0 or 1 (which is not allowed in the NANP), the MIN can have a unique address space that is distinct from all other NANP-based MIN values (see Fig. 16.16).

The best solution to the international numbering problem is to expand the length of the MIN to at least 12 digits to accommodate a country code value. This would enable true seamless automatic roaming using existing internationally standardized number formats. The

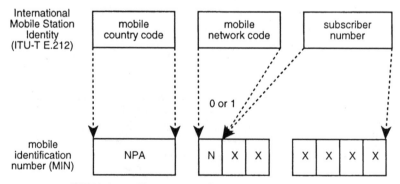

Figure 16.16 TSB29-A specifies a unique format for the MIN that can be recognized as international for roaming purposes.

problem is that this requires major changes throughout the cellular networks, and there are tens of millions of existing subscribers with mobile stations containing 10-digit MINs.

Another solution that has been proposed is known as HLR *double dipping*. This means that when an MS registers with a MIN that may be ambiguous, sequential queries are made from the serving system to the two HLRs that both maintain the ambiguous MIN. This solution has drawbacks, though, since extraneous signaling traffic is never desirable.

International call delivery problems

Another problem with international roaming involves the call delivery procedures used to deliver mobile-terminated calls to roaming subscribers. In IS-41-A and IS-41-B systems, the temporary local directory number (TLDN) used to redirect a call from an originating MSC to a serving MSC is commonly implemented as a 10-digit number (IS-41-A and IS-41-B do not specify minimum and maximum lengths for the TLDN digit string). This poses a problem when calls need to be redirected from the subscriber's home system to another country where the serving system resides. A 10-digit TLDN may not clearly identify the country for the call to be delivered to. Another problem for the originating MSC is knowing when to add an international dialing prefix code to the TLDN to properly deliver the call via an international gateway. Note that in IS-41-C the TLDN is explicitly stated to be between 0 and 15 digits long. This solves the problem as long as all serving systems providing call delivery are compliant with this revision.

The best long-term solution to this and other numbering problems involving international roaming is the implementation and deployment of the IMSI (see Chap. 15). This value would decouple the subscriber's dialable directory number from a unique internationally used identification number which can be used to route signaling information.

IS-41/GSM Interworking

With the growth of personal communications service (PCS) systems in North America, competing mobile networking and radio technologies for systems based on cellular technology have emerged. In North America, the primary cellular telecommunications networking technology is IS-41, supporting AMPS, NAMPS, TDMA, and CDMA radio technologies. The advent of PCS has provided the opportunity for the North American standardization of the Global System for Mobile

(GSM) communications, an alternative international cellular technology standard. GSM comprises both networking and radio technologies that are not directly compatible with IS-41 and its associated radio technologies.

GSM was originally standardized in Europe by the European Telecommunications Standards Institute (ETSI). The following versions of GSM have been standardized:

1. GSM 900 (GSM at 900 MHz, international)

2. DCS 1800 (GSM Digital Cellular System at 1800 MHz, international)

3. PCS 1900 (GSM Digital Cellular System at 1900 MHz, network and radio technology for North America)

For North American PCS, the PCS 1900 network and radio technology has been selected for use by many PCS service providers. However, IS-41 technology along with its associated radio technologies has also been selected for use by many other PCS service providers. IS-41 networking for PCS has the advantage of being fully compatible with existing North American cellular networks providing nationwide cellular coverage. The GSM application-layer MAP protocols to provide mobility management, radio system management, and call processing are incompatible with the IS-41 MAP protocol providing those functions. The radio technologies are also quite different; although PCS 1900 is a type of time-division multiple access, it is incompatible with North American TDMA.

PCS service providers implementing new GSM technology initially will have limited geographic coverage of their systems due to the time and cost of deploying equipment nationwide. Because of this lack of immediate nationwide coverage, GSM service providers are finding it desirable to interwork with IS-41 networks to provide full North American coverage for their subscribers (see Fig. 16.17 and 16.18).

Introduction to interworking

There are two separate areas of system interworking that apply to GSM and IS-41 interoperation: radio technology interworking and network interworking. *Radio technology interworking* refers to the capability of a mobile station (MS) to interoperate between PCS 1900 radio technology and one or more of the North American radio technologies (i.e., AMPS, NAMPS, TDMA, or CDMA) to access either a PCS 1900-based or IS-41-based network to receive service. This interworking is not feasible; however, a reasonable solution is a dual-mode MS similar to dual-mode AMPS/TDMA mobile stations used today.

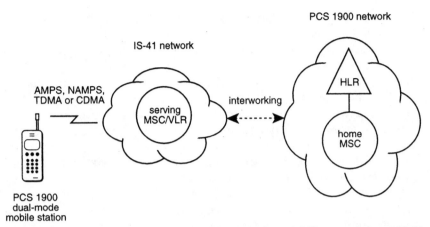

Figure 16.17   An interworking scenario where a PCS 1900 MS roams into an IS-41 network service area. Interworking is required to support the intersystem signaling operations between the two networks to support MS service qualification, location management, and many other functions.

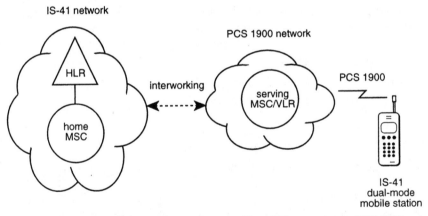

Figure 16.18   An interworking scenario where an IS-41 MS roams into a PCS 1900 network service area. Analogous interworking is required to that shown in Fig. 16.17.

The term *dual-mode* implies a "split-personality" MS that would be capable of operating in either of two modes based on the type of network access required. A dual-mode PCS 1900/AMPS MS would enable a single MS to access PCS 1900 networks in those service areas and IS-41 networks where PCS 1900 is not available. In fact, even multimode MSs have been studied to provide access for PCS/1900, AMPS, and either TDMA or CDMA. Multiple radio technologies can coexist (if not interwork), provided that the network technology adopted can simultaneously support the many network access technologies.

Network interworking introduces an entirely different set of problems to be solved. If two networking technologies are highly incompatible, it is difficult for a network service provider to offer the primary service need for their subscribers: seamless automatic roaming.

It is somewhat obvious that the best solution to incompatible networking protocols is to not require any interworking at all! A single MAP protocol supporting the many network access protocols (i.e., radio technologies) is truly the optimal solution to seamless automatic roaming. However, business issues often times outweigh the optimal engineering solutions, and these solutions are not always the most desirable in a fast-moving, high-growth industry like wireless telecommunications.

There are many examples of network protocol interworking, such as TCP/IP-to-X.25 and X.25-to-SS7. The problem in mobile networks is the complexity of the interworking; it is not simply the mapping of data packets and addresses to provide transport, although this is part of it. The primary problem is the interworking of the application-layer MAP protocols, along with the handling of the mobility management, radio system management, and call processing functions.

The interworking problem

The best method for approaching the interworking problem is to break it down into the following three steps:

1. Define interworking.

2. Identify the functions requiring interworking.

3. Design the methods to provide interworking for these functions.

Definition of interworking. The initial step to providing interworking between GSM-based and IS-41-based systems is to define *interworking* and the types of interworking to be supported. For the purposes of this discussion, we will address the problem of MAP interworking between PCS 1900 and IS-41. There is no standard definition for the term *interworking*. Instead of attempting to define such an all encompassing term, defining what an interworking solution needs to provide for this specific problem is more appropriate.

We will use the following assumptions to address the MAP interworking problem:

1. *Interworking implies the successful communication between PCS 1900 and IS-41 MAP protocols.* A more specific definition might limit a feasible solution to be called interworking.

2. *Interworking provides a degree of seamlessness to subscribers capable of transiting between the PCS 1900 and IS-41 networks.* Subscribers capable of roaming between the two networks should be able to access basic service seamlessly (i.e., originate and receive calls as usual). Since many call features are not compatible, their use may not be supported while roaming.

3. *It is not feasible to provide an interworking solution enabling full compatibility between PCS 1900 and IS-41 networks.* Since IS-41 and PCS 1900 networks are so different, it is not reasonable to provide interworking of all the functions of both networks to each other.

4. *Transmission facilities between PCS 1900 and IS-41 networks are compatible.* This is a reasonable assumption based on North American transmission standards.

5. *Signaling protocols providing transport between PCS 1900 and IS-41 networks are compatible.* This is a reasonable assumption based on North American signaling standards.

6. *Dual-mode PCS 1900/AMPS mobile stations are used to access PCS 1900 and IS-41 networks.* Without a dual-mode MS, it is not feasible to provide access to both network types with a single compatible interface.

7. *The subscriber has a single subscription allowing access to both networks (rather than two subscriptions, one to each network, which requires no interworking).* From a subscriber perspective, it is unreasonable to support two different subscriptions from two different networks.

These assumptions simplify the basic problem and permit a basic solution to be implemented that can evolve so that these limitations can be removed.

Functions requiring interworking. The next step toward solving the interworking problem involves identifying the set of operational functions to be provided across the two networks. The most necessary functions fall within the scope of mobility management, radio system management, and call processing. These functions include:

1. Mobile station service qualification

2. Mobile station location management

3. Mobile station state management

4. Authentication

5. Intersystem handoff

6. Call origination

7. Call delivery

8. Call features and control

Since PCS 1900 and IS-41 are such different systems, and interworking to provide some functions is so complex, an approach that prioritizes these functions into deployment phases can be used. Nearly all these functions require primarily HLR procedures to be performed. One of the functions that requires more processing and greater interaction between functional entities is intersystem handoff. This is due to the complexity of handing off a call between two different radio technologies and the changes required to support those technologies at the anchor and target MSCs. If a dual-mode MS is used, each mode can operate independently to provide the listed functions except for intersystem handoff.

In high-population cellular coverage areas, it is likely that both PCS 1900 and IS-41 networks will be deployed. Since an intersystem handoff between the two networks would be an uncommon event (i.e., only where the coverage areas of the two networks do not intersect), intersystem handoff can be considered a lower-priority function.

The other low-priority functions are the call features that PCS 1900 and IS-41 do not support in common. Implementation of these features also involves radical changes in many functional entities. The features that PCS 1900 and IS-41 do support in common are call forwarding (i.e., unconditional, busy, and no answer), call waiting, and three-way calling.

Although total compatibility between PCS 1900 and IS-41 is lacking, there is still benefit to providing basic mobile functionality for subscribers who can only receive nationwide coverage, at least in the short term, through interworking.

Methods to provide interworking. To provide interworking between PCS 1900 and IS-41 MAP protocols, three areas need to be addressed: protocol conversion, database mapping, and transaction management. Protocol conversion provides the translation of messages and parameters from one protocol to the other. Database mapping provides the translation and management of information elements that allow each of the application protocols to provide user service (e.g., subscriber identification, location, and status information). Transaction management enables the completion of queries between the two networks (e.g., reoriginating and maintaining queries and responses from one network to the other).

Interworking solutions

The basic model of a solution is a logical functional entity that performs the appropriate protocol conversion, database mapping, and transaction management to support the mobility management, call origination, and call delivery functions. This interworking function

(IWF) provides the logical interface between PCS 1900 and IS-41 networks (see Fig. 16.19).

There are many implementations that can be derived from this logical reference model. Note that this model does not necessarily imply the placement of the IWF outside both networks (see Fig. 16.20). The IWF can support interworking between any network elements where there is desired interworking functionality. All physical implementa-

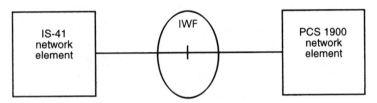

Figure 16.19 A logical interworking function (IWF) can support intersystem operations required between the IS-41 and PCS 1900 networks.

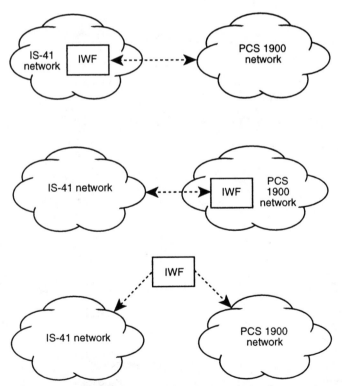

Figure 16.20 The IWF can reside in either the PCS 1900 or IS-41 networks or between the networks.

tions of the IWF need to provide the three basic functions of protocol conversion, database mapping, and transaction management.

Interworking between PCS 1900 and IS-41 networks to provide the identified mobility management, call origination, and call delivery functions involves the serving MSC/VLR and the HLR (see Fig. 16.21). Signaling is required between the HLR in one network and the MSC/VLR of the other to support all the functions identified for interworking. Trunk connections are required between the MSC/VLR and the PSTN to support call origination, call delivery, and the call features and control.

The most basic physical implementation of the IWF is known as the *dual-mode HLR*. This network node is essentially a complete IS-41-based HLR conjoined with a complete PCS 1900 HLR. Each "side" of the HLR adequately serves that "side" of the network. Internally, the protocol conversion, database mapping, and transaction management functions can be performed to support roaming between the two networks. However, the dual-mode HLR solution may be considered

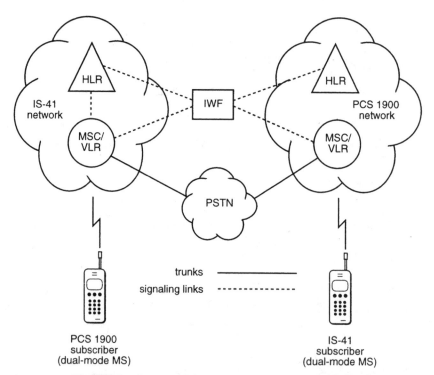

Figure 16.21 The IWF involves signaling connections to the serving MSC/VLR and the HLR of both networks to interwork the intersystem operations. Trunk connections to the PSTN are required for call origination and call delivery.

"overqualified" as an IWF. It is essentially the implementation of a PCS 1900 HLR inside an IS-41 HLR or vice versa. Although conceptually this seems to be a solution that intrudes the least on both networks, it is a very expensive implementation and can take many years to design, develop, test, and deploy (see Fig. 16.22). It also requires the replacement of existing HLRs in each network with this new dual-mode design.

An alternative—and in some cases preferable—solution is to implement protocol conversion, database mapping, and transaction management for all the functions supported between the networks as a single separate network element, as shown in Fig. 16.21. It provides a basic IWF solution that requires no changes to existing IS-41 or GSM networks and that can reside external or internal to either of the networks. The IWF, in this case, appears as the HLR to the network serving the roaming subscriber, and it also appears as the serving VLR to the subscriber's home network. Intersystem operations can then be routed to and translated by the IWF and sent to the "real" elements in the other network.

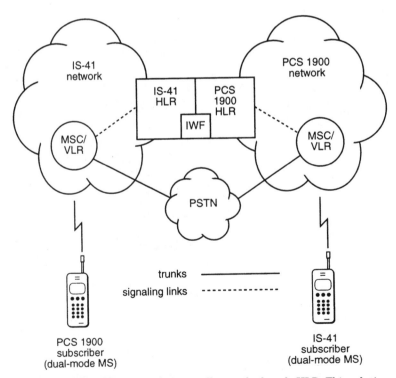

Figure 16.22 The IWF can reside internally to a dual-mode HLR. This solution tends to be prohibitively expensive and can take many years to develop.

Generally, interworking between PCS 1900 and IS-41 networks does not provide the best of what cellular technology and PCS potentially have to offer. There is limited functionality that can reasonably be performed between the two networks. However, it is a method of supporting nationwide roaming to subscribers of PCS 1900 networks until those networks provide nationwide coverage themselves.

17

IS-41 Implementations

The IS-41 standard does not specify physical implementations. It does, however, provide the intersystem operations that make many implementations possible. These implementations are generally based upon particular network architectures and the means for routing signaling messages between the network elements. Although a variety of implementations are possible, this chapter describes the more common and interesting implementations of IS-41 to provide mobile telecommunications services to subscribers.

Use of the Mobile Identification Number

A common misunderstanding about the MIN is that it must also be the directory number (DN), or phone number of the MS. The IS-41 standard evolved from many concepts applied to wire line networks. For wire line networks, the DN of a subscriber is also representative of the line number used to identify the line from the central office to the telephone. The cellular network in North America evolved from the same principle; thus the DN of the subscriber was implemented to represent the unique number used by the network to identify the subscriber. In practice today, the MIN and the DN are the same number; however, IS-41 never specifies the MIN as a DN. Implementations that provide for a separate DN from the MIN are not at all precluded by the IS-41 standard.

Directory numbers are phone numbers that can be dialed from anywhere. They follow the North American Numbering Plan (NANP) which is a subset of the international *ITU-T E.164* numbering plan. The DN is a 10-digit decimal number following the format NXX-NXX-XXXX, where N is any number from 2 to 9 and X is any number from

0 to 9. This format also indicates a geographic address. The first three digits (NXX) indicate the area code, the next three digits (NXX) indicate the exchange or office code, and the last four digits (XXXX) specify an individual subscriber.

DNs are used to call cellular mobile stations, but they can have other uses. If the DN is separate from the MIN, it is used in call delivery by the originating MSC to route a *LocationRequest* message to the correct HLR. The HLR can then find the subscriber's MIN based on the DN (i.e., the DN is a key field in the subscriber's service profile record) and perform subsequent operations based on the MIN rather than the DN.

The MIN is also a 10-digit decimal number. It is a programmable value that is entered into the MS by the subscriber or an authorized dealer of the service provider. The programmable MIN and the permanent electronic serial number (ESN) together represent a subscription to the network. The MIN-ESN pair also represents a key field that uniquely identifies the subscriber's service profile record in the HLR. The MIN identifies an MS over the radio interface and is used to route IS-41 messages from the serving MSC to the HLR and vice versa (e.g., *RegistrationNotification*).

In public cellular implementations, cellular service providers obtain blocks of directory numbers from the NANP administrator. If the DNs are used as MINs, they are distributed to cellular subscribers when they become authorized for service. There are two primary advantages when the MIN and DN are identical:

1. The DN does not need to be allocated and managed as a separate information element in the HLR.

2. Subscribers are only aware of a single number.

If the MIN and DN are different numbers, there are many other advantages, most due to the fact that the MIN is not necessarily representative of a geographic location:

1. Multiple DNs can be assigned to a MIN, allowing different treatment based on the called number.

2. Multiple MINs can be assigned to a DN, allowing multiple MSs to be called with the same number.

3. The directory number can be portable among service providers; i.e., the phone number does not have to change if the subscriber receives a new MIN with a different service provider.

4. Directory numbers are a limited resource. Separate MINs allow for numbers that are not limited to the NANP format, providing billions of additional numbers that are distinct from NANP numbers (i.e., numbers where the first and fourth digits can be 0 or 1 in the NXX-NXX-XXXX format).

5. Since MINs would not be as limited a resource as DNs, multiple MINs can be implemented in a single MS. Separate MINs could make it easier to route IS-41 messages from the MS to different locations based on different routing of the MIN (e.g., short messages routed directly to a message center).

6. Mobile phones could easily be preprogrammed with the MINs and sold off the shelf in retail stores in any location. This preprogramming eliminates the geographic tie of a programmed phone to the location indicated by a directory number.

Some cellular service providers today provide private network services using nondialable MIN values programmed into the MSs. However, a directory number must still be assigned to support call delivery to these MSs. Special procedures need to be implemented in the HLR to translate the DN to the MIN to enable the IS-41 intersystem operations to perform correctly.

IS-41-C provides a new parameter, known as the *mobile directory number* (MDN). This parameter is designed for implementations where the MIN does not also represent the directory number. There are no procedures in IS-41-C that prescribe a use for the MDN. It was added to the standard in support of implementations where the DN and MIN are separate values.

Use of Signaling System No. 7 (SS7)

SS7 is the primary signaling protocol used to support the IS-41 mobile application part (MAP). SS7 provides the transaction capabilities, message transport, and reliable network route management capabilities to support the IS-41 intersystem operations.

IS-41 Revisions A and B specify the initial 1988 version of ANSI SS7 as the preferred data transport protocol for intersystem operations. Subsequent versions of ANSI SS7 were published in 1992 and later. Although IS-41-C was published in February 1996, it still specifies the initial 1988 version of ANSI SS7 for signaling. This was done for three reasons:

1. Later versions of ANSI SS7 are not completely backward-compatible.

2. Later versions of SS7 specified the intermediate signaling network identifier (ISNI) function, enabling constrained network routing. This was deemed an undesirable feature by the TIA.

3. Nearly all the features of the later versions of ANSI SS7 fail to be implemented in the public SS7 networks.

Currently, the use of an older version of SS7 has not been a problem. However, many of the features of the later versions of SS7 may eventually be desired, and subsequent versions of IS-41 may specify the use of these versions.

SS7 protocols supported

IS-41-C specifies the use of the following SS7 protocol levels (see Fig. 17.1):

- Message Transfer Part (MTP) level 1
- Message Transfer Part (MTP) level 2
- Message Transfer Part (MTP) level 3
- Signaling Connection Control Part (SCCP) class 0 service
- Transaction Capabilities Application Part (TCAP)

MTP level 1 provides the physical transmission layer of the protocol, based on DS0 channels. MTP level 2 provides the data link layer of the protocol, providing point-to-point link reliability between SS7 signaling points. MTP level 3 provides the network address and route management layer of the protocol. SCCP class 0 service provides

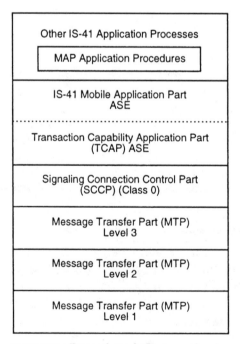

Figure 17.1 Protocol stack diagram showing the SS7 protocols used for IS-41 applications.

basic connectionless transport service, along with an enhanced addressing technique known as *global title* addressing. TCAP provides the application-layer communications protocol supporting transaction management and remote operation capabilities between two IS-41 functional entities.

An IS-41 message is encapsulated within a TCAP operation message to either invoke a remote mobile operation or return some result of an operation. The TCAP operation message is encapsulated within an SCCP unit data (UDT) message for transport. The UDT message is encapsulated within an MTP level 3 message signal unit (MSU), the basic SS7 packet routed through the network via physical node addresses (see Fig. 17.2).

Although SCCP provides three other classes of service, class 0 is the only one specified for use. It provides unsequenced connectionless data packets that are routed through the SS7 network independently of one another.

SS7 supports multiple TCAP component operations per transaction (i.e., per message). Although IS-41-C does not explicitly prohibit this practice, existing implementations generally support only one component per transaction. This is due to simpler designs and because the use of multiple components could exceed the maximum message length for an MSU based on potential IS-41-C message sizes.

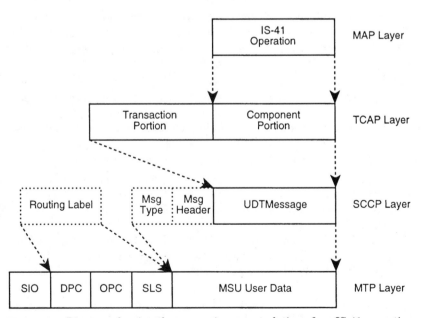

Figure 17.2 Diagram showing the successive encapsulation of an IS-41 operation into levels of the SS7 protocol.

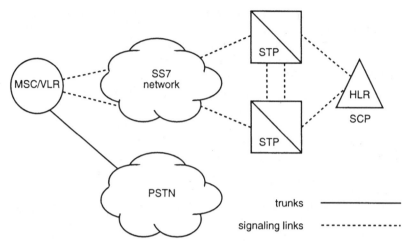

Figure 17.3 Network architecture showing the connectivity of the serving MSC/VLR to a subscriber's HLR via the SS7 network. The MSC is an end SP, and the HLR is an SCP.

SS7 routing and addressing

In IS-41 networks that use SS7, the MSC is an end signaling point (SP) in the SS7 network. The HLR is sometimes considered a service control point (SCP) (see Fig. 17.3). SS7 routing is based on *point codes.* Each node in the SS7 network has a unique point code address. MSUs that are sent between SS7 signaling points contain an originating point code (OPC) address and a destination point code (DPC) address. Routing of MSUs through the SS7 network based on point code addresses only is known as *point code routing.*

IS-41 supports point code routing as well as *global title routing.* A global title (GT) is an address that, by itself, does not explicitly allow routing in the network. This GT address needs to be translated somewhere in the network (usually by STPs) to a point code, *subsystem number* (SSN), or both. The SSN is an address of an SS7 application residing at a signaling point represented by a point code. This is known as *global title translation* (GTT). This addressing mechanism makes network routing of MSUs to and from end signaling points much simpler than point code routing. The end SPs do not need to maintain tables of each point code to properly route messages. They can supply a GT address, and the network will perform the GTT function to route the MSU to the correct SP.

Routing from the MSC to the HLR. The MSC/VLR currently serving the subscriber routes IS-41 signaling messages to the HLR based on the subscriber's MIN. The serving MSC/VLR must contain tables that enable the routing of the messages to the proper STP mated pair that is serving the HLR.

A group of related nodes such as an STP pair can be identified by a single point code called an *alias*. The alias point code address is used by network nodes that send MSUs to the related nodes via a combined link set (i.e., A links). The nodes are assigned the alias point code in addition to their individual point codes.

If the serving MSC and HLR reside within an SS7 network that does not support the GTT function, it must perform point code routing of IS-41 signaling messages destined to the HLR. The MTP routing tables within the MSC are required to maintain the point code address of the HLR. The serving MSC (which has a roaming agreement with the HLR service provider) maintains a table mapping ranges of MIN values to the point code of the HLR serving those subscribers. If the HLR is deployed as a mated-pair configuration for redundancy, the ranges of MIN values are mapped to an alias point code representing either of the HLRs. In this configuration, one HLR is considered active and the other is standby. The serving MSC is never aware of which one is active, and it has no need to know. Figures 17.4 and 17.5 show the point code routing of IS-41 messages (within MSUs) to the HLR.

MINs are mapped to the alias SS7 point code and SSN of the serving HLR. The serving MSC enters the MIN into the called-party address field of the SCCP UDT message. The MSC also enters the alias point code of the HLR mated pair into the DPC of the message. When the MSU is delivered to either of the STPs of the mated pair,

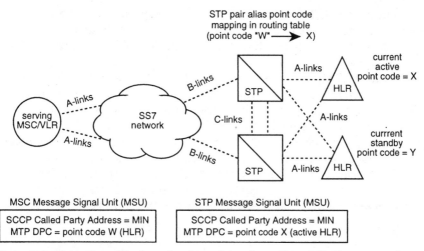

Figure 17.4 Network architecture showing point code routing of IS-41 messages from the serving MSC to the subscriber's HLR.

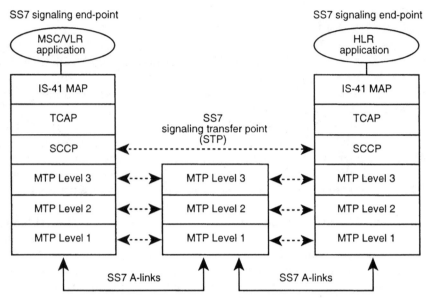

SS7 signaling end-point   SS7 signaling end-point

Figure 17.5 Protocol model of point code routing from the serving MSC to the subscriber's HLR.

the STP maps the actual point code of the active HLR into the DPC field of the MSU and delivers the message across the appropriate link set. In the case of failure in one of the HLRs, all IS-41 messages are routed to the other HLR.

If the serving MSC and HLR reside within an SS7 network that supports the GTT function, a translation is performed on the MIN to route the messages to the HLR. The serving MSC (which has a roaming agreement with the HLR) maintains a table mapping ranges of MIN values to the alias point code of the mated STP pair. Figures 17.6 and 17.7 show the GTT routing of IS-41 messages to the HLR.

MINs are mapped to the alias SS7 point code of the STP mated pair serving the HLR. The serving MSC enters the MIN into the called-party address field of the SCCP UDT message. The MSC also enters the alias point code of the STP mated pair into the DPC of the message. When the MSU is delivered to either of the STPs of the mated pair, the STP performs GTT and routes the MSU to the appropriate HLR.

The serving MSCs are required to contain routing table information to support routing of IS-41 transactions to the subscriber's HLR. Each MSC requires two databases:

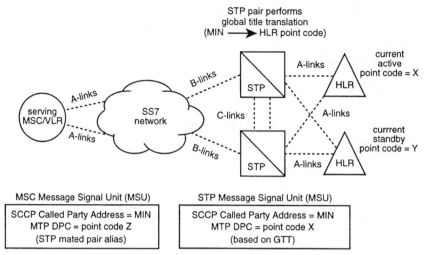

Figure 17.6 Network architecture showing global title routing of IS-41 messages from the serving MSC to the subscriber's HLR.

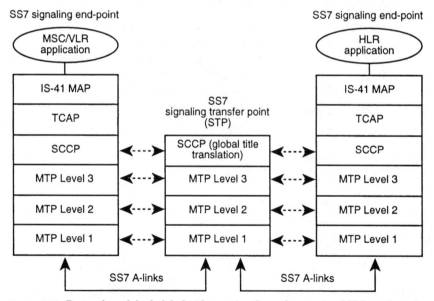

Figure 17.7 Protocol model of global title routing from the serving MSC to the subscriber's HLR.

1. The *number-range database* maps ranges of subscriber numbers (MINs in the form of NPA-NXX codes) onto MSC IDs (i.e., the identification of the HLR which serves the associated subscriber).

2. The *current network address database* identifies the DPC for the associated MSC ID.

If GTT is used, the current network address database can contain merely the DPC for the first STP which performs GTT, as opposed to "hard-coding" HLR network addresses into each MSC.

In most cases, the STP provides SS7 message screening based on the following parameters:

- MTP OPC
- MTP DPC
- SCCP calling party address (before and after GTT)
- SCCP called-party address (before and after GTT)

This screening provides the security mechanism to prevent unauthorized network access to the HLR. It further validates the identity of a calling party address by restricting it to preapproved link sets.

Point code/SSN as system identifier. Point code addresses are managed by the MTP level 3 portion of the SS7 protocol. They are unique across SS7 networks and represent the physical node address of each signaling point. Subsystem numbers (SSNs) are managed by the SCCP portion of SS7. They represent the address of an application at an SS7 signaling point. The following SSNs are used for IS-41:

- 5—mobile application part (MAP)
- 6—home location register (HLR)
- 7—visitor location register (VLR)
- 8—mobile switching center (MSC)
- 9—equipment identity register (EIR)
- 10—authentication center (AC)
- 11—short message service (SMS)

An SSN value of 5, referring to the MAP as an application of the IS-41 protocol as a whole, is an artifact of older revisions of IS-41. IS-41 Revision A specified a single SSN, representing the entire MAP application. IS-41 Revision B added the other SSNs when the need to differentiate IS-41 applications (e.g., HLR, MSC, AC) was discovered. IS-41 functional entities should be able to accept messages containing the SSN value for the MAP so that they are backward compatible with functional entities that have not implemented the more specific SSN values.

Generally, the MSCID parameter is sent in IS-41 application messages to indicate the identity of the serving MSC (or VLR) originating the message. This MSCID parameter is validated by the HLR when messages are received to ensure that the subscriber's HLR has a roaming agreement with that MSC (or VLR).

Many of the IS-41 intersystem operations optionally contain the PC_SSN parameter. This parameter is used as an alternative address to the MSCID for the IS-41 application layer. This PC_SSN parameter is used to more clearly identify a serving MSC. That is, there are situations that can occur where the MSCID value may not always be indicative of an exact point code address. The MSCID identifies a system and market, including all the cell sites associated with that system. A point code address identifies the actual physical node that sends or receives SS7 signaling messages.

Of course, there already are the separate SS7 originating point code (OPC) and SSN addresses sent with the IS-41 messages for use at the MTP and SCCP portions of the SS7 protocol. But the PC_SSN parameter for use in IS-41 applications is indicative of the originating address of the transaction, which may not be the same as the OPC and SSN sent along with an individual MSU containing the IS-41 message. This can occur, for instance, when the PC_SSN parameter represents a serving MSC but the message is forwarded to the HLR by a separate VLR that is also an SS7 signaling point. In this case the OPC and SSN values together indicate the VLR and not the MSC originating the message.

Alternate MAP Addressing

Although IS-41 specifies the use of SS7 and its addressing techniques, there are other network addressing techniques afforded by IS-41. These techniques are based upon the needs of individual service providers. The two additional address types specified by IS-41-C are:

1. MSCIdentificationNumber parameter (not to be confused with MSCID)

2. SenderIdentificationNumber parameter

These two parameters represent addresses that can be used in place of lower-layer address parameters, such as PC_SSN, if they are supported by the network. In contrast, the MSCID parameter identifies a system and market, including all the cell sites associated with that system. The MSCIdentificationNumber identifies an address that can be used to represent an MSC system in a manner divorced from physical considerations.

The SenderIdentificationNumber identifies an address of a functional entity such as a VLR or HLR similar to the way that a MIN identifies a particular mobile station. The use of these addresses facilitates routing of IS-41 messages over X.25 networks or between SS7 and X.25 networks. The format of these identifiers is defined so that they can be supported by global title translation functions.

Procedures for the use of these parameters are not defined in IS-41. Their use would have to be supported by systems receiving these addresses to know exactly what they identify. Note that use of these addresses could also facilitate routing between GSM and IS-41 systems, since GSM network elements are all identified by unique *ITU-T E.164* addresses. An E.164 address for an IS-41 network element can be supported via the MSCIdentificationNumber and SenderIdentificationNumber address parameters.

Short Message Service Architecture

SMS is a service that can have many, many implementation options. Aside from a standardized bearer service capability and some teleservice options, few other aspects of SMS have been standardized in IS-41-C. The reason is to allow a variety of implementation-dependent options that can be made compatible based on the bearer service and specified addressing techniques.

With the growth of Internet use, it can be very advantageous to implement SMS based on the internet protocol (IP) as the networking and transport mechanisms to deliver short messages. Also, the IS-41-C network reference model supports connectivity between message centers (MCs) and short message entities (SMEs). Figure 17.8 shows a sample generic network architecture using IP as the transport method for providing short message service.

The MC in the IS-41 network can provide protocol conversion between IP and SS7 for short messages to and from the serving MSC. If the MSC provides an interworking function supporting IP data transfer, short messages can be transferred directly between the MSC and the MC using IP. The architecture allows IS-41 network access to external IP applications (i.e., SMEs) as well as inexpensive IP access between MCs belonging to different networks. This allows for delivery of a short message to a local MC where a subscriber is roaming, or between MCs supporting mobile-to-mobile short messages.

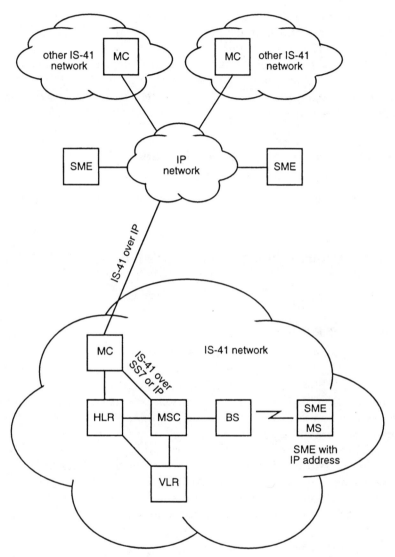

Figure 17.8 A generic SMS architecture based on the internet protocol (IP). IP allows inexpensive access to applications (SMEs) and is an easy protocol to interconnect with other message centers.

Home Location Register

The HLR is the primary repository for information about a given set of subscribers. In contrast, the VLR is a temporary storage function for those subscribers currently visiting the MSC or MSCs served by that VLR. When a subscriber is served by an MSC, the subscriber's characteristics are retrieved from the HLR and stored in the VLR. The MSC retrieves those characteristics from the VLR as needed based upon subscriber call activity.

The HLR stores both semipermanent and transient data about subscribers. These data enable subscriber mobility between MSCs of the same service provider or between MSCs of different service providers. The following list provides the types of subscriber information stored at an HLR:

- Current subscriber location and activity status

- Subscriber identification information

- Subscription information such as the subscribed features and privileges

- Service restrictions

- Subscriber feature information including feature activation status and feature registration information such as forward-to numbers

- Data enabling calls to be delivered to subscribers

HLR functional philosophy

The HLR acts as a controller for the IS-41 network. IS-41-C provides many procedures that enable the HLR to be the primary controlling mobile network element in an intelligent network-like architecture. The procedures enable the control of all features at the HLR so that serving systems must always obtain the rights and privileges of subscribers before providing any services.

The HLR resides in a fixed network location and is queried to find the location of subscribers who can move throughout and between networks. These queries enable visiting networks to acquire the rights and privileges of roaming subscribers. The HLR controls a subscriber's features so that those features follow that subscriber while roaming between MSC serving systems.

As a centralized node for subscriber signaling traffic, the HLR is able to control certain features for a subscriber on a network-wide basis. All feature control requests from throughout the network are directed to the specific HLR serving a particular subscriber. A mobile subscriber will be served by only one HLR.

The HLR is essentially a transaction processor handling network queries and operations. It is administered and controlled by the service provider with whom the subscribers have a service agreement. The service provider can retain control over the following functionality:

- Subscriber validation

- Subscriber features to maintain a uniform subscriber interface
- Signaling network access
- Delivery of incoming calls to a mobile subscriber regardless of location
- Subscriber roaming
- Strategic market information, such as subscriber activity and number of database accesses
- Fraud protection, investigation, and subscriber shutdown (based on authentication)
- Service offerings

The HLR is the most important network element in the mobile telecommunications network. It can be implemented as a stand-alone function, such as a service control point (SCP) application, or it can be implemented as an integrated function within an MSC.

HLR databases

The HLR holds all subscriber service profile records which define the features that have been authorized for each subscriber. The HLR is responsible for validating a subscriber to allow it to incur charges while roaming on another system. It also tracks the location of all its subscribers so that it can direct incoming calls to the MSC currently serving the subscriber. The HLR application is able to perform these functions because of the information it stores in its databases (see Fig. 17.9).

The HLR supports the following database tables:

- Subscriber service profile record table
- Subscriber feature option table
- Served MIN table
- Roamer agreement table

The HLR supports other tables, but these include the most necessary information for the HLR application functionality.

Subscriber service profile record table. The subscriber service profile record table is the HLR database containing the records established for each subscriber. This table generally includes the following subscriber information:

1. Identification data:
    - MIN

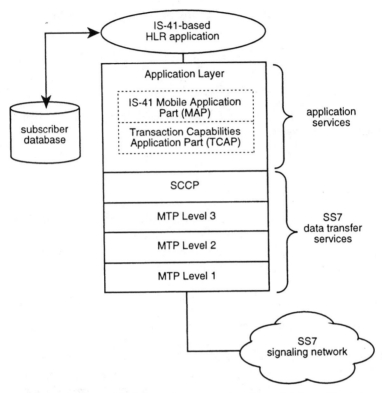

Figure 17.9 The HLR application has access to a large database. The interface to the database is proprietary.

- ■ ESN
- ■ MDN (if applicable)
2. Call profile data:
   - ■ Credit status (e.g., paid-up, delinquent account)
   - ■ Call origination indicator (e.g., local calls only, single NPA-NXX-XXXX only)
   - ■ Call restriction digits (e.g., by NPA only, NPA-NXX only, or NPA-NXX-XXXX)
   - ■ Call termination restriction (e.g., none, all)
   - ■ Preferred interexchange carrier
3. Qualification data (closely related to the authentication procedures)
   - ■ Authorization period
   - ■ Call history count

- Carrier-requested temporary disconnect (e.g., for delinquent accounts)
- Subscriber-requested temporary disconnect (e.g., for vacation)

4. Feature activations (for each feature available)
   - Feature authorized/activated
   - Feature states
   - Feature timers

5. Feature registrations (one example)
   - Call-forwarding forward-to-number

6. Location and activity status
   - VLR identification information (e.g., MSCID, PC_SSN)
   - Registration status (i.e., active, inactive)

7. Fraud protection (e.g., subscriber behavior information)

Subscriber feature option table. The subscriber feature option table maintains the configurations required for certain features to operate. An example of this database information is

1. Allowed feature code values

2. Number of subscriber modifications

3. Feature timer values (e.g., message-waiting notification)

4. Short message service options

Served MIN table. The served MIN table defines MIN ranges assigned to a particular HLR. The MINs are defined in terms of office codes, thousands groups, or hundreds groups. If numbers are allocated individually, a subscriber record is usually created for each number in the list and marked as an "unassigned" MIN if it is not yet allocated. The basic structure of the served MIN table is as follows:

1. MIN block specification, one of the following:
   - Full office code
   - Thousands group
   - Hundreds group

2. Allocated individually (i.e., true or false)

Roamer agreement table. The roamer agreement table defines other systems (i.e., MSCs/VLRs) which have a service agreement with the current HLR. The table reflects individual privileges granted to each system. The basic structure of the roamer agreement table is as follows:

1. MSCID (primary key)

2. Point code (alternate key)

3. Subsystem number

4. Node type (e.g., HLR, VLR, MSC, AC)

5. Allowed screening type:

   ▪ Match point code

   ▪ Match subsystem number

   ▪ Match SID

6. Messages allowed from specified subsystem, node, or network

7. HLR revision level of node:

   ▪ IS-41 Revision 0

   ▪ IS-41 Revision A

   ▪ IS-41 Revision B

   ▪ IS-41 Revision C

Visitor Location Register

The VLR is the temporary storage function for those subscribers currently visiting the MSC or MSCs served by that VLR. Subscribers are dynamically added and removed from the VLR database based upon their registration status.

In most cases, the serving VLR is physically integrated with the MSC currently serving a given subscriber. This is primarily due to the close functional relationship required between the MSC and VLR. However, stand-alone VLR configurations are not precluded by the IS-41 protocol.

An interesting implementation option is that a subscriber currently being served in the home area can be served by the VLR associated with the serving MSC. Although the subscriber is not roaming, signaling transactions can be reduced by not serving the subscriber directly from the HLR.

Authentication Center

The authentication function is very closely related to the individual subscriber functions provided by the HLR. The AC can be a separate network element or can be integrated as an application function within the HLR; the latter is more common. The HLR interface to the AC is based on IS-41 as an application-layer interface. Lower-layer protocol interfaces between them are specified as SS7 or X.25 by IS-41, but may be implementation-dependent, of course.

The AC provides the function of executing the authentication algorithm, based on information provided by the HLR about a subscriber. However, the AC can optionally maintain authentication information about subscribers that is separate from the HLR.

IS-41 does not preclude the implementation of a centralized AC as a separate and distinct network node; however, there is no separate SS7 global title translation (GTT) type to address the AC as an SS7 node separate from the HLR. All IS-41 AC transactions are required to pass through the HLR. IS-41 does, however, support a separate subsystem number (SSN) for the AC.

A-key management

A primary issue with the implementation of IS-41-based authentication is A-key management. A-key management consists of control and distribution of the A-key. The basic requirements of A-key management are the following:

1. The A-key must remain unknown to the customer and the distributor of the MS.

2. The HLR/AC must be very secure.

3. The A-key must not be readable by any party except the manufacturer.

4. Updates to the A-key in the MS and HLR or AC must be performed securely.

The service provider typically controls distribution of the initial and subsequently reprogrammed A-key. There is currently no standard method to distribute the A-keys for MSs; however, *TIA/EIA TSB50* specifies a user programming procedure to enter the A-key into the MS. This specification is not currently considered by the industry to be an adequate method of installing the A-key into the MS.

Authentication is a function that has, until recently, not been widely implemented. The importance of the function did not become pressing to cellular service providers until enormous revenue losses became more important than the implementation costs of a non-revenue-generating function. The service providers deploying digital radio systems (i.e., TDMA and CDMA) plan to deploy authentication. The service providers currently providing AMPS-based systems are beginning to deploy authentication today. The problem is that most MSs currently in use are not authentication-capable.

A Interface

The A interface has few impacts on the functioning of the IS-41 network. It is interesting though, to see why there are different standardized A-interface protocols as well as the potential impacts of these protocols on the rest of the cellular network.

Most A-interface protocols deployed today are proprietary. This stems from the original concept in early systems where mobile switches were sold along with the radio systems as a single bundled system. The equipment manufacturers had no need to support a standard A interface since their base station systems were sold with their MSCs. With the tremendous growth of cellular and new PCS systems, service providers began to desire open standards, allowing them to deploy switches that fit their needs with other vendors' radio systems that fit their needs.

There are three standardized A-interface protocols: SS7, ISDN, and frame relay. The use of SS7 for the A interface is a concept originally specified by GSM-based systems. SS7 allows reliable transport of the signaling messages between the base station and the MSC. It is interesting to note that SS7 is not a protocol designed for point-to-point transport (as is the case with the A interface). Most of the functions that SS7 provides are based on packet-switched adaptive routing techniques and the ability to transfer messages to different links for rerouting. It is a commonly available protocol, though, and the extraneous functionality can be removed for the A-interface implementation.

ISDN was chosen to be standardized specifically for PCS systems. ISDN provides a good user-network interface that can connect to many wire line telecommunications switches today. In fact, almost no MSCs support ISDN. This protocol was desired primarily by service providers implementing the Bellcore Wireless Access Communications System (WACS) and its cousin, the Personal Access Communications System (PACS). These systems were designed to allow wire line ISDN-based switches to provide network service to small PCS service providers that did not have their own network. ISDN allows these small PCS service providers to use the intelligent network infrastructure provided by the large wire line carriers.

Frame relay is the other A-interface protocol standardized for use in cellular-based networks. Frame relay enables fast packet-switching technology over very reliable transmission facilities. Frame relay provides some interesting implications for the way that radio systems access the cellular network (see Fig. 17.10). Frame relay allows for a network of radio systems that can communicate with one another and access one or more MSCs. With this configuration, radio systems can

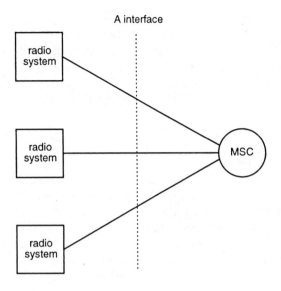

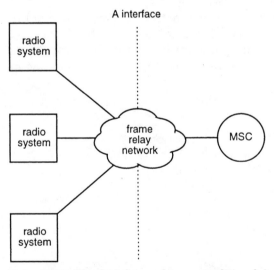

Figure 17.10 The first figure shows a traditional A interface architecture connecting radio systems to the MSC. The second figure shows the implications of a frame relay-based A interface.

be added at very little cost since they only need access to an available frame relay network. Handoffs between radio systems can also be performed without MSC intervention. The radio systems can be easily assigned to different MSC serving areas depending on the changing traffic patterns and capacity of adjacent MSCs.

Authorization and Call Routing Equipment (ACRE)

The ACRE is a logical functional entity employing IS-41 to provide an interface from a personal base station to the public cellular network. A personal base station is a small radio access system that allows mobile telecommunications in a small geographic area (such as an office building) which may or may not interface with the public cellular network. These personal base stations sometimes interface to a PBX that can act as a local MSC for the small radio systems (see Fig. 17.11).

The ACRE communicates with personal base stations through the PSTN to authorize, authenticate, and support call routing to mobile stations accessing those personal base stations. The ACRE also communicates with the cellular network via IS-41-C to inform the HLR that the subscriber is now being served by the personal base station. The technique requires no special agreements between the PSTN and the cellular network since the personal base station simply dials through the PSTN to reach the cellular network. The combination of

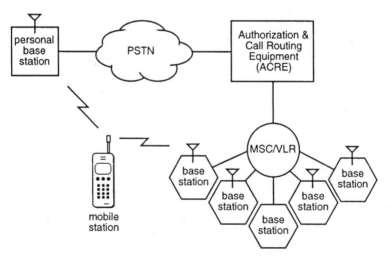

Figure 17.11 Network architecture diagram showing the authorization and call routing equipment (ACRE) and its relationship to wire line and wireless networks.

the ACRE and the personal base station enables a standard cellular mobile station to access the cellular networks as well as any areas covered by the personal base stations.

When subscribers move from the IS-41 public cellular network to an area covered by a personal base station, the ACRE system informs the HLR that the subscriber is now being served by the personal base station. In this sense, the ACRE acts as a "modified VLR." Subscribers can then receive cellular calls in the environment covered by the personal base station, whether it be an office environment or home.

IS-41 Deployment

IS-41 has been deployed by many service providers in North America for many years. As IS-41 has been revised and becomes more sophisticated, more service providers have deployed the protocol to provide enhanced services to subscribers. Examples of the services provided

Figure 17.12 AMPS technology is extensively deployed around the world and supported by IS-41 in many cases. (*Reproduced under written permission of Cellular Networking Perspectives.*)

by IS-41 that are now taken for granted include automatic call delivery, automatic roaming, and features such as call forwarding and call waiting.

The Cellular Telecommunications Industry Association (CTIA) tracks the implementation and deployment of IS-41 networks. The status has been changing so rapidly over the past few years that it is not worth listing statistics in this book, since by the time of publication those statistics will be outdated. Needless to say, IS-41-C Revision A and Revision B are currently deployed in *most* networks across the United States and Canada. Nearly all the service providers have worked with the equipment manufacturers to implement IS-41 for well over a year before its official publication in February 1996. Deployment schedules begin sometime around mid-1997.

Mexico also uses IS-41 as the signaling protocol for its cellular network infrastructure; however, it lags well behind the United States and Canada in keeping up with revision levels.

In the rest of the world, many, many countries have cellular systems based on the IS-41 protocol, and the list is growing. The International Telecommunications Union—Radio (ITU-R) group has officially approved IS-41-C as an international network standard for support of CDMA-based radio systems.

Wherever AMPS, NAMPS, TDMA, or CDMA radio technologies are chosen for use, IS-41 is the networking technology to support them. Figure 17.12 shows where AMPS is deployed around the world, in many cases supporting IS-41 today, and potentially supporting IS-41 in the future.

Standards and Bulletins Related to IS-41

This appendix contains the standards, bulletins, and specifications closely related to IS-41. The latest versions listed are valid as of the publication date of this book and are subject to change. Currently, IS-41-C has been elevated to a national standard, numbered *ANSI/TIA/EIA-41*.

Note that there are many other standards and bulletins not listed here. These standards pertain to other aspects of cellular systems not directly related to IS-41, such as the ancillary air interface standards that belong to the AMPS, TDMA, and CDMA families of standards.

Name: ANSI/TIA/EIA-553

Title: Mobile Station–Land Station Compatibility Specification

Summary: Provides the original standard analog AMPS air interface supported by IS-41 networks. Note that there are other standards and TSBs closely associated with this air interface standard, but they do not impact IS-41.

Latest version: ANSI/TIA/EIA-553 Revision A (1996)

Name: TIA/EIA IS-41

Title: Cellular Radio-Telecommunications Intersystem Operations

Summary: Provides standard intersystem operations among MSCs,

VLRs, and HLRs for cellular networks to support subscriber mobility.

Latest version: Revision C (1996) (ANSI/TIA/EIA-41)

Name: TIA/EIA IS-52

Title: Uniform Dialing Procedures and Call Processing Treatment for Use in Cellular Radio Telecommunications

Summary: Provides the standard subscriber dialing plan for cellular networks and the network treatment of the call types dialed. ANSI/TIA/EIA-660 is the full national standard of IS-52-A and incorporated only minor editorial changes as a modification to IS-52-A.

Latest version: ANSI/TIA/EIA-660 (1996)

Name: TIA/EIA IS-53

Title: Cellular Features Description

Summary: Provides the standard stage 1 descriptions (from the subscriber's perspective) of the cellular supplementary services supported by IS-41. IS-53 Revision 0 provides service descriptions for IS-41 Revision B. IS-53 Revision A provides service descriptions for IS-41 Revision C. ANSI/TIA/EIA-664 is the full national standard of IS-53-A and is identical to IS-53-A.

Latest version: ANSI/TIA/EIA-664 (1996)

Name: TIA/EIA IS-54

Title: Cellular System Dual-Mode Mobile Station–Base Station Compatibility Standard

Summary: Provides the standard dual-mode AMPS-TDMA air interface protocol supported by IS-41. Note that there are other standards and TSBs closely associated with this air interface standard that do not impact IS-41.

Latest version: Revision B (1992); also revised and updated as TIA/EIA IS-136

Name: TIA/EIA IS-88

Title: Mobile Station–Land Station Compatibility Standard for Dual-Mode Narrowband Analog Cellular Technology

Summary: Provides the standard analog narrowband AMPS (NAMPS) air interface protocol supported by IS-41. Note that there are other standards and TSBs closely associated with this air interface standard that do not impact IS-41.

Latest version: Revision 0 (1993); also revised as TIA/EIA IS-91

Name:	TIA/EIA IS-91
Title:	Mobile Station–Base Station Compatibility Standard for 800 MHz Analog Cellular
Summary:	Provides the standard analog (AMPS and NAMPS) air interface protocol supported by IS-41. This standard provides modifications of ANSI/TIA/EIA-553 (AMPS) and TIA/EIA IS-88 (NAMPS). Note that there are other standards and TSBs closely associated with this air interface standard that do not impact IS-41.
Latest version:	TIA/EIA IS-691 (1996)

Name:	TIA/EIA IS-93
Title:	Cellular Radio Telecommunications A_i-D_i Interfaces Standard
Summary:	Provides the standard protocol interfaces from a cellular MSC to wire line switches in the PSTN/ISDN.
Latest version:	Revision 0 (1993)

Name:	TIA/EIA IS-95
Title:	Mobile Station–Base Station Compatibility Standard for Dual-Mode Wideband Spread Spectrum Cellular Systems
Summary:	Provides the standard dual-mode AMPS-CDMA air interface protocol supported by IS-41. Note that there are other standards and TSBs closely associated with this air interface standard that do not impact IS-41.
Latest version:	Revision A (1995)

Name:	TIA/EIA IS-124
Title:	Cellular Radio Telecommunications Intersystem Non-Signalling Data Communications (DMH)
Summary:	Provides the standard protocol for the transport of near real-time call detail records in support of billing and fraud management for IS-41 networks.
Latest version:	Revision A (1997)

Name:	TIA/EIA IS-136
Title:	800 MHz TDMA Cellular–Radio Interface–Mobile Station–Base Station Compatibility
Summary:	Provides the standard dual-mode AMPS-TDMA air interface protocol supported by IS-41. This standard defines the use of a digital control channel for enhanced services. Note that there are other standards and TSBs closely associated with this air interface standard that do not impact IS-41.
Latest version:	Revision A (1996)

Name: TIA/EIA IS-634

Title: MSC-BS Interface for Public 800 MHz

Summary: Provides the standard A-interface transport protocol for cellular systems based on both SS7 and frame relay and supported by IS-41.

Latest version: Revision A (1996)

Name: TIA/EIA IS-653

Title: ISDN-Based A-Interface for 1800 MHz PCS

Summary: Provides the standard A-interface transport protocol for PCS systems based on ISDN and supported by IS-41.

Latest version: Revision A (1996)

Name: TIA/EIA TSB29

Title: International Implementation of Cellular Radiotelephone Systems Compliant with ANSI/TIA/EIA-553

Summary: Provides recommended guidelines for the implementation, assignment, and administration of MIN and SID values for international implementations of AMPS and IS-41.

Latest Version: Revision B (1996)

Name: TIA/EIA TSB41

Title: IS-41-B Technical Notes

Summary: Provides recommended modifications to IS-41 Revision B to clarify ambiguities and correct errors in the intersystem operations.

Latest version: Revision 0 (1994); also revised and added to TIA/EIA IS-41 Revision C

Name: TIA/EIA TSB50

Title: User Interface for Authentication Key Entry

Summary: Provides recommended user procedures for programming the A-key required for authentication into the mobile station.

Latest version: Revision 0 (1993)

Name: TIA/EIA TSB51

Title: Cellular Radiotelecommunications: Authentication, Signaling Message Encryption and Voice Privacy

Summary: Provides recommended algorithms and intersystem operations for authentication, encryption, and voice privacy for cellular networks based on IS-41 Revision B.

Latest version:	Revision 0 (1993); also revised and incorporated into TIA/EIA IS-41-C
Name:	TIA/EIA TSB55
Title:	IS-41-A Forward Compatibility Rules
Summary:	Provides recommended forward-compatibility rules for implementations of IS-41 Revision A to be compatible with subsequent versions of IS-41.
Latest version:	Revision 0 (1994); also revised and incorporated into TIA/EIA IS-41-C
Name:	TIA/EIA TSB56
Title:	Cellular Application Level Testing for IS-41 Revision B, TSB51 and IS-53
Summary:	Provides recommended test scenarios to verify intersystem signaling and cellular feature compatibility with IS-41 Revision B, IS-53 Revision 0 and TSB51.
Latest version:	Revision A (1994)
Name:	TIA/EIA TSB64
Title:	IS-41-B Support for Dual-Mode Wideband Spread Spectrum Mobile Stations
Summary:	Provides recommended handoff operations for CDMA-based intersystem handoffs as modifications to intersystem handoff as described in IS-41 Revision B.
Latest version:	Revision 0 (1994); also revised and incorporated into TIA/EIA IS-41-C
Name:	TIA/EIA TSB65
Title:	IS-41-B Mobile Border System Problems
Summary:	Provides recommended solutions to cellular border cell problems identified in IS-41 Revision B.
Latest Version:	Revision 0 (1994); also revised and incorporated into TIA/EIA IS-41-C
Name:	TIA/EIA TSB68
Title:	PCN to PCN Intersystem Operations—IS-41 Based
Summary:	Provides recommended implementation modifications for the use of IS-41 for personal communications service (PCS) networks.
Latest version:	Revision 0 (1995)
Name:	TIA/EIA TSB76
Title:	PCS Multi-band

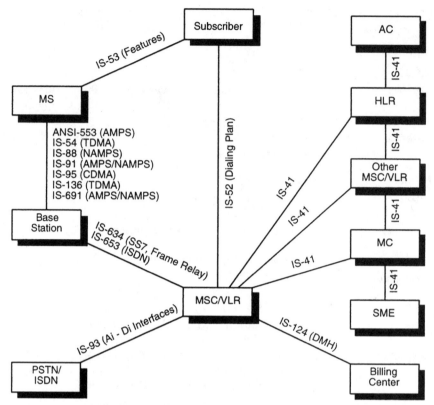

Figure A.1 Relationship of the TIA/EIA standards to the interfaces using those standards. (*Reproduced under written permission of Cellular Networking Perspectives*)

Summary: Defines modifications to IS-41 messages and procedures
                      to enable interworking between cellular and PCS sys-
                      tems.

Latest version: Revision 0 (1996)

B

The Standards
Creation Process

Telecommunications Industry Association (TIA) Interim Standards and specifications are created to define and promote compatibility among telecommunications systems equipment.

The process begins when a representative of a member company introduces a contribution to an engineering committee or subcommittee (i.e., a standard-formulating group) requesting the creation of a new standard specification for a particular technology. If a large enough consensus is reached within the formulating group to develop the specification, a project request letter along with a project initiation notice form is submitted to the TIA Technical Standards Subcommittee (TSSC) for approval.

If the TSSC approves the project request, the TIA Standards Secretariat will assign a project number (PN) to the specification and inform the formulating body of the approval. The standards formulating group then solicits written contributions to begin creating the technical standard. The formulating group assigns an editor to develop a draft of the standard based on the contributions, or portions of the contributions, that have been accepted by consensus. Typically, contributions continue to be added to the draft over a period of several months. The draft specification is continually revised and rewritten based on accepted contributions. Eventually, when enough core material has been agreed upon, the draft specification becomes *baseline text*. The advance to baseline text is simply an affirmation that the material in the draft is in a relatively stable state.

Written contributions continue to be added to the baseline text until it is decided by consensus that the technical material in the specification is complete, accurate, and adequate for the standard. The baseline text is then *frozen,* and no more new technical changes are allowed to be made to the text, unless, of course, an error has been found. Editorial changes are always allowed, up until publication, so that clarifications can be made as well as grammatical corrections.

After the technical information is frozen, the specification may be submitted to a process known as *verification and validation* (V&V). The V&V process is meticulous, since the baseline text is carefully analyzed, line by line, for correctness and accuracy. Once the V&V process is complete, the specification can be submitted to the TSSC to become a *standards proposal* (SP). This SP and a ballot are distributed to the TIA committee membership for vote. Ballots that are returned with a negative vote or with technical comments must be resolved before the SP can be published as a standard. If enough significant technical changes are required to the SP, it may need to be reballoted. Once all the comments and negative votes have been resolved, the SP can be submitted to the TIA for publication as a standard.

TIA standards are considered satisfactory for a maximum of 5 years. Within this time, the standards-formulating group must review the standard and take one of the following three actions:

1. Reaffirm the standard; i.e., determine that the technical content is still valid and requires no changes.

2. Revise the standard; i.e., determine that the technical content is still valid, but requires changes.

3. Rescind the standard; i.e., determine that the technical content no longer has value and declare the standard to no longer be in force.

The IS-41 standard is one of the few TIA standards to be officially revised at least four times for technical modifications and enhancements.

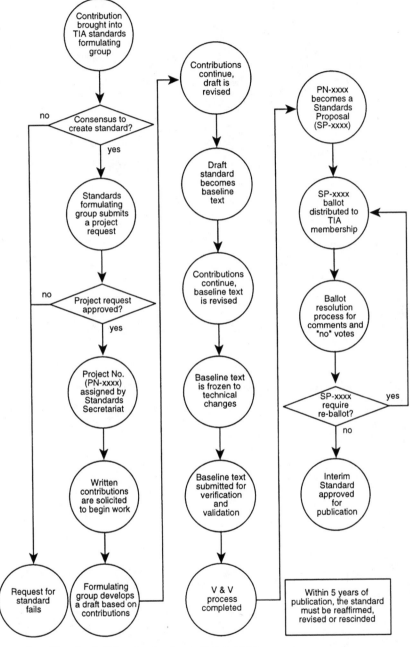

Figure B.1   Flowchart of the standards creation process.

Bibliography

American National Standards Institute (ANSI) T1 Committee Standards

American National Standards Institute, Inc., American National Standard for Telecommunications, *Signaling System Number 7 (SS7)—General Information*, Exchange Carriers Standards Association Committee T1, T1.110, 1988.

American National Standards Institute, Inc., American National Standard for Telecommunications, *Signaling System Number 7 (SS7)—Message Transfer Part (MTP)*, Exchange Carriers Standards Association Committee T1, T1.111, 1988.

American National Standards Institute, Inc., American National Standard for Telecommunications, *Signaling System Number 7 (SS7)—Signaling Connection Control Part (SCCP)*, Exchange Carriers Standards Association Committee T1, T1.112, 1988.

American National Standards Institute, Inc., American National Standard for Telecommunications, *Signaling System Number 7 (SS7)—Transaction Capabilities Application Part (TCAP)*, Exchange Carriers Standards Association Committee T1, T1.114, 1988.

American National Standards Institute, Inc., American National Standard for Telecommunications, *Signaling System Number 7 (SS7)—Supplementary Services for Non-ISDN Subscribers*, Exchange Carriers Standards Association Committee T1, T1.611, 1991.

American National Standards Institute, Inc., American National Standard for Telecommunications, *Operations, Administration, Maintenance and Provisioning (OAM&P)—Network Tones and Announcements*, Exchange Carriers Standards Association Committee T1, T1.209, 1989.

American National Standards Institute, Inc., EIA/TIA American National Standard for Telecommunications, *EIA/TIA 553, Mobile Station–Land Station Compatibility Specification*, Exchange Carriers Standards Association, September, 1989.

International Organization for Standardization (ISO) Standards

International Organization for Standardization (ISO), ISO-7776, *Information Technology—Telecommunication and Information Exchange between Systems—High Level Data Link Control Procedures—Description of the X.25 LAPB-Compatible DTE Data Link Procedures*, September 1994.

International Organization for Standardization, ISO-8208, *Information Technology—Data Communications—X.25 Packet Layer Protocol for Data Terminal Equipment*, April 1993.

International Organization for Standardization, ISO-8878, *Information Technology—Telecommunication and Information Exchange between Systems—Use of X.25 to Provide the OSI Connection-Mode Network Service*, October 1992.

International Telegraph and Telephone Consultative Committee (CCITT) Standards and International Telecommunications Union— Telecommunication (ITU-T) Standards

CCITT, vol. II, fascicle II.2, *Telephone Network and ISDN—Operation, Numbering, Routing and Mobile Service*, Recommendations E.100–E.333, 1988.

CCITT, vol. III, fascicle III.7, *Integrated Services Digital Network (ISDN) General Structure and Service Capabilities*, Recommendations I.110–I.257, 1988.

CCITT, vol. VI, fascicle VI.1, *General Recommendations on Telephone Switching and Signaling, Functions and Information Flows for Services in the ISDN, Supplements*, Recommendation Q.65, 1988.

CCITT, vol. VI, fascicle VI.13, *Public Land Mobile Network—Mobile Application Part and Interfaces*, Recommendations Q.1051–Q.1063, 1988.

CCITT, vol. VI, fascicle VI.7, *Specifications of Signalling System No. 7*, Recommendations Q.700–Q.716, 1988.

CCITT, vol. VI, fascicle VI.8, *Specifications of Signalling System No. 7*, Recommendations Q.721–Q.766, 1988.

CCITT, vol. VI, fascicle VI.9, *Specifications of Signalling System No. 7*, Recommendations Q.771–Q.795, 1988.

CCITT, vol. VII, fascicle VII.3, *International Reference Alphabet (IRA)* [formerly International Alphabet No. 5 (IA5)], Recommendation T.50, 1988.

CCITT, vol. VIII, fascicle VIII.2, *Data Communication Networks: Services and Facilities, Interfaces*, Recommendations X.1–X.32, 1988.

CCITT, vol. VIII, fascicle VIII.3, *Data Communication Networks: Transmission, Signalling and Switching, Network Aspects, Maintenance and Administrative Arrangements*, Recommendations X.40–X.181, 1988.

CCITT, vol. VIII, fascicle VIII.4, *Data Communication Networks: Open Systems Interconnection (OSI)—Model and Notation, Service Definition*, Recommendations X.200–X.219, 1988.

CCITT, vol. VIII, fascicle VIII.5, *Data Communication Networks: Open Systems Interconnection (OSI)—Protocol Specifications, Conformance Testing*, Recommendations X.220–X.290, 1988.

ITU-T, *Numbering Plan for the ISDN Era*, Recommendation E.164, 1991.

Cellular Telecommunications Industry Association (CTIA) Specifications

CTIA, *Network Reference Model Definition*, Revision 1.1, April 12, 1993.

Telecommunications Industry Association (TIA)* Standards

TIA, *Standards and Technology Annual Report* (STAR), 1994.

TIA, *Standards and Technology Annual Report* (STAR), 1995.

TIA, *Engineering Manual*, 1995.

TIA, TR45.0.A, *Common Cryptographic Algorithms*, Revision B, May 9, 1995.

TIA/EIA, Interim Standard IS-41, *Cellular Radio-Telecommunications Intersystem Operations*, February 1988.

TIA/EIA, Interim Standard IS-41-A, *Cellular Radio-Telecommunications Intersystem Operations*, January 1991.

*To purchase the complete text of any TIA document, call Global Engineering Documents at + 1 (800) 854-7179 or send a facsimile to + 1 (303) 397-2740.

TIA/EIA, Interim Standard IS-41-B, *Cellular Radio-Telecommunications Intersystem Operations,* December 1991.

TIA/EIA, Interim Standard IS-41-C, *Cellular Radio-Telecommunications Intersystem Operations,* February 1996.

TIA/EIA, Interim Standard IS-52, *Uniform Dialing Procedures and Call Processing Treatment for Use in Cellular Radio Telecommunications,* November 1989.

TIA/EIA, Interim Standard IS-52-A, *Uniform Dialing Procedures and Call Processing Treatment for Use in Cellular Radio Telecommunications,* March 1995.

TIA/EIA, Interim Standard IS-53, *Cellular Features Description,* August 1991.

TIA/EIA, Interim Standard IS-53-A, *Cellular Features Description,* May 1995.

TIA/EIA, Interim Standard IS-54-B, *Cellular System Dual-Mode Mobile Station—Base Station Compatibility Standard,* April 1992.

TIA/EIA, Interim Standard IS-88, *Mobile Station–Land Station Compatibility Standard for Dual-Mode Narrowband Analog Cellular Technology,* January 1993.

TIA/EIA, Interim Standard IS-91, *Mobile Station–Base Station Compatibility Standard for 800 MHz Analog Cellular,* October 1994.

TIA/EIA, Interim Standard IS-93, *Cellular Radio Telecommunications A_i -D_i Interfaces Standard,* November 1993.

TIA/EIA, Interim Standard IS-95-A, *Mobile Station–Base Station Compatibility Standard for Dual-Mode Wideband Spread Spectrum Cellular Systems,* May 1995.

TIA/EIA, Interim Standard IS-124, *Cellular Radio Telecommunications Intersystem Non-Signalling Data Communications (DMH),* November 1993.

TIA/EIA, Interim Standard IS-136, *800 MHz TDMA Cellular–Radio Interface–Mobile Station–Base Station Compatibility,* December 1994.

TIA/EIA, Interim Standard IS-634-A, *MSC-BS Interface for Public 800 MHz,* 1996.

TIA/EIA, Interim Standard IS-653, *ISDN-Based A-Interface (Radio System—PCSC) for 1800 MHz Personal Communications Systems,* May 1996.

TIA/EIA, Interim Standard IS-680, *Personal Base Station—Authorization and Call Routing Equipment Compatibility Standard,* May 1996.

TIA/EIA, Telecommunications Systems Bulletin TSB29-A, *International Implementation of Cellular Radiotelephone Systems Compliant with ANSI/TIA/EIA2D553,* September 1992.

TIA/EIA, Telecommunications Systems Bulletin TSB41, *IS-41B Technical Notes,* November 1994.

TIA/EIA, Telecommunications Systems Bulletin TSB50, *User Interface for Authentication Key Entry,* March 1993.

TIA/EIA, Telecommunications Systems Bulletin TSB51, *Cellular Radiotelecommunications: Authentication, Signaling Message Encryption and Voice Privacy,* May 1993.

TIA/EIA, Telecommunications Systems Bulletin TSB55, *IS-41-A Forward Compatibility Rules,* May 1994.

TIA/EIA, Telecommunications Systems Bulletin TSB56-A, *Cellular Application Level Testing for IS-41 Revision B, TSB51 and IS-53,* June 1994.

TIA/EIA, Telecommunications Systems Bulletin TSB64, *IS-41-B Support for Dual-Mode Wideband Spread Spectrum Mobile Stations,* January 1994.

TIA/EIA, Telecommunications Systems Bulletin TSB65, *IS-41-B Mobile Border System Problems,* April 1994.

TIA/EIA, Telecommunications Systems Bulletin TSB68, *PCN to PCN Intersystem Operations—IS-41 Based,* June 1995.

Other Sources

Bell Communications Research (Bellcore), *Compatibility Information for Interconnection of a Wireless Service Provider and Local Exchange Carrier Network,* Technical Advisory TA-NPL-000145, Issue 2, August 1993.

Cellular Networking Perspectives, David Crowe. Calgary, Canada: Cellular Networking Perspectives Ltd., 1992–1996.*

CIBERNET Corporation, *Cellular Intercarrier Billing Exchange Roamer Record,* Release 2, April 1, 1994.

Suggested Readings

Bates, Bud: *Wireless Networked Communications.* New York: McGraw-Hill, 1994.

Lee, William C. Y.: *Mobile Cellular Telecommunications Systems: Analog and Digital Systems.* New York: McGraw-Hill, 1995.

Russell, Travis: *Signaling System #7.* New York: McGraw-Hill, 1995.

*To purchase a subscription to *Cellular Networking Perspectives,* call + 1 (800) 633-5514 or send a facsimile to + 1 (403) 289-6658.

Glossary of Acronyms

3WC	three-way calling
AC	authentication center
ACRE	authorization and call routing equipment
AFREPORT	IS-41-C AuthenticationFailureReport operation
A-key	authentication key
AMPS	Advanced Mobile Phone System
ANI	automatic number identification
ANSI	American National Standards Institute
APDU	application protocol data unit
ASE	application service element
ASI	application service interface
ASP	application service part
ASREPORT	IS-41-C AuthenticationStatusReport operation
ATIS	Alliance for Telecommunications Industry Solutions
AUTHDIR	IS-41-C AuthenticationDirective operation
AUTHDIRFWD	IS-41-C AuthenticationDirectiveForward operation
AUTHREQ	IS-41-C AuthenticationRequest operation
BCD	binary-coded decimal
Bellcore	Bell Communications Research, Inc.
BID	billing identification
BLK	IS-41-C blocking operation
BS	base station
BSC	base station controller
BSCHALL	IS-41-C BaseStationChallenge operation
BTA	basic trading area
BTS	base transceiver system
BULKDEREG	IS-41-C BulkDeregistration operation
CAVE	cellular authentication and voice encryption
CC	conference calling or country code
CCI	conference call indicator
CCITT	International Telegraph and Telephone Consultative Committee (currently ITU-T)

CD	call delivery
CDMA	code division multiple access
CFB	call forwarding—busy
CFD	call forwarding—default
CFNA	call forwarding—no answer
CFU	call forwarding—unconditional
CGSA	cellular geographic service area
CIBER	Cellular Intercarrier Billing Exchange Roamer
CMRS	commercial mobile radio service
CMT	cellular messaging teleservice
CNI	calling number identification
CNIP	calling number identification presentation
CNIR	calling number identification restriction
COUNTREQ	IS-41-C CountRequest operation
CPN	calling party number
CPT	cellular paging teleservice
CSS	cellular subscriber station (MS preferred)
CT	call transfer
CTIA	Cellular Telecommunications Industry Association
CW	call waiting
D-AMPS	digital AMPS
DCE	data circuit-terminating equipment
DECT	Digital European Cordless Telephone
DN	directory number
DND	do not disturb
DS0	digital signal level 0
DS1	digital signal level 1
DTE	data terminal equipment
DTMF	dual-tone multifrequency
ECSA	Exchange Carrier Standards Association
EIA	Electronic Industries Association
EIR	equipment identity register
ESN	electronic serial number
FA	flexible alerting
FACDIR	IS-41-C FacilitiesDirective operation
FACDIR2	IS-41-C FacilitiesDirective2 operation
FACREL	IS-41-C FacilitiesRelease operation
FC	feature code

FCC	Federal Communications Commission
FCS	feature code string
FE	functional entity
FEATREQ	IS-41-C FeatureRequest operation
FLASHREQ	IS-41-C FlashRequest operation
FM	frequency modulation
HANDBACK	IS-41-C HandoffBack operation
HANDBACK2	IS-41-C HandoffBack2 operation
HANDMREQ	IS-41-C HandoffMeasurementRequest operation
HANDMREQ2	IS-41-C HandoffMeasurementRequest2 operation
HANDTHIRD	IS-41-C HandoffToThird operation
HANDTHIRD2	IS-41-C HandoffToThird2 operation
HLR	home location register
IA5	International Alphabet No. 5
ID	identification or identifier
IEC	International Electrotechnical Commission
IEEE	Institute of Electrical and Electronic Engineers
IMSI	international mobile station identification
INFOBACK	IS-41-C InformationBackward operation
INFODIR	IS-41-C InformationDirective operation
INFOFWD	IS-41-C InformationForward operation
IS	Interim Standard
ISANSWER	IS-41-C InterSystemAnswer operation
ISDN	Integrated Services Digital Network
ISO	International Organization for Standardization
ISP	intermediate service part
ISPAGE	IS-41-C InterSystemPage operation
ISPAGE2	IS-41-C InterSystemPage2 operation
ISSETUP	IS-41-C InterSystemSetup operation
ISUP	ISDN User Part
ITU	International Telecommunications Union
ITU-T	International Telecommunications Union—Telecommunications (formerly CCITT)
IVR	interactive voice response
IXC	interexchange carrier
kbits/s	kilobits per second
kHz	kilohertz
LEC	local exchange carrier

LOCREQ	IS-41-C LocationRequest operation
MAH	mobile access hunting
MAHO	mobile station-assisted handoff
MAP	mobile application part
Mbits/s	megabits per second
MC	message center
MCC	mobile country code
MDN	mobile directory number
MF	multifrequency
MFJ	Modification of the Final Judgement
MHz	megahertz
MIN	mobile identification number
MNC	mobile network code
MS	mobile station
MSA	metropolitan statistical area
MSC	mobile switching center
MSIN	mobile station identification number
MSINACT	IS-41-C MSInactive operation
MSONCH	IS-41-C MobileOnChannel operation
MTA	major trading area
MTP	Message Transfer Part
MTSO	mobile telephone switching office
MWN	message-waiting notification
NAM	number assignment module or numeric address module
NAMPS	Narrowband Advanced Mobile Phone System
NANP	North American Numbering Plan
NANPA	North American Numbering Plan Administration
NDC	national destination code
NIST	National Institute of Standards and Technology
NIUF	National ISDN Users Forum
NPA	numbering plan area
NSN	national significant number
OA&M	operations, administration, and maintenance
OAM&P	operations, administration, maintenance, and provisioning
OMT	overhead message train
ORREQ	IS-41-C OriginationRequest operation
OSI	Open Systems Interconnection
OTA	over the air activation

PACA	priority access and channel assignment
PACS	Personal Access Communications System
PBX	private branch exchange
PCA	password call acceptance
PCIA	Personal Communications Industry Association
PCN	personal communications network
PCS	personal communications services
PCW	priority call waiting
PIN	personal identification number
PL	preferred language
PN	project number
POTS	plain old telephone service
PSAP	public safety answering point
PSPDN	public switched packet data network
PSTN	public switched telephone network
QUALDIR	IS-41-C QualificationDirective operation
QUALREQ	IS-41-C QualificationRequest operation
RANDREQ	IS-41-C RandomVariableRequest operation
RBOC	Regional Bell Operating Company
REDDIR	IS-41-C RedirectionDirective operation
REDREQ	IS-41-C RedirectionRequest operation
REGCANC	IS-41-C RegistrationCancellation operation
REGNOT	IS-41-C RegistrationNotification operation
RESET	IS-41-C ResetCircuit operation
RF	radio frequency
RFC	remote feature control
ROUTREQ	IS-41-C RoutingRequest operation
RSA	rural service area
RUIDIR	IS-41-C RemoteUserInteractionDirective operation
SCA	selective call acceptance
SCCP	Signaling Connection Control Part
SCP	service control point
SDL	Specification and Description Language
SID	system identifier
SMDBACK	IS-41-C SMSDeliveryBackward operation
SMDFWD	IS-41-C SMSDeliveryForward operation
SMDPP	IS-41-C SMSDeliveryPointToPoint operation
SME	short message entity or signaling message encryption

SMS	short message service
SMSNOT	IS-41-C SMSNotification operation
SMSREQ	IS-41-C SMSRequest operation
SN	subscriber number or station number
SP	signaling point or standards proposal
SPINA	subscriber PIN access
SPINI	subscriber PIN intercept
SS#7	Signaling System No. 7 (ITU-T Standard)
SS7	Signaling System No. 7 (ANSI Standard)
SSD	shared secret data
SSD_A	shared secret data A
SSD_B	shared secret data B
STP	signaling transfer point
SWNO	switch number
TC	transaction capabilities
TCAP	Transaction Capabilities Application Part
TDMA	time-division multiple access
TIA	Telecommunications Industry Association
TID	transaction identifier
TMSI	temporary mobile station identity
TRANUMREQ	IS-41-C TransferToNumberRequest operation
TSB	Telecommunications Systems Bulletin
TSSC	Technical Standards Subcommittee
TTEST	IS-41-C TrunkTest operation
TTESTDISC	IS-41-C TrunkTestDisconnect operation
UBLK	IS-41-C Unblocking operation
UNRELDIR	IS-41-C UnreliableRoamerDataDirective operation
UNSOLRES	IS-41-C UnsolicitedResponse operation
V&V	verification and validation
VLR	visitor location register
VMR	voice message retrieval
VP	voice privacy
WG	working group
WZ1	world zone 1

Glossary of Definitions

access signaling   A type of signaling used to manage communications between a network user and an access point into the network.

adaptive routing   In packet-switched networks, the ability of traffic to be dynamically routed around congested or inaccessible switches and trunks, enabling communications to continue over alternate network paths.

Advanced Mobile Phone System (AMPS)   The analog cellular telecommunications system used in North America. It is characterized by analog radio frequencies between 800 and 900 MHz.

alerting   The function that provides an indication to the MS that a mobile-terminated call is being delivered or that some other feature has been invoked. Alerts are typically in the form of audible tones, visual signals, or vibrations.

alternate procedures   These are optional substitute procedures that may be used under certain circumstances for feature operation. They may be used in addition to the normal specified procedures, but not in place of them.

American National Standards Institute (ANSI)   The primary U.S. national public standards-making group responsible for creating and accrediting standards related to telecommunications systems and services.

anchor MSC   The role that an MSC assumes when it is the initial serving MSC for the call, i.e., the MSC that controls the base station that is first to assign an air interface traffic channel to an MS-originated or MS-terminated call. It is called the *anchor* MSC because it serves as the *fixed* MSC—with special processing responsibilities—for the duration of the call, even if the serving MSC changes as a result of call handoff.

antenna system   A system that converts electric signals from a radio transmitter to electromagnetic waves that comprise the radio transmission signals and vice versa.

application service element (ASE)   The portion of the application layer (layer 7) of the OSI reference model that performs a specific task for the application, such as specialized communications and transaction-based operations. Many ASEs can serve a single application.

application service interface (ASI)   The interface between the MAP application processes and the application service elements serving the MAP.

A side   One of the two sets of frequency spectrum originally licensed to cellular service providers, generally characterized by mobile station transmit frequencies between 824 and 836 MHz and receive frequencies between 869 and 880 MHz.

authentication   A procedure used to validate a mobile station's or subscriber's identity.

authentication center (AC)   A functional entity in the IS-41 network that manages the authentication information related to a mobile station or subscriber.

authentication controller   This term is not used in IS-41-C; we use it to refer to the entity controlling various authentication processes. This can be either the VLR or the AC, depending on whether SSD is shared with the serving system.

authentication key (A-key)   The primary private 64-bit key used in the CAVE algorithm. Private key encryption means that the same key is used to encrypt and decrypt the information; therefore, the key must be shared by at least two network elements, the authentication center and the mobile station. The A-key is only used to generate the temporary private key, called the shared secret data.

automatic number identification (ANI)   The number that provides the charge number of the line or trunk that originated a call. It is a number used to support exchange access and billing. This number may or may not be identical to the calling party number.

automatic roaming   The ability of a visited cellular network to support roaming of a mobile subscriber automatically, without additional actions being taken by the subscriber or a party calling the subscriber.

autonomous registration   Another name for timer-based registration, whereby the mobile station periodically registers with the network without the subscriber's taking any actions.

bandwidth   The specific range of frequencies that a given signal can occupy.

base station (BS)   The functional entity that represents all of the functions that terminate radio communications at the network side of the MS interface to the cellular network. It controls radio resources and manages network information that is required to provide telecommunications services to the MS. The BS essentially consists of the radio transceiver and radio transceiver controller equipment serving one or more cells.

basic trading area (BTA)   One of the 493 personal communications services (PCS) areas defined by Rand McNally & Company and adopted by the FCC that overlays part of the area covered by a major trading area (MTA) and is significantly smaller than an MTA.

bearer information   The necessary information that enables communications between end users of a bearer service to occur.

bearer service   A communications service defined by the basic capabilities of the transmission medium over which the communications traverse. In SMS, the transport protocol along with basic addressing and identification functions defines the bearer service.

Bell Communications Research, Inc. (Bellcore)   The name for the central service organization providing technical and management services for the Regional Bell Operating Companies (RBOCs).

Bell System   The name of the predivestiture conglomerate of AT&T, Western Electric, Bell Laboratories, and the Regional Bell Operating Companies.

border MSC   This term is synonymous with *neighbor MSC,* although its use is generally limited to discussions of *problems* that occur between neighboring systems.

border system   This term generally refers to an MSC (or MSC and associated VLR) that serves cells that are in close proximity to those of the current serving system, i.e., either adjacent to or sufficiently close that signals intended for one system are overheard by the other. The use of this term is limited to discussions of problems that occur between neighboring systems.

B side   One of the two sets of frequency spectrum originally licensed to cellular service providers, generally characterized by mobile station transmit frequencies between 837 and 849 MHz and receive frequencies between 881 and 894 MHz.

call   A temporary communication between telecommunications end users for the purpose of exchanging information. A call includes the sequence of events that allocate and assign resources and signaling channels required to establish a communications connection.

call control   See **call processing.**

called party   The intended recipient of a telephone call. For mobile-terminated calls, the MS is the called party. For mobile-originated calls, any valid telecommunications termination point can be the called party.

call history count   The value of a 6-bit counter maintained by both the AC (and the serving system if SSD is shared) and the MS. The MS counter is incremented under command of the AC (or serving system) and is used to detect the existence of an illegal MS.

calling party   The party originating a call. For mobile-originated calls, the MS is the calling party. For mobile-terminated calls, any valid telecommunications terminal can be the calling party.

calling party number   A set of digits and related indications that provides numbering information related to a calling party.

call processing   A set of functions that establishes, maintains, and releases calls to and from the mobile subscriber.

candidate MSC   A neighbor MSC that the current serving MSC is considering for the role of the new serving MSC before intersystem handoff occurs.

cell   An individual geographic area that is the topological component of a cellular or personal communications services system. The area is defined by the telecommunications coverage of the radio equipment located at the cell site.

cell sector   A geographic portion of a cell (typically one-third) that is served by directional antennas. The aggregate of the cell sectors that form an entire cell increases capacity by providing frequency reuse among the sectors.

cell site   The physical location of a cell's radio equipment and supporting systems. This term also refers to the equipment located at the cell site.

cellular authentication and voice encryption (CAVE) The common term for the collection of algorithms specified for the authentication processes. The CAVE algorithms make use of the MIN, ESN, A-key, and SSD values to generate authentication results that determine the legitimacy of a mobile station.

cellular carrier The cellular network operator that provides the network infrastructure and cellular services.

cellular digital packet data (CDPD) A technology that provides digital packet-switched data services through standard analog cellular channels. CDPD consists of a radio portion (the air link) and a network that employs standard OSI protocols to support end-to-end transfer of packetized messages.

cellular geographic service area (CGSA) A cellular market coverage area defined by the U.S. government. Two cellular network operators are licensed in each of these areas (i.e., A side and B side). A CGSA is further defined as either a metropolitan statistical area (MSA) or a rural service area (RSA). There are 733 CGSAs.

Cellular Intercarrier Billing Exchange Roamer (CIBER) The call detail record format defined by CIBERNET Corp. that is specified for cellular calls.

cellular service provider A company, organization, business, etc. authorized to provide cellular radio-telephone service. It is the organization which sells, administers, maintains, and charges for the cellular service. The service provider may or may not be the provider and operator of the network.

cellular subscriber station (CSS) See **mobile station.**

cellular system A type of wireless system characterized by high capacity, frequency reuse, and mobility management that controls communications to and from specific geographic "cells."

Cellular Telecommunications Industry Association (CTIA) An international organization of the wireless communications industry including both wireless carriers and manufacturers formed in 1984. The membership of the association includes all commercial mobile radio service providers, including cellular, personal communications services, enhanced specialized mobile radio, and mobile satellite services.

central office An exchange carrier switching system that terminates station loops and connects the loops to each other and to trunks. Sometimes called **end office.**

channel A single unidirectional or bidirectional path for transmitting or receiving electric signals between two entities.

channel-associated signaling A type of signaling in which the signals necessary for the traffic carried by a single channel are transmitted in the channel itself or in a channel permanently associated with it.

charge number See **automatic number identification.**

circuit An electromagnetic path between two or more points capable of providing a number of channels to support communications.

circuit-switched connection A circuit connection that is established and maintained, usually on demand, between two or more stations to allow the exclusive use of the circuit until the connection is released.

code-division multiple access (CDMA) A direct sequence spread-spectrum radio technique that digitally codes a base signal and employs different signal encoding patterns for each channel to redistribute that base signal across a broad bandwidth. Channels for the signal are distinguished from one another through the use of unique code sequences.

common carrier A company that provides public telecommunications facilities at nondiscriminatory rates. Common carriers are regulated by the FCC.

common-channel signaling A type of signaling in which signaling information relating to many circuits or channels is conveyed over a single channel.

conversation The portion of a call where the calling and called parties may engage in a spoken exchange.

data circuit-terminating equipment (DCE) The interface equipment that provides all the functions that are required to establish, maintain, and terminate a connection. It couples the data terminal equipment to a transmission circuit and vice versa.

data terminal equipment (DTE) The equipment comprising the data source, data termination, or both. It converts user information to data signals for transmission or reconverts received data signals to user information.

digital AMPS (D-AMPS) An informal term for the dual-mode TDMA/AMPS technology standardized by the Telecommunications Industry Association in North America (i.e., TIA/EIA IS-136).

digital signal level 0 (DS0) A 64 kbits/s digital pulse-code-modulated signal (or channel) derived from a 4-kHz voice-band analog signal.

digital signal level 1 (DS1) A time-division multiplexed digital signal comprised of 24 DS0 level signals (or channels) at a composite rate of 1.544 Mbits/s.

directory number (DN) The dialable number used to call a mobile station or telecommunications terminal. It is the actual number used by a calling party to reach a called party.

distributed network A type of network characterized by network resources, such as switching equipment and databases, that are distributed throughout the geographic area being served by the network.

dual-tone multifrequency (DTMF) A method for signaling digits inband over a voice channel by adding two multifrequency signal tones.

E-1 carrier A digital transmission system used in Europe that multiplexes 32 separate 64 kbits/s digital channels onto one circuit at a composite rate of 2.048 Mbits/s. Only 30 of the channels are used for voice and data. The other two are used for frame synchronization and signaling.

electronic serial number (ESN) A 32-bit number permanently stored in North American mobile stations that uniquely identifies the mobile station equipment.

end user   The user or subscriber of wireless network services. The end user is either the calling party for mobile-originated calls or the called party for mobile-terminated calls.

equipment identity register (EIR)   The functional entity in the network that provides the database repository for mobile equipment–related data.

feature   A bearer service, teleservice, or supplementary service provided to a subscriber, such as call waiting, call forwarding, or short message service.

feature activation   An action taken by the service provider, subscriber, or network to enable a feature operation to be invoked. While active, a feature may be either *quiescent* or *operative* according to whether the system is actually providing the feature at a given time. For example, call forwarding can be active and quiescent while no mobile-terminated calls are made. It becomes operative when a call is actually made and subsequently invoked.

feature authorization   The administrative process of making the feature available to a subscriber, involving the subscriber and the service provider. Subscription options may be associated with the authorization. A feature can be made available to all subscribers or to an individual subscriber by pre-arrangement with the service provider.

feature code   A special sequence of digits entered by a subscriber into the MS to perform some action within the cellular network.

feature interactions   The rules to follow when multiple features are invoked simultaneously.

feature invocation   An action or event that triggers the feature to be performed. It may be an action taken by the subscriber (e.g., by pressing a specific key), taken automatically by the network, or taken automatically by the terminal as a result of a particular condition that occurs.

feature registration   The process of programming, by the subscriber or service provider, information related to the feature (e.g., *registering* the forward-to number for call forwarding).

Federal Communications Commission (FCC)   The U.S. federal regulatory body consisting of a board of seven commissioners appointed by the President. The FCC has the power to regulate all interstate and foreign electric communications systems originating in the United States, including radio, television, facsimile, telephone, and cable systems.

flash   A message sent on a voice channel from an MS to the MSC used to invoke some special processing. Typically, the pressing of the Send key by the subscriber during a call is considered a flash from the MS.

frequency modulation (FM)   The modulation of the frequency of an electromagnetic carrier wave, with another wave serving as the modulating signal, such that the frequency excursions of the carrier wave are proportional to the modulating signal bearing the intelligence to be transmitted.

frequency reuse   A method of simultaneously transmitting the same channel frequency at many cells at specific power levels to expand the capacity of the system (provided that the distance between cell sites using the same frequency is great enough to avoid interference).

functional entity (FE)   A logical entity in the network consisting of one or more logical functions. The relationship between logical functional entities and physical network nodes can be one-to-one or many-to-one.

gateway MSC   This term is not used in IS-41-C (except for a very brief description in IS-41.1) but is usually acknowledged to mean an MSC that is assigned the MS's directory number, but does not provide radio access services (i.e., does not control any base stations).

handoff   The seamless transfer of a wireless call from one base station to another as a mobile unit crosses cell boundaries. (Also see **intersystem handoff, intrasystem handoff, mobile station–controlled handoff, mobile station–assisted handoff, network-controlled handoff, hard handoff, soft handoff,** and **softer handoff.**)

handoff chain   As a call is handed off from the first serving MSC (the anchor MSC) to another serving MSC and so on, a handoff chain develops; i.e., it is the sequence of MSCs, from the anchor to the current serving MSC, actively involved in the call at a given time.

hard handoff   A "break-before-make" form of call handoff between radio channels, whereby the MS temporarily disconnects from the network as it changes channels. This is the only type of handoff that requires IS-41 signaling support (as of IS-41-C); specifically, IS-41 intersystem operations are required when the handoff is between radio channels served by different MSCs. AMPS and TDMA MSs are limited to hard handoff, whereas CDMA MSs can perform soft handoff.

home location register (HLR)   The functional entity in the network that provides the primary database repository of subscriber information used to provide control and intelligence in cellular and wireless networks. The HLR contains a record for each home subscriber that includes location information, subscriber statuses, subscribed features, and directory numbers.

home message center   The message center belonging to the home service provider that stores and forwards short messages for a particular subscriber.

home MSC   The MSC associated with an MS that broadcasts the system identifier (SID) that matches one programmed into the MS's memory. The home MSC is assigned the MS's directory number (i.e., calls to the MS's directory number will terminate on the home MSC).

home system   The home MSC and its associated home location register (HLR). These two functional entities are sometimes combined in the same physical equipment; however, this need not be the case.

identification   A process to identify a network functional entity, telecommunications end user, telecommunications terminal, or telecommunications service provider.

implementation-dependent   A specification of functionality in the network that is determined or influenced by a particular method of implementation and is thus limited to that implementation.

implementation-independent   A specification of functionality in the network that is not influenced by a particular method of implementation and is thus independent of that implementation.

in-band signaling A type of signaling in which the frequencies or time slots used to carry the signals are within the bandwidth of the information channel.

incoming call A call described from the mobile station perspective that is mobile-terminated and originated from some calling party.

initiating MSC The MSC that initiates the release of an inter-MSC trunk.

Integrated Services Digital Network (ISDN) An integrated digital network in which the same time-division multiplexing switches and transmission routes are used to establish connections for different types of services, such as telephone, data, electronic mail, or facsimile services.

Integrated Services Digital Network User Part (ISUP) The functional part of the SS7 protocol which is responsible for providing call control signaling functions required to support basic bearer services and supplementary services for voice and nonvoice applications in an ISDN.

intelligent network A type of distributed network characterized by the capabilities of the network nodes to provide and interpret information via local processing and functionality at those nodes.

interexchange carrier (IXC) A carrier authorized to provide interexchange telecommunications services using the North American Numbering Plan. IXCs typically carry traffic between local exchange areas.

interface The boundary or point of connection between two adjacent physical or logical entities. This boundary establishes the technical interface, the test points, and the points of operational responsibility of the two entities. The boundary is defined in communications systems by functional, interconnection, and signaling characteristics.

Interim Standard The name for a standard that has been approved by the Telecommunications Industry Association membership, but not yet by the entire ANSI membership as a full national standard. Interim Standards are not considered permanent standards; rather they are an interim step toward a permanent national ANSI standard.

inter-MSC trunks The dedicated voice circuits between MSCs required for call handoff purposes.

international roaming The ability of mobile stations to roam between two or more countries. This type of roaming is usually difficult to achieve seamlessly due to differences in protocol implementations in different countries.

intersystem handoff A type of handoff where the MS is moving between two cells subtending two different MSCs.

intersystem operations Operations that are performed between cellular systems that enable subscriber mobility in and between those systems.

intrasystem handoff A type of handoff where the MS is moving between two cells subtending the same MSC.

IS-41 network A cellular network based on the TIA/EIA IS-41 standard. This term is very generic and describes any network that uses IS-41 intersystem operations to provide mobility to subscribers.

local exchange carrier (LEC) Any exchange carrier that provides telecommunication exchange and exchange access service. An LEC serves subscribers directly and provides local service inside its service area.

logical channel A means of two-way simultaneous transmission across a communications link that comprises associated transmit and receive channels. A number of logical channels may be derived from a common physical channel by multiplexing the logical channels onto the physical channel.

major trading area (MTA) One of the 51 personal communications service (PCS) areas defined by Rand McNally & Company and adopted by the FCC that cover large geographic portions of the United States.

message center (MC) The functional entity in the network that provides the store-and-forward function for short message services (SMS).

Message Transfer Part (MTP) The portion of the Signaling System No. 7 protocol which is responsible for the transfer of signaling messages throughout the SS7 network as required by the application user parts. The MTP comprises levels 1, 2, and 3 of the SS7 protocol.

metropolitan statistical area (MSA) One of the 305 cellular geographic service areas (CGSAs) defined by the U.S. government that cover the most populated urban cellular markets.

mobile application part (MAP) An application-layer signaling protocol (layer 7) mainly providing the mobility management function supporting wireless network operations.

mobile directory number (MDN) The mobile station's directory number which may be different from its mobile identification number.

mobile identification number (MIN) The 10-digit number that represents a mobile station's identity, directory number, or both.

mobile radio-telephone See **mobile station.**

mobile station (MS) The mobile or portable subscriber radio-telephone equipment.

mobile station–assisted handoff (MAHO) A type of handoff where the network requires the MS to measure signals from surrounding cells and report those measurements back to the network. The network then uses these measurements to determine where a handoff is required and to which channel.

mobile station–controlled handoff A type of handoff where the MS continuously monitors the radio signal's strength and quality. When predefined criteria are met, the MS checks the best candidate cell for an available traffic channel and requests the handoff to occur.

mobile switching center (MSC) The functional entity in the network that provides the cellular radio-telephone switching function.

mobile telephone switching office (MTSO) The wireless switching office including the mobile switching equipment, the hardware interfaces, and the real estate where the switching system resides.

mobility management   A general set of network application functions that enable the mobility of cellular subscribers in a cellular network.

Modification of the Final Judgement (MFJ)   The 1984 settlement (associated with the 1982 Consent Decree) between AT&T and the U.S. government involving the separation of the Regional Bell Operating Companies, Western Electric, AT&T Long Lines, and Bell Laboratories.

multifrequency (MF)   A trunk signaling method in which a combination of audio frequencies is used to convey address and control signaling.

narrowband advanced mobile phone system (NAMPS)   A narrowband version of the analog cellular telecommunications system used in North America and patented by Motorola, Inc. It is characterized by dividing the 30-kHz analog radio channels used for AMPS into three 10-kHz channels, thereby increasing capacity 3 times.

neighbor MSC   An MSC that serves cells that are adjacent to those of the current serving MSC. See **border MSC.**

network   The telecommunications equipment which has any part in processing a call or a supplementary service for the subscriber referred to. It may include local exchanges and transit exchanges, but does not include the mobile station and is not limited to the "public network" or any other particular set of equipment.

network architecture   The design principles, physical structure, functional organization, data formats, operational procedures, and other features used as the basis for the design, development, and operation of a network.

network-controlled handoff   A type of handoff where the radio base station, MSC, or both monitor the radio signal of the MS. When the signal's strength and quality deteriorate below a predefined threshold, the network arranges for a handoff to another channel.

network operator   See **cellular carrier** and **cellular service provider.**

network reference model   A diagram that depicts the entities of a network and the interfaces between those entities. The model is used as a graphical representation of the mobile telecommunications system as a whole and does not depict a specific network implementation.

network signaling   A type of signaling used between network nodes to operate, manage, and control the network to support certain types of functionality (e.g., mobility, voice traffic).

North American Numbering Plan (NANP)   A plan for the allocation of unique 10-digit directory numbers. The numbers consist of a three-digit area (numbering plan area) code, a three-digit office code, and a four-digit line number. The plan also extends to format variations (e.g., three-digit and seven-digit numbers), prefixes (for example, 1, 0, 01, and 011), and special code applications (e.g., service access codes like 800 numbers).

number assignment module (NAM)   The mobile station's electronic memory module where the MIN and other subscriber specific parameters are stored. MSs that have multi-NAM features offer subscribers the option of obtaining multiple subscriptions with different service providers simultaneously with the same MS.

operations, administration, and maintenance (OA&M) A set of functions that enable the network operator to monitor and control the network. OA&M functions are primarily used to observe and record operational characteristics of the network.

originating MSC This is the MSC that initiates the mobile-terminated call delivery procedures for an MS. Generally, this would be the home system for the MS, but in some network architectures a gateway MSC provides this function.

outgoing call A call described from the mobile station perspective that is mobile-originated and destined toward some called party.

out-of-band signaling A type of signaling in which the frequencies or time slots used to carry the signals are outside the bandwidth of the information channel.

over-the-air activation (OTA) A feature that provides initial activation of a mobile station via programming of the MS by the service provider remotely over the air interface.

packet-switched connection A logical connection that is established between two or more stations to allow the routing and transfer of data in the form of packets. The channel is occupied during the transmission of the packet only. Upon completion of the transmission, the channel is made available for the transmission of other packets.

packet-switched network A network in which logical connections are established between two or more network nodes that allow the routing and transfer of data in the form of packets so that a channel is occupied during the transmission of the packet only. Upon completion, the channel is made available for the transmission of other packets.

paging The function performed at the cell site to seek an MS for a mobile-terminated call.

party Any participant in a mobile telephone call.

path minimization The process of removing the second MSC from the handoff chain when an MS is handed off to a third MSC during the same call.

personal communications services (PCS) A family of mobile or portable radio communications services which provides services to individuals and businesses and is integrated with a variety of other networks. In North America, PCS is characterized by the use of radio frequencies between 1800 and 2200 MHz.

physical channel A single unidirectional or bidirectional path for transmitting or receiving electric signals between two entities using a particular encoding and modulation scheme.

point-to-point A service designed to convey data between two well-defined entities. In SMS, the short message entities (SMEs) represent the communication points that can originate and receive short messages.

protocol   A set of rules or conventions governing the interactions of process-es, applications, and components in a communications system.

public switched telephone network (PSTN)   The telecommunications net-work commonly accessed by ordinary telephones, key telephone systems, pri-vate branch exchange trunks, and data transmission equipment that pro-vides service to the general public.

radio frequency (RF)   The frequency of the electromagnetic spectrum nor-mally associated with radio-wave propagation.

radio frequency (RF) channel   A single allocation of contiguous spectrum (that is, 30 kHz in AMPS, 10 kHz in NAMPS, 30 kHz in TDMA, and 1.25 MHz in CDMA).

radio system   The portion of a mobile telecommunications network which includes antenna systems, transceivers, and transceiver controllers that is responsible for managing and controlling the radio resources in the network.

radio system management   A set of functions that manage the radio resources, connections, and transmission paths between the MS and the net-work, before, during, and after a call.

radio-telephone   See **mobile station.**

receiving MSC   The MSC that receives the request for inter-MSC trunk release.

registration   A set of messages and functions that enable the network to become aware of a subscriber in a particular location. Registration encom-passes the location management, service qualification, and mobile station state management functions.

reseller   A wireless service provider that buys service in bulk from another wireless network operator and "resells" that service to subscribers.

revertive call   A call to oneself. In the case of an MS, it is a call to the MS's own mobile directory number.

Revision 0   An informal term used to refer to the initial version of a specifi-cation published by the Telecommunications Industry Association. This revi-sion number is not officially listed as an actual revision number by the TIA.

roaming   The act of operating a mobile station in a cellular system or net-work other than the one from which service was subscribed.

rural service area (RSA)   One of the 428 cellular geographic service areas (CGSAs) defined by the U.S. government that cover the less-populated rural cellular markets.

seamless roaming   A type of roaming whereby the subscriber obtains mobile telecommunications services in the exact same way that they are provided in the home service area (i.e., no apparent *seams* between network service areas).

service   Any function that can be performed or supported in the wireless network. This term is sometimes used to refer to a **feature,** which is a specif-ic type of service.

service control point (SCP)   A network element in the SS7 network which uses transaction capabilities (i.e., the TCAP protocol) to directly access application databases.

service switching point (SSP)   A network element in the SS7 network (usually an end signaling point) which uses transaction capabilities (i.e., the TCAP protocol) to communicate with an SCP that can directly access application databases.

serving MSC   The MSC that, at any given time, is "serving" the mobile station (MS); i.e., in the case of an active call, the serving MSC controls the base station through which the connection is made between the MS and the other party to the call.

serving system   The serving MSC and its associated VLR. In general, the serving system maintains the most current location and status of the MS. These two functional entities are often combined in the same physical equipment; however, this need not be the case.

serving VLR   The visitor location register associated with the serving MSC.

shared secret data (SSD)   The temporary private 128-bit key that is calculated and stored in the MS and known by the network. The SSD is a concatenation of two 64-bit subsets: SSD_A, which is used to support the authentication procedures, and SSD_B, which is used to support the voice encryption procedures.

short message   A message, represented as a single packet of data, no more than 200 octets in length, that is the user traffic sent and received via the short message service.

short message entity (SME)   A functional entity in the network that can originate short messages, terminate short messages, or do both.

short message service (SMS)   A packet-switched messaging service that provides store-and-forward functions for the handling of short messages destined to or originated from wireless service subscribers.

signal   The intelligence, operation, message, or control function that is to be conveyed over a communication system.

signaling   The interchange of signals for the purpose of operating, controlling, managing, supervising, or maintaining a communication system.

Signaling Connection Control Part (SCCP)   The portion of the SS7 protocol that is responsible for connectionless and connection-oriented network data transfer including the global title translation function.

signaling link   In SS7, a bidirectional transmission path for signaling, consisting of two data channels operating together in opposite directions at the same data rate. A signaling link consists of a physical signaling data link together with its transfer control functions. The signaling link provides reliable transfer of signaling messages between nodes in a signaling network.

signaling point (SP)   A node in an SS7 signaling network that originates or receives signaling messages, transfers signaling messages from one signaling link to another, or does both.

Signaling System No. 7 (SS7)   An out-of-band common-channel signaling protocol standard that is designed to be used over a variety of digital telecommunication switching networks. It is optimized to provide a reliable means for information transfer for call control, remote network management, and maintenance.

signaling transfer point (STP)   A signaling point at which a message is received on a signaling link and is transferred to another link. It is neither the source nor the final destination of the message.

soft handoff   A "make-before-break" form of call handoff between radio channels, whereby the MS temporarily communicates with both the serving cell and one or more target cells before being directed to release all but the selected target radio channel. Soft handoff generally refers to the case where all the cells—serving and target—are controlled by a single MSC; therefore, this type of handoff does not require IS-41 signaling. However, future revisions of IS-41 may support soft handoff *between* MSCs. Currently, soft handoff is supported only for handoffs based on the CDMA air interface.

softer handoff   A soft handoff between two different parts or sectors of a cell controlled by a single base station. This type of handoff does not require IS-41 signaling and is currently supported by CDMA mobile stations only.

Specification and Description Language (SDL)   An internationally standardized flowchart language used to specify the behavior and general parameters of a telecommunications system.

spectrum   A continuous range of frequencies of electromagnetic waves that have common characteristics.

spread spectrum   A communication technique in which an information signal is transmitted across an entire bandwidth of frequencies based upon a special coding algorithm.

SSD sharing   Network signaling traffic between the serving system and the AC can be reduced if some authentication tasks are off-loaded from the AC to the serving system. To do this, the AC must *share* the SSD with the serving system.

stage 1   This stage is part of the overall method used to characterize telecommunication services. Stage 1 defines the service aspects of a capability. Specifically, stage 1 provides a service description of a telecommunication service from the subscriber's perspective.

stage 2   This stage is part of the overall method used to characterize telecommunication services. Stage 2 defines the functional network aspects of a capability. Specifically, stage 2 provides a description of the functions at the user-network interface and inside the network between network elements.

stage 3   This stage is part of the overall method used to characterize telecommunication services. Stage 3 defines the network implementation aspects of a capability. Specifically, stage 3 provides a description of the actual protocols and formats used to develop the telecommunication service.

standard   A specification that establishes engineering and technical requirements for processes, procedures, practices, and methods that have been decreed by authority or adopted by consensus.

store-and-forward   A method of transmitting data to their destination recipient and the storage of those data if the recipient is unavailable to receive them. The data are stored until it is convenient for them to be sent to their destination.

subscriber   A person authorized for a wireless feature or service.

switch   A function that transfers connection, contact, or signal continuity from one circuit or channel to another.

switching system   An electronic system that performs the selection and switching of telecommunication circuits.

system identifier (SID)   A 15-bit value used to uniquely identify a licensed cellular network operator within a cellular geographic service area for cellular systems and a trading area for PCS systems.

T-1 carrier   A digital transmission system used in North America that multiplexes 24 separate 64 kbits/s digital channels onto one circuit at a composite rate of 1.544 Mbits/s.

tandem MSC   An MSC in the handoff chain that is neither the anchor MSC nor the current serving MSC.

target MSC   The candidate MSC that is chosen by the serving MSC as the most appropriate system to take on the serving MSC's responsibilities.

Telecommunications Industry Association (TIA)   A U.S. public standards-making group responsible for accrediting other standards groups and approving standards for use in the United States.

Telecommunications Systems Bulletin (TSB)   The name for a publication that is not a standard but has significant value to the wireless telecommunications industry. TSBs are used as addenda to existing standards or to publicize material that is of industry importance.

teleservice   A type of telecommunication service that provides the complete capability, including terminal equipment functions, for communication between end users according to protocols established between administrations. In SMS, the applications that use the bearer service are considered the teleservices. These include application protocol elements such as supported character sets, priorities, and time stamps.

terminal mobility   The ability of the terminal to access telecommunication services from different locations while in motion, and the capability of the network to identify and locate that terminal.

time-division multiple access (TDMA)   The use of time interleaving to provide multiple, and apparently simultaneous, transmission in a single radio transceiver with a minimum of interference.

traffic   The aggregate of all user requests and communications being serviced by the network.

transaction   A controlled exchange of information between two functional entities in a network.

Transaction Capabilities (TC)   A set of functions that enable transaction processing between two functional entities in a network.

Transaction Capabilities Application Part (TCAP) The portion of the SS7 protocol which is responsible for information transfer between two or more nodes in the signaling network. The TCAP consists of transaction capabilities that manage remote operations via SCCP message transfer.

transcoding The conversion of digital signals from one coding plan to another.

transmission facilities management A set of functions that manage the physical means of providing voice and data communications services to cellular subscribers. These functions manage the trunk and line types and support the physical rates and digital encoding of the trunks and lines associated with the switching system.

trigger An event, action, or occurrence of a condition that is used to initiate a call processing function.

trunk A transmission channel connected between two network elements in a telecommunications system.

trunk group A group of trunks connected between two elements in a telecommunications system.

trunk maintenance A set of functions that control and maintain individual trunks.

user information Information that is transferred across the functional interface from a source user to a communication system for ultimate delivery to a destination user.

user part The portion of the SS7 protocol that transfers signaling messages via the Message Transfer Part (MTP). Different types of user parts exist, each of which specifies a particular use of the signaling system.

virtual circuit A logical circuit or channel that is established between source and destination packet switches and usually requires some form of setup prior to data transfer. This type of circuit may not be visible or obvious to a user.

visited system This is a serving system for an MS that is broadcasting an SID that is different from any other system programmed into the MS's memory. Since IS-41's primary application is in the case where the serving system is a visited system (not the home system), the terms *visited system* and *serving system* can be used interchangeably.

visitor location register (VLR) A local database function that maintains temporary records associated with individual network subscribers. The VLR contains subscriber location and service information. The local network switch accesses the VLR to retrieve information for the handling of calls to and from subscribers.

wireless system Any network system that is accessed via the propagation of electromagnetic waves through the air (i.e., instead of wires) supporting system communications.

wire line system Any network system that is accessed via the propagation of electromagnetic waves through a physical wire path to provide connections supporting system communications.

world zone 1 (WZ1)   The group of countries in the World Numbering Plan that are identified by the single-digit country code 1. WZ1 is defined in ITU-T Recommendation E.164. WZ1 generally includes the United States, Canada, and the Bahamas.

X.25   An ITU-T specification for a packet-switched protocol specified for use on public data networks between data terminal equipment and data circuit-terminating equipment.

Index

A interface, 45, 46, 48, 49, 376
A side, 5
Access signaling, 25, 118
Active/standby platforms, 50, 51, 52
Adaptive routing, 26
Addressing:
  alternate techniques, 367
Advanced digital features, 323
Advanced Mobile Phone System (AMPS),
  4, 5, 64, 72, 98, 139, 323, 379
A-key, 185
A-key management, 187, 375
A_i interface, 45, 48, 49, 334
Alert pip tone, 239
Alliance for Telecommunications Industry
  Solutions (ATIS), 17
American National Standards Institute
  (ANSI), 16, 46
Anchor MSC's role in feature processing,
  164
ANSI T1.111, 120
ANSI T1.112, 120
ANSI T1.114, 121, 124, 126, 127
ANSI/TIA/EIA-41, xiii, 321, 381, 382
ANSI/TIA/EIA-553, 49, 381, 383, 384
ANSI/TIA/EIA-660, 382
ANSI/TIA/EIA-664, 382
Antenna, 4, 38
Antenna system, 72
Application process, 111, 123, 128–132
Application service element (ASE),
  127–132
Application service interface, 131
Application services, 111, 112, 121
AUTH bit, 198
Authentication, 68, 71, 79, 184, 351, 371

Authentication center (AC), 12, 40, 184,
  187
  implementation, 374
  in network reference model, 44, 104
Authentication controller, 187, 205, 216,
  247
Authentication policy, 185, 201, 203, 205
Authentication reporting, 183, 219–222
Authentication-capable, 192, 193, 221
Authorization and call routing equipment
  (ACRE), 378, 379
Automatic number identification (ANI),
  79, 334
Automatic roaming, 31, 62, 68, 106,
  161–165
Autonomous registration, 163, 170, 171,
  177, 185, 293

B interface, 45, 49
B side, 5
Backward compatibility, 97, 319
Base station, 4, 10, 199, 208, 211, 216, 247
  in network reference model, 42
Base station challenge, 188, 213, 215
Base station controller (BSC), 38, 72
Base transceiver system (BTS), 38, 72
Baseline text, 387
Basic call origination (see IS-41-C voice
  services)
Basic trading area (BTA), 67
Bearer information, 25
Bell System, 17
Bellcore, 17, 334, 376
Billing, 56, 98, 163, 332
Billing ID (BID), 66, 67

ABOUT THE AUTHORS

MICHAEL D. GALLAGHER is Director of Systems Engineering for Synacom Technology, Inc., in San Jose, California. His experience spans hardware design, software design, and development in wireline and wireless telecommunications, and he was a major contributor to the development of the IS-41 Revision C cellular networking protocol. Mr. Gallagher is a graduate of the University of Ottawa and Rensselaer Polytechnic Institute.

RANDALL A. SNYDER is Director of Consulting Services and Principal Engineer for Synacom Technology, Inc., in San Jose, California. He has extensive experience in systems engineering and network design, as well as development of SS7, IS-41, and GSM networks. Mr. Snyder is a graduate of Franklin & Marshall College.